Cyberspace: First Steps

Cyberspace: First Steps

edited by
Michael Benedikt

The MIT Press
Cambridge, Massachusetts
London, England

Second printing, 1992

© 1991 Massachusetts Institute of Technology

This book was set in Stone Serif and Stone Sans by The MIT Press and printed and bound in the United States of America.

Library of Congress Cataloging-in-Publication Data

Cyberspace: first steps / edited by Michael Benedikt.
     p. cm.
   Includes bibliographical references.
   ISBN 0-262-02327-X
   1. Space and time.   2. Cybernetics.   3. Artificial intelligence.
  4. Metaphysics.   I. Benedikt, Michael.
  QC173.59.S65C93   1991
  003'.3—dc20   91-27372
                                          CIP

Contents

Cyberspace: First Steps

1 *Introduction*

Michael Benedikt

Cyberspace: A word from the pen of William Gibson, science fiction writer, circa 1984. An unhappy word, perhaps, if it remains tied to the desperate, dystopic vision of the near future found in the pages of *Neuromancer* (1984) and *Count Zero* (1987)—visions of corporate hegemony and urban decay, of neural implants, of a life in paranoia and pain—but a word, in fact, that gives a name to a new stage, a new and irresistible development in the elaboration of human culture and business under the sign of technology.

Cyberspace: A new universe, a parallel universe created and sustained by the world's computers and communication lines. A world in which the global traffic of knowledge, secrets, measurements, indicators, entertainments, and alter-human agency takes on form: sights, sounds, presences never seen on the surface of the earth blossoming in a vast electronic night.

Cyberspace: Accessed through any computer linked into the system; a place, one place, limitless; entered equally from a basement in Vancouver, a boat in Port-au-Prince, a cab in New York, a garage in Texas City, an apartment in Rome, an office in Hong Kong, a bar in Kyoto, a cafe in Kinshasa, a laboratory on the Moon.

Cyberspace: The tablet become a page become a screen become a world, a virtual world. Everywhere and nowhere, a place where nothing is forgotten and yet everything changes.

Cyberspace: A common mental geography, built, in turn, by consensus and revolution, canon and experiment; a territory swarming with data and lies, with mind stuff and memories of nature, with a million voices and two million eyes in a silent, invisible concert of enquiry, deal-making, dream sharing, and simple beholding.

Cyberspace: Its corridors form wherever electricity runs with intelligence. Its chambers bloom wherever data gathers and is stored. Its depths increase with every image or word or number, with every addition, every contribution, of fact or thought. Its horizons recede in every direction; it breathes larger, it complexifies, it embraces and involves. Billowing, glittering, humming, coursing, a Borgesian library, a city; intimate, immense, firm, liquid, recognizable and unrecognizable at once.

Cyberspace: Through its myriad, unblinking video eyes, distant places and faces, real or unreal, actual or long gone, can be summoned to presence. From vast databases that constitute the culture's deposited wealth, every document is available, every recording is playable, and every picture is viewable. Around every participant, this: a laboratory, an instrumented bridge; taking no space, a home presiding over a world . . . and a dog under the table.

Cyberspace: Beneath their plaster shells on the city streets, behind their potted plants and easy smiles, organizations are seen as the organisms they are—or as they would have us believe them be: money flowing in rivers and capillaries; obligations, contracts, accumulating (and the shadow of the IRS passes over). On the surface, small meetings are held in rooms, but they proceed in virtual rooms, larger, face to electronic face. On the surface, the building knows where you are. And who.

Cyberspace: From simple economic survival through the establishment of security and legitimacy, from trade in tokens of approval and confidence and liberty to the pursuit of influence, knowledge, and entertainment for their own sakes, everything informational and important to the life of individuals—and organizations—will be found for sale, or for the taking, in cyberspace.

Cyberspace: The realm of pure information, filling like a lake, siphoning the jangle of messages transfiguring the physical world, decontaminating the natural and urban landscapes, redeeming them, saving them from the chain-dragging bulldozers of the paper industry, from the diesel smoke of courier and post office trucks, from jet fuel fumes and clogged airports, from billboards, trashy and pretentious architecture, hour-long freeway commutes, ticket lines, and choked subways...from all the inefficiencies, pollutions (chemical and informational), and corruptions attendant to the process of moving information attached to *things*—from paper to brains—across, over, and under the vast and bumpy surface of the earth rather than letting it fly free in the soft hail of electrons that is cyberspace.

Cyberspace as just described—and, for the most part, as described in this book—does not exist.

But this states a truth too simply. Like Shangri-la, like mathematics, like every story ever told or sung, a mental geography of sorts has existed in the living mind of every culture, a collective memory or hallucination, an agreed-upon territory of mythical figures, symbols, rules, and truths, owned and traversable by all who learned its ways, and yet free of the bounds of physical space and time. What is so galvanizing today is that technologically advanced cultures—such as those of Japan, Western Europe, and North America—stand at the threshold of making that ancient space both uniquely visible and the object of interactive democracy.

Sir Karl Popper, one of this century's greatest philosophers of science, sketched the framework in 1972. The world as a whole, he wrote, consists of three, interconnected worlds. *World 1*, he identified with the objective world of material, natural things and their physical properties—with their energy and weight and motion and rest; *World 2* he identified with the subjective world of consciousness—with intentions, calculations, feelings, thoughts, dreams, memories, and so on, in individual minds. *World 3*, he said, is the world of objective, real, and public structures which are the not-necessarily-intentional products of the minds of living creatures, interacting with each other and with the natural *World 1*. Anthills, birds' nests, beavers' dams, and similar, highly complicated structures built by animals to deal with the environment, are forerunners. But many *World 3* structures, Popper noted, are abstract; that is, they are purely informational: forms of social organization, for example, or patterns of communication. These abstract structures have always equaled, and often surpassed, the *World 3* physical structures in their complexity, beauty, and importance to life. Language, mathematics, law, religion, philosophy, arts, the sciences, and institutions of all kinds, these are all edifices of a sort, like the libraries we build, physically, to store their operating instructions, their "programs." Man's developing belief in, and effective behavior with respect to, the objective existence of *World 3* entities and spaces meant that he could examine them, evaluate, criticize, extend, explore, and indeed make discoveries in them, in public, and in ways that could be expected to bear on the lives of all. They could evolve just as natural things do, or in ways closely analogous. Man's creations in this abstract realm create their own, autonomous problems too, said Popper: witness the continual evolution of the legal system, scientific and medical practice, the art world, or for that matter, the computer and entertainment industries. And always these *World 3* structures feed back into and guide happenings in *Worlds 1* and *2*.

For Popper, in short, temples, cathedrals, marketplaces, courts, libraries, theatres or amphitheaters, letters, book pages, movie reels, videotapes, CDs, newspapers, hard discs, performances, art shows . . . are all physical manifestations—or, should one say, the physical components of—objects that exist more wholly in *World 3*. They are "objects," that is, which are patterns of ideas, images, sounds, stories, data . . . patterns of pure information. And cyberspace, we might now see, is nothing more, or less, than the latest stage in the evolution of *World 3*, with the ballast of materiality cast away—cast away again, and perhaps finally.

This book explores the consequences and limits of such a development. But let it be said that, in accordance with the laws of evolution, and no matter how far it is developed, cyberspace will not *re*place the earlier elements of *World 3*. It will not *re*place but *dis*place them, finding, defining, its own niche and causing the earlier elements more closely to define theirs too. This has been the history of *World 3* thus far. Nor will virtual reality replace "real reality." Indeed, real reality—the air, the human body, nature, books, streets . . . who could finish such a list?—in all its exquisite design, history, quiddity, and meaningfulness may benefit from both our renewed appreciation and our no longer asking it to do what is better done "elsewhere."

I have introduced Popper's rather broad analysis to set the stage for a closer examination of the origins and nature of our subject, cyberspace. I discern four threads within the evolution of *World 3*. These intertwine.

Thread One   This, the oldest thread, begins in language, and perhaps before language, with a commonness-of-mind among members of a tribe or social group. Untested by dialogue—not yet brought out "into the open" in this way—this commonness-of-mind is tested and effective nonetheless in the coordinated behavior of the group around a set of beliefs held simply to be "the case:" beliefs about the environment, about the magnitude and location of its dangers and rewards, what is wise and foolhardy, and about what lies beyond; about the past, the future, about what lies within opaque things, over the horizon, under the earth, or above the sky. The answers to all these questions, always "wrong," and always pictured in some way, are common property before they are privately internalized and critiqued. (The group mind, one might say, precedes the individual mind, and consensus precedes critical exception, as Mead and Vygotsky pointed out.) With language and pictorial representation, established some ten to twenty thousand years ago, fully entering the artifactual world, *World 3*, these ideas blossom and elaborate at a rapid pace. Variations develop on the common themes of life and death, the whys and wherefores, origins and ends of all things, and these coalesce ecologically into the more or less coherent systems of narratives, characters, scenes, laws, and lessons that we now recognize, and sometimes disparage, as *myth*.

One does not need to be a student of Carl Jung or Joseph Campbell to acknowledge how vital ancient mythological themes continue to be in our advanced technological cultures. They inform not only our arts of fantasy, but, in a very real way, the way we understand each other, test ourselves, and shape our lives. Myths both reflect the "human condition" and create it.

Now, the segment of our population most visibly susceptible to myth and most productive in this regard are those who are "coming of age," the young. Thrust inexorably into a complex and rule-bound world that, it begins to dawn on them, they did not make and that, further, they do not understand, adolescents are apt to reach with some anger and some confusion into their culture's "collective unconscious"—a world they already possess—for anchorage, guidance, and a base for resistance. The boundary between fiction and fact, between wish and reality, between possibility and probability, seems to them forceable; and the archetypes of the pure, the ideal, the just, the good, and the evil, archetypes delivered to them in children's books and movies, become now, in their struggle towards adulthood, both magnified and twisted. It is no surprise that adolescents, and in particular adolescent males, almost solely support the comic book, science fiction, and videogame industries (and, to a significant extent, the music and movie industries too). These "media" are alive with myth and lore and objectified transcriptions of life's more complex and invisible dynamics. And it is no surprise that young males, with their cultural bent—indeed mission—to master new technology, are today's computer hackers and so populate the on-line communities and newsgroups. Indeed, just as "cyberspace" was announced in the pages of a science fiction novel, so the young programmers of on-line "MUDs" (Multi-User Dungeons) and their slightly older cousins hacking networked videogames after midnight in the laboratories of MIT's Media Lab, NASA, computer science departments, and a hundred tiny software companies are, in a very real sense, by their very activity, creating cyberspace.

This is not to say that cyberspace is for kids, even less is it to say that it is for boys: only that cyberspace's inherent immateriality and malleability of content provides the most tempting stage for the acting out of mythic realities, realities once "confined" to drug-enhanced ritual, to theater, painting, books, and to such media that are always, in themselves, somehow less than what they reach for, mere gateways. Cyberspace can be seen as an extension, some might say an inevitable extension, of our age-old capacity and need to dwell in fiction, to dwell empowered or enlightened on other, mythic planes, if only periodically, as well as this earthly one. Even without appeal to sorcery and otherworldy virtual worlds, it is not too farfetched to claim that already a great deal of the attraction of the networked personal computer in general—once it is no longer feared as a usurper of consciousness on the one hand, nor denigrated as a toy or adding machine on the other—is due to its lightning-fast enactment of special "magical" words, instruments, and acts, including those of induction and mastery, and the instant connection they provide to distant realms and buried resources. For the mature as well as the young, then, and for the purposes of art and self-definition as well as rational communications and business, it is likely that cyberspace will retain a good measure of *mytho-logic*, the exact manifestation(s) of which, at this point, no one can predict.

Three of the authors in this book—Michael Heim, Allucquere Rosanne Stone, and David Tomas—take up the cultural-anthropological theme, the latter two with special reference to the changing meaning of the "technophilic" physical body. Chip Morningstar and F. Randall Farmer describe their experiences with on-line games, in particular, LucasFilm's Habitat. William Gibson's short piece also makes its contribution at this level, if more directly, as an allegorical work of fiction itself.

Thread Two   Convolved with the history of myth is the thread of the history of media technology as such, that is, the history of the technical means by which absent and/or abstract entities—events, experiences, ideas—become symbolically represented, "fixed" into an accepting material, and thus conserved through time as well as space. Again, this a fairly familiar story, one whose detailed treatment is far beyond the scope of this introduction and this book. Nevertheless it is one worth rehearsing. It is also a topic that is extremely deep, for the secret of life itself is wrapped up in the mystery of genetic encoding and the replication and motility of molecules that orchestrate each other's activity. Genes are information; molecules are media as well as motors, so to speak . . .

But we cannot begin here, where the interests of computation theorists and biologists coincide. Our story best begins with evolved man's conscious co-option of the physical environment, specifically those parts, blank themselves, that best receive markings—such as sand, wood, bark, bone, stone, and the human body—for the purpose of preserving and delivering messages: signs, not unlike spoors, tracks, or tell-tale colors of vegetation or sky, but now intentional, between man and man, and man and his descendants. What a graceful and inspired step it was, then, to begin to *produce* the medium, to create smooth plastered walls, thin tablets, and papyrus, and to reduce the labor of marking—carving, chiseling—to the deft movement of a pigmented brush or stylus. As society elaborated itself and as the need to keep records and to educate grew, how much more efficient it was to shrink and conventionalize the symbols themselves, then to crowd them into rows and layers, "paper-thin" scrolls and stacks.

At this early stage already, the double movement towards the dematerialization of media on the one hand and the reification of meanings on the other is well underway. Against the ravages of time, nonetheless, and to impress the illiterate masses, only massive sculptures, friezes, and reliefs in stone would do. These are what we see today; these are what survive of ancient cultures and impress us still. But it would be wrong therefore to underestimate the traffic of information in more ephemeral media that must have sustained day-to-day life: the scratched clay tablets, the bark shards, graffitied walls, counters, papyri, diagrams in the sand, banners in the wind, gestures, demonstrations, performances, and, of course, the babble of song, gossip, rumor, and instruction that continuously filled the air. Every designed and every made thing was also the story of its use and its ownership, of its making and its maker.

This world sounds strangely idyllic. Many of its components, in only slightly updated forms, survive today. It was a period perhaps four thousand years long when objects, even pure icons and symbols, were not empty or ignorable but were real and meaningful, when craftsmanship, consensus, and time were involved in every thing and its physical passage through society. But first, with the development of writing and counting and modes of graphic representation, and then, centuries later, with the invention of the printing press and the spread of literacy beyond the communities of religious scholars and noblemen, the din of ephemeral communications came to be recorded at an unprecedented scale. More important for our story, these "records" came to be easily duplicable, transportable, and broadcastable.

Life would never be the same. The implications of the print revolution and the establishment of what Marshall McLuhan called the "Gutenberg galaxy" (in his book of the same name) for the structure and function of technologically advancing societies can hardly be overestimated. Not the least of these implications were (1) the steady, de facto, democratization of the means of idea production and dissemination, (2) the exponential growth of that objective body of scientific knowledge, diverse cultural practices, dreams, arguments, and documented histories called *World 3*, and (3) the fact that this body, containing both orthodoxies and heresies, could neither be located at any one place, nor be entirely controlled.

However, our double movement did not stop there, as we are all witness today. Although "printed matter" from proclamations to bibles to newspapers could, in principle, be taken everywhere a donkey, a truck, a boat, or an airplane could physically go, there was a limit, namely, *time*. No news could be fresh days or weeks later. The coordination of goods transportation in particular was a limiting case, for if no message could travel faster than that whose imminent arrival it was to announce . . . then of what value the message? Hence the telegraph, that first "medium" after semaphore, smoke signals, and light-flashing, to connect distant "stations" on the notion of a permanent *network*.

Another related limit was expense: the sheer expenditure of energy required to convey even paper across substantial terrain. The kind of flexible common-mindedness made possible in small communities by the near-simultaneity and zero-expense of natural voice communications, or even rumor and leaflets, collapses at the larger scale. Social cohesion is a function of ideational consensus, and without constant update and interaction, such cohesion depends crucially on early, and strict, education in—and memory of—the architectures, as it were, of *World 3*.

With the introduction of the telephone, both the problem of speed and the problem of expense were largely eliminated. Once wired, energy expenditure was trivial to relay a message, and it was soon widely realized (interestingly only in the 1930s and 40s) that the telephone need not be used like a "voice-telegraph," which is to say, sparingly and for serious matters only. Rather, it could be used also as an open channel for constant, meaningful, community-creating and business-running interchanges; "one-on-one" interchanges, to be sure, but "many-to-many" over a period of time. Here was a medium, here *is* a medium, whose communicational limits are still being tested, and these quite apart from what can be accomplished using the telephone system for computer networks.

Of course, the major step being taken here, technologically, is the transition, wherever advantageous, from information transported physically, and thus against inertia and friction, to information transported electrically along wires, and thus effectively without resistance or delay. Add to this the ability to *store* information electromagnetically (the first tape recorder was demonstrated commercially in 1935), and we see yet another significant and evolutionary step in dematerializing the medium and conquering—as they say—space and time.

But this was paralleled by a perhaps more significant development: *wire-less* broadcasting, that is, radio and television. Soon, encoded words, sounds, and pictures from tens of thousands of sources could invisibly saturate the world's "airwaves," every square millimeter and without barrier. What poured forth from every radio was the very sound of life itself, and from every television set the very sight of it: car chases, wars, laughing faces, oceans, volcanos, crying faces, tennis matches, perfume bottles, singing faces, accidents, diamond rings, faces, steaming food, more faces . . . images, ultimately, of a life not really lived anywhere but arranged for the viewing. Critic and writer Horace Newcomb (1976) calls television less a medium of *communication* than a medium of *communion,* a place and occasion where nightly the British, the French, the Germans, the Americans, the Russians, the Japanese . . . settle down by the million to watch and ratify their respective national mythologies: nightly variations on a handful of dreams being played out, over and over, with addicting, tireless intensity. Here are McLuhan's acoustically structured global villages (though he wished there to be only one), and support for the notion that the electronic media, and in particular television, provide a medium not unlike the air itself—surrounding, permeating, cycling, invisible, without memory or the demand for it, conciliating otherwise disparate and perhaps antagonistic individuals and regional cultures.

With cordless and then private cellular telephones, and "remote controls" and then hand-held computers communicating across the airwaves too, the very significance of geographical location at all scales begins to be questioned.

We are turned into nomads . . . who are always in touch.

All the while, material, print-based media were and are growing more sophisticated too: "vinyl" sound recording (a kind of micro-embossing), color photography, offset lithography, cinematography, and so on . . . the list is long. They became not only more sophisticated but more egalitarian as the general public not only "consumed" ever greater quantities of magazines, billboards, comic books, newspapers, and movies but also gained access to the means of production: to copying machines, cameras, movie cameras, record players, and the rest, each of which soon had its electronic/digital counterpart as well as a variety of hybrids, extensions, and cross-marriages: national newspapers printed regionally from satellite-transmitted electronic data, facsimile transmission, digital preprint and recording, and so on.

The end of our second narrative thread is almost at hand.

With the advent of fast personal computers, digital television, and high bandwidth cable and radio-frequency networks, so-called post-industrial societies stand ready for a yet deeper voyage into the "permanently ephemeral" (by which I mean, as the reader is well aware, cyberspace). As a number of chapters in this book observe, so-called on-line community, electronic mail, and information services (USENET, the Well, Compuserve, and scores of others) already form a technological and behavioral beginning. But the significance of this voyage is perhaps best gauged by the almost irrational enthusiasm that today surrounds the topic of *virtual reality*.

Envisaged by science fiction writer/promoter Hugo Gernsback as long ago as 1963 (see Stashower 1990) and explored experimentally by Ivan Sutherland (1968), the technology of virtual reality (VR) stands at the edge of practicality and at the current limit of the effort to create a communication/communion medium that is both phenomenologically engulfing and yet all but invisible. By mounting a pair of small video monitors with the appropriate optics directly to the head, a stereoscopic image is formed before the "user's" eyes. This image is continuously updated and adjusted by a computer to respond to head movements. Thus, the user finds himself entirely surrounded by a stable, three-dimensional visual world. Wherever he looks he sees what he would see were the world real and around him. This *virtual* world is either generated in real time by the computer, or it is preprocessed and stored, or it exists physically elsewhere and is "videographed" and transmitted in stereo, digital form. (In the last two cases the technique is apt to be named *telepresence* rather than virtual reality.) In addition, the user may be wearing stereo headphones. Tracked for head movements, a complete acoustic sensorium is thus added to the visual one. Finally, the user may wear special gloves, and even a whole body suit, wired with position and motion transducers to transmit to others—and to represent to himself—the shape and activity of his body in the virtual world. There is work underway also to provide some form of force-feedback to the glove or suit so that the user will actually feel the presence of virtual "solid" objects—their weight, texture, and perhaps even temperature (see Stewart 1991a for a recent survey, and Rheingold 1991). With a wishful eye cast towards such fictional technologies as the Holodeck, portrayed in the television series "Star Trek, the Next Generation," devices sketched in such films as *Total Recall* and *Brainstorm,* and, certainly, the direct neural connections spoken of in Gibson's novels, virtual reality/telepresence technology is as close as one can come in reality to entering a totally synthetic sensorium, to immersion in a totally artificial and/or remote world.

Much turns on the question of whether this is done alone or in the company of others; and if the latter, of how many, and how. Most of the chapters in this book tackle the question in one form or another. For, engineering questions aside, as the population of a virtual world increases, with it comes the need for consensus of behavior, iconic language, modes of representation, object "physics," protocols, and design—in a word, the need for *cyberspace* as such, seen as a general, singular-at-some-level, public, consistent, and democratic "virtual world." Herein lies the very power of the concept. In this volume, the chapters by Wendy A. Kellogg, John M. Carroll, and John T. Richards, by Steve Pruitt and Tom Barrett, by Meredith Bricken, and, again, by Michael Heim look specifically at the remarkable phenomenon of telepresence or "virtuality" as a prime component of the experience of cyberspace. Other authors in this volume imagine a viable cyberspace operating with less completely immersive techniques, although these nonetheless are thought of as considerably advanced over today's rather simple, low-resolution, two-dimensional graphical and textual interfaces.

Thread Two, then, is drawn from the history of communication media. The broad historical movement from a universal, preliterate actuality of *physical doing,* to an education-stratified, literate reality of *symbolic doing* loops back, we find. With movies, television, multimedia computing, and now VR, it loops back to the beginning with the promise of a *post*literate era, if such can be said; the promise, that is, of "post-symbolic communication" to put it in VR pioneer Jaron Lanier's words (Lanier 1989, Stewart 1991b). In such an era, language-bound descriptions and semantic games will no longer be required to communicate personal viewpoints, historical events, or technical information. Rather, direct—if "virtual"—demonstration and interactive experience of the "original" material will prevail, or at least be a universal possibility. We would become again "as children," but this time with the power of summoning worlds at will and impressing speedily upon others the particulars of our experience.

In future computer-mediated environments, whether or not this kind of literal, experiential sharing of worlds will supersede the symbolic, ideational, and implicit sharing of worlds embodied in the traditional mechanisms of text and representation remains to be seen. While pure VR will find its unique uses, it seems likely that cyberspace, in full flower, will employ all modes.

Thread Three   Another narrative, this one is spun out of the history of architecture.

The reader may remember that Popper saw architecture as belonging to *World 3*. This it surely does, for although shelter, beauty, and meaning can be found in "unspoiled" nature, it is only with architecture that nature, as habitat, becomes co-opted, modified, and codified.

Architecture, in fact, begins with displacement and exile: exile from the temperate and fertile plains of Africa two million years ago—from Eden, if you will, where neither architecture nor clothing was required—and displacment through emigration from a world of plentiful food, few competitors, and no more kin than the earth would provide for. Rapid climatic change, increasing competition, and exponential population growth was to change early man's condition irreversibly. To this day, architecture is thus steeped in nostalgia, one might say; or in defiance.

Architecture begins with the creative response to climatic stress, with the choosing of advantageous sites for settlements (and the need to defend these), and the internal development of social structures to meet population and resource pressure, to wit: with the mechanisms of privacy, property, legitimation, task specialization, ceremony, and so on. All this had to be carried out in terms of the availability of time, materials, and design and construction expertise. Added to these were the constraints and conventions manufactured by the culture up to that point. These were often arbitrary and inefficient. But always, even as conventions and constraints transformed, and as man passed from hunting and gathering to agrarianism to urbanism, the theme of return to Eden endured, the idea of return to a time of (presumptive) innocence and tribal/familial/national oneness, with each other and with nature.

I bring up this theme not because it "explains" architecture, but because it is a principle theme driving architecture's self-dematerialization. *De*materialization? The reader may be surprised. What *is* architecture, after all, if not the creation of durable physical worlds that can orient generations of men, women, and children, that can locate them in their own history, protect them always from prying eyes, rain, wind, hail, and projectiles . . . durable worlds, and in them, permanent monuments to everything that should last or be remembered?

Indeed these are some of architecture's most fundamental charges; and most sacred among them, as I have argued elsewhere (Benedikt 1987), is architecture's standard bearing, along with nature, for our sense of what we mean by "reality." But this should not blind us to a significant countercurrent, one fed by a resentment of quotidian architecture's bruteness and claustrophobia, which itself is a spilling-over of the resentment we feel for our own bodies' cloddishness, limitations, and final treachery: their mortality. Reality is death. If only we could, we would wander the earth and never leave home; we would enjoy triumphs without risks, eat of the Tree and not be punished, consort daily with angels, enter heaven now and not die. In the name of these unreasonable desires we revere finery and illumination, and reward bravery, goodness, and learning with the assurance of eternal life. As though *we* could grow wings! As though we could grow wings, we erect gravity-defying cathedrals resplendent with colored windows and niches crowded with allegorical life, create paradisiacal gardens such as those at Alhambra, Versailles, the Taj Mahal, Roan-Ji, erect stadia for games, create magnificent libraries, labyrinths, and observatories, build on sacred mountain tops, make enormous, air conditioned greenhouses with amazing flying-saucer elevators, leap from hillsides strapped to kites, dazzle with gold, chandeliers, and eternally running streams; we scrub and polish and whiten . . . all in a universal, cross-cultural act of reaching beyond brute nature's grip in the here and now. And this with the very materials nature offers us.

In counterpoint to the earthly *garden* Eden (and even to that walled garden, Paradise) then, floats the image of the Heavenly City, the new Jerusalem of the book of Revelation. Like a bejeweled, weightless palace it comes down out of heaven itself "its radiance like a most rare jewel, like jasper, transparent" (Revelation 21:9). Never seen, we know its geometry to be wonderfully complex and clear, its twelves and fours and sevens each assigned a set of complementary cosmic meanings. A city with streets of crystalline gold, gates of solid pearl, and no need for sunlight or moonlight to shine upon it for "the glory of God is its light."

In fact, all images of the Heavenly City—East and West—have common features: weightlessness, radiance, numerological complexity, palaces upon palaces, peace and harmony through rule by the good and wise, utter cleanliness, transcendence of nature and of crude beginnings, the availability of all things pleasurable and cultured. And the effort at describing these places, far from a mere exercise in superlatives by medieval monks and painters, continues to this day on the covers and in the pages of innumerable science fiction novels and films. (Think of the mother ship in *Close Encounters of the Third Kind*.) Here is what it means to be "advanced," they all say.

From Hollywood Hills to Tibet, one could hardly begin to list the buildings actually built and projects begun in serious pursuit of realizing the dream of the Heavenly City. If the history of architecture is replete with visionary projects of this kind, however, these should be seen not as naive products of the fevered imagination, but as hopeful fragments. They are attempts at physically realizing what is properly a cultural archetype, something belonging to no one and yet everyone, an image of what would adequately compensate for, and in some way ultimately justify, our symbolic and collective expulsion from Eden. They represent the creation of a place where we might *re-enter* God's graces. Consider: Where Eden (before the Fall) stands for our state of innocence, indeed ignorance, the Heavenly City stands for our state of wisdom, and knowledge; where Eden stands for our intimate contact with material nature, the Heavenly City stands for our transcendence of both materiality and nature; where Eden stands for the world of unsymbolized, asocial reality, the Heavenly City stands for the world of enlightened human interaction, form and information. In Eden the sun rose and set, there were days and nights, wind and shadow, leaf and stone, and all perfumed. The Heavenly City, though it may contain gardens, breathes the crystalline gleam of its own lights, sparkling, insubstantial, laid out like a beautiful equation. Thus, while the biblical Eden may be imaginary, the Heavenly City is *doubly* imaginary: once, in the conventional sense, because it is not actual, but once again because even if it became actual, because it *is* information, it could come into existence only as a virtual reality, which is to say, fully, only "in the imagination." The image of The Heavenly City, in fact, is an image of *World 3* become whole and holy. And a religious vision of cyberspace.

I must now return briefly to the history of architecture, specifically in modern times. After a century of the Industrial Revolution, the turn of the twentieth century saw the invention of high-tensile steels, of steel-reinforced concrete, and of high-strength glass. Very quickly, and under economic pressure to do more with less, architects seized and celebrated the new vocabulary of lightness. Gone were to be the ponderous piers, the small wooden windows, the painstaking ornament, the draughty chimneys and lanes, the chipping and smoothing and laying! Instead: daring cantilevers, walls reduced to reflective skins, openness, light, swiftness of assembly, chromium. Gone the stairs, the horse-droppings in the street, and the cobbles. Instead, the highway, the bulletlike car, the elevator, the escalator. Gone the immovable monument, instead the demountable exhibition; gone the Acropolis, instead the World's Fair. In 1924, the great architect Le Corbusier proposed razing half of Paris and replacing it with *La Ville Radieuse*, the Radiant City, an exercise in soaring geometry, rationality, and enlightened planning, unequaled since. A Heavenly City.

By the late 1960s, however, it was clear that the modern city was more than a collection of buildings and streets, no matter how clearly laid out, no matter how lofty its structures or green its parks. The city became seen as an immense node of communications, a messy nexus of messages, storage and transportation facilities, a massive education machine of its own complexity, involving equally all media, *including buildings*. To no one was this more apparent than to a group of architects in England calling themselves Archigram. Their dream was of a city that built itself unpredictably, cybernetically, and of buildings that did not resist television and telephones and air conditioning and cars and advertising but accommodated and played with them; inflatable buildings, buildings on rails, buildings like giant experimental theaters with video cameras gliding like sharks through a sea of information, buildings bedecked in neon, projections, lasers beams . . . These were described in a series of poster-sized drawings called *archi*tectural tele*grams*, which were themselves, perhaps not incidentally, early examples of what multimedia computer screens might look like tomorrow (Cook 1973). Although the group built nothing themselves, they were and are, nonetheless, very influential in the world of architecture.

Now, a complete treatment of the signs of the ephemeralization of architecture and its continuing capitulation to media is outside the scope of this introduction. It occurs on many fronts, from the wild "Disneyfication" of form, to the overly meek accommodation of services. Most interesting, however, is a thread that arises from thinking of architecture itself as an abstraction, a thread that has a tradition reaching back to ancient Egypt and Greece and the coincidence of mathematical knowledge with geometry and hence correct architecture. As late as the eighteenth century, architects were also scientists and mathematicians; witness Andrea Palladio, Sir Christopher Wren, and before them, of course, Leonardo da Vinci and Leon Battista Alberti. From the 1920s till the 1960s, the whole notion that architecture is about the experiential modulation of *space* and *time*—that it is "four dimensional"—captivated architectural theory, just as it had captivated a generation of artists in the 20s and 30s (Henderson 1983). This was something conceptually far beyond the simple mathematics of good proportions, even of structural engineering. It is an idea that still has force.

Then too there is the tradition of architecture seen for its symbolic content; that is, for not only the way it shapes and paces information fields in general (the emanations of faces, voices, paintings, exit signs, etc.) but the way buildings carry meaning in their anatomy, so to speak, and in what they "look like." After five thousand years, the tradition is very much alive as part of society's internal message system. In recent years, however, the architectural "message system" has taken on a life of its own. Not only have architectural drawings generated an art market in their own right—as illustrated conceptual art, if you will—but buildings themselves have begun to be considered as arguments in an architectural discourse about architecture, as propositions, narratives, and inquiries that happen, also, to be inhabitable. In its most current avant-garde guise, the movement goes by the name of Deconstructivism, or Post-Structuralism (quite explicitly related to the similarly named movements in philosophy and literary criticism). Its interests are neither in the building as an object of inhabitation nor as an object of beauty, but as an object of information, a collection of ciphers and "moves," junctions and disjunctions, reversals and iterations, metaphorical woundings and healings, and so on, all to be "read." This would be of little interest to us here were it not an indication of how far architecture can go towards attempting to become pure demonstration, and intellectual process, and were it not fully a part of the larger movement I have been describing. (And we should remember that, as a rule, today's avant-garde informs tomorrow's practice. See Betsky 1990.)

But there is a limit to how far notions of dematerialization and abstraction can go and still help produce useful and interesting, real architecture. That limit has probably been reached, if not overshot (Benedikt 1987). And yet the impetus toward the Heavenly City remains. It is to be respected; indeed, it can usefully flourish . . . in cyberspace.

The door to cyberspace is open, and I believe that poetically and scientifically minded architects can and will step through it in significant numbers. For cyberspace will require constant planning and organization. The structures proliferating within it will require *design*, and the people who design these structures will be called *cyberspace architects*. Schooled in computer science and programming (the equivalent of "construction"), in graphics, and in abstract design, schooled also along with their brethren "real-space" architects, cyberspace architects will design electronic edifices that are fully as complex, functional, unique, involving, and beautiful as their physical counterparts if not more so. Theirs will be the task of visualizing the intrinsically nonphysical and giving inhabitable visible form to society's most intricate abstractions, processes, and organisms of information. And all the while such designers will be *re*realizing in a virtual world many vital aspects of the physical world, in particular those orderings and pleasures that have always belonged to architecture.

Two chapters in this volume "come out of" architecture, my own and Marcos Novak's. My chapter attempts to discuss cyberspace in terms of certain basic design principles and then show some visualized examples; Novak discusses the idea of cyberspace as a poetic medium that, among other things, creates a "liquid architecture," an architecture of information, being less a proposition about designing buildings, of course, than a prelude as to how we might evolve legible forms in the context of a user-driven and self-organizing cyberspace system.

Thread Four   This thread is drawn from the larger history of mathematics. It is the line of arguments and insights that revolve around

(1) the propositions of geometry and space, (2) the spatialization of arithmetical/algebraic operations, and (3) reconsideration of the nature of space in the light of (2).

Since Artistotle, operating alongside this "spatial-geometrical" thread in mathematics has been a complementary one, that is, the development of symbolic logic, algebraic notation, calculus, finite mathematics, and so on, to modern programming languages. I say "complementary" because these last-named subjects could (and can still) proceed purely symbolically, with little or no geometrical, spatial interpretation; algebra, number theory, computation theory, logic . . . these are symbolic operations upon symbolic operations and have a life of their own.

In practice, of course, *diagrams*, which are spatial and geometrical, and *symbol strings* (mathematical notation, language) are accepted as mutually illuminating representations and are considered together. But the distinction between them, and the tension, still remain. There are those who think most easily and naturally in symbolic sequences, and linear operations upon them; there are those who think most easily and naturally in shapes, actions, and spaces. Apparently more than one type of intelligence is involved here (West 1991, Gardner 1983, Hadamard 1945). Be this as it may, cyberspace clearly is premised upon the desirability of spatialization per se for the understanding of information. Certainly, it extends the current paradigm in computing of "graphic user interfaces" into higher dimensions and more involving experiences, and it extends current interest, as evidenced by the popularity of Edward Tufte's books (1983, 1990), in "data cartography" in general and in the field of scientific visualization. But, more fundamentally, cyberspace revivifies and then extends some of the more basic techniques and questions having to do with the spatial nature of mathematical entities, and the mathematical nature of spatial entities, that lie at the heart of what we consider both real and measurable.

Rigorous reasoning with shape—deductive geometry—began, as we all know, in ancient Greece with Thales around 600 B.C., continuing through 225 B.C. with Pythagoras, Euclid, and Apollonius. The subject was twin: (1) the nature (and methods of construction) of the idealized forms studied—basically lines, circles, regular polygons and polyhedra, although Apollonius began work on conic sections—and (2) the nature of perfect reasoning itself, which the specifiability and universality of geometrical operations seemed to exemplify. The results of geometrical study had practical use in building and road construction, land surveying, and what we today call mechanical engineering. Its perfection and universality also supported the casting of astrological/cosmological models along geometrical lines.

The science and art of geometry has developed sporadically since, receiving its last major "boost" of renewed interest—after Kepler and Newton—in the late nineteenth century, with Bolyai and Lobatchevsky's discovery of non-Euclidean geometry. Soon, however, with the concept of pure topology and the discovery of consistent geometries of higher dimensionality than three, first Euclidean geometry and then geometry in general began to lose something of its luster as a science wherein significant new discoveries could be made. *All* statements of visual geometrical insight, it seemed, could be studied more generally and accurately in the symbolic/algebraic language of analytical mathematics—final fruit of Descartes' project in *La Géométrie*, which was precisely to show how the theorems of geometry could be transcribed into analytical (algebraic) form.

Of course the linkage, once made, between geometry and algebra, space and symbol, form and argument, is a two-way one. Descartes had both "algebraized" geometry and "geometrized" algebra. (And it is this second movement that is of most interest to us here.) With one profound invention, he had built the conceptual bridge we today call the Cartesian coordinate system. Here was the insight: just as the positions of points in natural, physical space could be encoded, specified, by a trio of numbers, each referring to a distance from a common but arbitrary origin in three mutually orthogonal directions, so too could the positions of points in a "mathematical space" where the "distances" are not physical distances but numerical values, derived algebraically, of the solution of equations of (up to) three variables. In this way, thousands of functions could accurately be "graphed" and made visible.

Today, procedures based on Descartes' insight are a commonplace, taught even at good elementary schools. But this should not mask the power of the implicit notion that *space itself* is something not necessarily physical: rather that it is a "field of play" for all information, only one of whose manifestations is the gravitational and electromagnetic field of play that we live in, and that we call the real world. Perhaps no examples are more vivid than the beautiful forms that emerge from simple recursive equations—the new science of "fractals"—and recent discoveries of "strange attractors," objects of coherent geometry and behavior that "exist" only in mathematical spaces (coordinate systems with specially chosen coordinates) and that economically map/describe/prescribe the behavior of complex, chaotic, physical systems. Which reality is the primary one? we might fairly ask.

Actually, why choose?

Modern physicists are sanguine: Minkowski had shown the utility of mapping time together with space, Hamiltonian mechanics lent themselves beautifully to visualizing the dynamics of a physical system in *n*-dimensional *state* or *phase space* where a single point represents the entire state of the system, and quantum mechanics seems to play itself out in the geometrical behavior of vectors in *Hilbert space*, in which one or more of the coordinates are "imaginary" (see Penrose 1989 for a recent explication).

In the meantime, the more common art of diagrams and charts proliferated—from old maps, schedules, and scientific treatises, to the pages of modern economics primers, advertisements, and boardroom "business graphics." Many of these representations are in fact hybrids, mixing physical, energic or spatiotemporal, coordinates with abstract, mathematical ones, mixing histories with geographies, simple intervallic scales with exponential ones, and so on. The practice of *diagramming* (surely one whose origins are earlier than writing) continues too, today enhanced by the mathematics of *graph theory* with its combinatorial and network techniques to analyze and optimize complex processes. What, we may ask, is the ontological status of such representations? All of them—from simple bar charts and organizational "trees" through matrices, networks, and "spreadsheets" to elaborate, multidimensional, computer-generated visualizations of invisible physical processes—all of these, and all abstract phase-, state-, and Hilbert-space entities, seem to exist in a geography, a space, borrowed from, but not identical with, the space of the piece of paper or computer screen on which we see them. All have a reality that is no mere picture of the natural, phenomenal world, and all display a physics, as it were, from elsewhere.

What are they, indeed? Neither discoveries nor inventions, they are of *World 3*, entities themselves evolved by our intelligence in the world of things and of each other. They represent first evidence of a continent about which we have hitherto communicated only in sign language, a continent "materializing," in a way. And at the same time they express a new etherealization of geography. It is as though, in becoming electronic, our beautiful old astrolabes, sextants, surveyor's compasses, observatories, orreries, slide rules, mechanical clocks, drawing instruments and formwork, maps and plans—physical things all, embodiments of the purest geometry, their sole work to make us at home in space—become environments themselves, the very framework of what they once only measured.

The contributions by Tim McFadden, Carl Tollander, and Alan Wexelblat are partially woven from this thread, as is a good part of my own. McFadden examines the idea of cyberspace as an informational Indra's Net, a universe of pointlike beads, infinite in number, each of which reflects all the others. Here cyberspace is an evolving, four-dimensional hologram of itself. (It was this ancient Hindu image of Indra's Net that also informed Leibniz's Monadology, as Heim discusses.) Tollander introduces Edelman's Neuronal Group Selection theory into the design of a noncentralized system of computational "engines" to create cyberspaces that can evolve in a "natural" way. (Novak also discusses this notion). Wexelblat examines the nature of coordinates in abstract spaces in general, in modern personal computing, and then, extrapolated, in terms of cyberspace specifically considered as an outgrowth of these.

My account of the intertwining "threads" that seem to lead to cyberspace is, of course, impressionistic and incomplete, and not just for lack of space in this introduction. Cyberspace itself is an elusive and future thing, and one can hardly be definitive at this early stage.

But it is also clear that the "threads" themselves are made of threads, and that there are others. For example, the history of art into modern times tells a related story, fully involving mythology, changing media, a relationship to architecture, logic, and so on. It is a thread I have not described, and yet the contribution of artists—visual, musical, cinematic—to the design of virtual worlds and cyberspace promises to be considerable, as Nicole Stenger, poet and animation artist, attests in this volume. Similarly, the story of progress in telecommunications and computing technology—the miniaturizations, speeds, and economies, the new materials, processes, interfaces and architectures—is a thread in its own right, with its own thrusts and interests in the coming-to-be of cyberspace. This story is well chronicled elsewhere (Rheingold 1985, Gilder 1988). Then there is the sociological story, and the economic one, the linguistic one, even the biological one . . . and one begins to realize that every discipline can have an interest in the enterprise of creating cyberspace, a contribution to make, and a historical narrative to justify both. How could it be otherwise? We are contemplating the arising shape of a new world, a world that must, in a multitude of ways, *begin,* at least, as both an extension and a transcription of the world as we know it and have built it thus far.

Another reason that my account is impressionistic and incomplete, however, is that the very metaphor of threads is too tidy and cannot support all that needs to be said. Scale aside, something deeper and more formless is going on. Consider: if information is the very *stuff* of space and time, what does it mean to manufacture information, and what does it mean to transfer it at ever higher rates between spatiotemporally distinct points, and thus dissolve their very distinctness? With mature cyberspaces and virtual reality technology, this kind of warpage, tunneling, and lesioning of the fabric of reality will become a perceptual, phenomenal fact at hundreds of thousands of locations, even as it falls short of complete, quantum level, physical achievement. Today intellectual, tomorrow practical, one can only guess at the implications.

Finally, my "narrative of threads" has not done justice to the authors represented in this volume. Each has their own perspective, expertise, and interest, and each draws inspiration from matters I have not mentioned, and stories I have not sketched or have only touched upon. Rather than extend this introduction with fuller discussion of each chapter, however, I recommend that the reader turn to them forthwith! Many are expanded and revised versions of presentations made at The First Conference on Cyberspace.[1] Others are written especially for the present collection.[2] All the authors address themselves to the topic with extraordinary seriousness, acumen, and enthusiasm, even though— and perhaps because—the varieties of cyberspace they imagine, describe, and sometimes criticize, do not yet exist. Indeed, the very definition of cyberspace may well be in their hands (or yours, dear reader). Of this much, one can be sure: the advent of cyberspace will have profound effects on so-called postindustrial culture, and the material and economic rewards for those who first and most properly conceive and implement cyberspace systems will be enormous.

But let us set aside talk of rewards. With this volume, with these "first steps," let us begin to face the perplexities involved in making the unimaginable imaginable and the imaginable real. Let the ancient project that is cyberspace continue.

Notes

1. Held on May 4 and 5, 1990, at The University of Texas at Austin. The Second (International) Conference on Cyberspace was held April 18–19, 1991, at The University of California at Santa Cruz.

2. Gibson, Tomas, Stone, and Wexelblat.

Selected Bibliography

Benedikt, Michael, *For an Architecture of Reality* (New York, Lumen Books, 1987) and *Deconstructing the Kimbell* (New York, Lumen Books, 1991).

Betsky, Aaron, *Violated Perfection* (New York, Rizzoli, 1990).

Cook, Peter, ed., *Archigram* (New York, Praeger, 1973).

Gardner, Howard, *Frames of Mind* (New York, Basic Books, 1983).

Gibson, William, *Neuromancer* (New York, Ace Books, 1984) and *Count Zero* (New York, Ace Books, 1987).

Gilder, George, *Microcosm* (New York, Simon and Schuster, 1988).

Hadamard, Jacques, *An Essay on the Psychology of Invention in the Mathematical Field.* (New York, Dover Publications, 1945).

Henderson, Linda D., *The Fourth Dimension and Non-Euclidean Geometry in Modern Art* (Princeton, N.J., Princeton University Press, 1983).

Lanier, Jaron, Interview in *Whole Earth Review,* Fall 1989, pp. 108ff.

McLuhan, Marshall, *Gutenberg Galaxy: The Making of Typographic Man* (Toronto, University of Toronto Press, 1962).

Masuda, Yoneji, *The Information Society as Post-industrial Society* (Washington, D.C., World Future Society, 1981).

Newcomb, Horace, *Television: The Critical View* (New York, Oxford University Press, 1976).

Penrose, Roger, *The Emperor's New Mind: Concerning Computers, Minds, and the Laws of Physics* (New York, Oxford University Press, 1989).

Popper, Karl, *Objective Knowledge: An Evolutionalry Approach* (Oxford University Press, 1972, revised 1979), pp. 106–152. See also Clifford Geertz, *The Interpretation of Cultures* (New York, Basic Books, 1973), pp. 55–83; Leslie A. White, "The Locus of Mathematical Reality: An Anthropological Footnote" in J. R. Newman, ed., *The World of Mathematics* (New York, Simon and Schuster, 1956), p. 2325; and Emile Durkheim, *The Elementary Forms of the Religious Life*. Translated by J. S. Swain (New York, Free Press,1965 [1915]).

Rheingold, Howard, *Tools for Thought* (New York, Simon and Schuster, 1985).

Rheingold, Howard, *Virtual Reality* (New York, Simon and Schuster, 1991).

Stashower, Daniel "A Dreamer Who Made Us Fall in Love with the Future," *Smithsonian,* vol. 21, #5 August 1990; p. 50. Contains reprint from *Life Magazine* (1963).

Sterling, Bruce, "Cyberspace (TM)," *Interzone*, vol. 2, November 1990.

Stewart, Doug, "Artificial Reality: Don't Stay Home Without It" *Smithsonian*, vol. 21, #10, January 1991a, pp. 36ff; "Interview: Jaron Lanier," *Omni* , vol. 13., #4, January 1991b, pp. 45ff.

Sutherland, I. E., "A Head-Mounted Three Dimensional Display," in *Fall Joint Computer Conference (*FJCC) (Washington, D.C., Thompson Books, 1968), pp. 757–764.

Tufte, Edward R., *Envisioning Information* (Cheshire, Conn. , Graphics Press, 1990) and *The Visual Display of Quantitative Information* (Cheshire, Conn., Graphics Press, 1983).

West, Thomas G., *In the Mind's Eye* (Buffalo, N.Y., Prometheus Books, 1991).

2 *Academy Leader*

William Gibson

"Ride music beams back to base."

He phases out on a vector of train whistles and the one particular steel-engraved slant of winter sun these manifestations favor, leaving the faintest tang of Players Navy Cut and opening piano bars of East St. Louis, this dangerous old literary gentleman who sent so many of us out, under sealed orders, years ago . . .

Inspector Lee taught a new angle—

Frequencies of silence; blank walls at street level. In the flat field. We became field operators. Decoding the lattices. Patrolling the deep faults. Under the lights. Machine Dreams. The crowds, swept with con . . . Shibuya Times Square Picadilly. A parked car, an arena of grass, a fountain filled with earth. In the hour of the halogen wolves . . . The hour remembered. In radio silence . . .

Just a chance operator in the gasoline crack of history, officer . . .

Assembled word *cyberspace* from small and readily available components of language. Neologic spasm: the primal act of pop poetics. Preceded any concept whatever. Slick and hollow—awaiting received meaning.

All I did: folded words as taught. Now other words accrete in the interstices.

"Gentlemen, that is not now nor will it ever be *my* concern . . . "

Not what I do.

I work the angle of transit. Vectors of neon plaza, licensed consumers, acts primal and undreamed of . . .

The architecture of virtual reality imagined as an accretion of dreams: tattoo parlors, shooting galleries, pinball arcades, dimly lit stalls stacked with damp-stained years of men's magazines, chili joints, premises of unlicensed denturists, of fireworks and cut bait, betting shops, sushi bars, purveyors of sexual appliances, pawnbrokers, wonton counters, love hotels, hotdog stands, tortilla factories, Chinese greengrocers, liquor stores, herbalists, chiropractors, barbers, bars.

These are dreams of commerce. Above them rise intricate barrios, zones of more private fantasy . . .

Angle of transit sets us down in front of this dusty cardtable in an underground mall in the Darwin Free Trade Zone, muzak-buzz of seroanalysis averages for California-Oregon, factoids on EBV mutation rates and specific translocations at the breakpoint near the c-myc oncogene . . .

Kelsey's second week in Australia and her brother is keeping stubbornly in-condo, doing television, looping *Gladiator Skull* and a new Japanese game called *Torture Garden*. She walks miles of mall that could as easily be Santa Barbara again or Singapore, buying British fashion magazines, shoplifting Italian eye-shadow; only the stars at night are different, Southern Cross, and the Chinese boys skim the plazas on carbon-fiber skateboards trimmed with neon.

She pauses in front of the unlicensed vendor, his face notched with pale scars of sun-cancer. He has a dozen cassettes laid out for sale, their plastic cases scratched and dusty. "Whole city in there," he says, "Kyoto, yours for a twenty." She sees the security man, tall and broad, Kevlar-vested, blue-eyed, homing in to throw the old man out, as she tosses the coin on impulse and snatches the thing up, whatever it is, and turns, smiling blankly, to swan past the guard. She's a licensed consumer, untouchable, and looking back she sees the vendor squinting, grinning his defiance, no sign of the $20 coin . . .

No sign of her brother when she returns to the condo. She puts on the glasses and the gloves and slots virtual Kyoto . . .

Once perfected, communication technologies rarely die out entirely; rather, they shrink to fit particular niches in the global info-structure. Crystal radios have been proposed as a means of conveying optimal seed-planting times to isolated agrarian tribes. The mimeograph, one of many recent dinosaurs of the urban office-place, still shines with undiminished *samisdat* potential in the century's backwaters, the Late Victorian answer to desktop publishing. Banks in uncounted Third World villages still crank the day's totals on black Burroughs adding machines, spooling out yards of faint indigo figures on long, oddly festive curls of paper, while the Soviet Union, not yet sold on throwaway new-tech fun, has become the last reliable source of vacuum tubes. The eight-track tape format survives in the truckstops of the Deep South, as a medium for country music and spoken-word pornography.

The Street finds its own uses for things—uses the manufacturers never imagined. The micro-tape recorder, originally intended for on-the-jump executive dictation, becomes the revolutionary medium of *magnetisdat,* allowing the covert spread of banned political speeches in Poland and China. The beeper and the cellular phone become economic tools in an increasingly competitive market in illicit drugs. Other technological artifacts unexpectedly become means of communication . . . The aerosol can gives birth to the urban graffitti-matrix. Soviet rockers press homemade flexidisks out of used chest x-rays . . .

Fifteen stones against white sand.

The sandals of a giant who was defeated by a dwarf.

A pavilion of gold, another of silver.

A waterfall where people pray . . .

Her mother removes the glasses. Her mother looks at the timer. Three hours. "But you don't like games, Kelsey . . . "

"It's not a game," tears in her eyes. "It's a city." Her mother puts on the glasses, moves her head from side to side, removes the glasses.

"I want to go there," Kelsey says.

"It's different now. Everything changes."

"I want to go there," Kelsey insists. She puts the glasses back on because the look in her mother's eyes frightens her.

The stones, the white sand: cloud-shrouded peaks, islands in the stream . . .

She wants to go there . . .

"The targeted numerals of the ACADEMY LEADER were hypnogogic sigils preceding the dreamstate of film."

3 Old Rituals for New Space: Rites de Passage *and William Gibson's* Cultural Model of Cyberspace

David Tomas

Across the communications landscape move the specters of sinister technologies and the dreams that money can buy.

—J. G. Ballard, *Crash*

The experience of subjective and intersubjective flow in ritual performance, whatever its sociobiological or personalogical concomitants may be, often convinces performers that the ritual situation *is* indeed informed with powers both transcendental and immanent.

—Victor Turner, "Social Dramas and Stories about Them"

The computer-generated interactive virtual environment of cyberspace has recently engaged the creative imaginations of a narrow spectrum of government, corporate, and academic researchers from various disciplines, as well as an assortment of other individuals including artists and science fiction writers. It has especially gained notoriety among cultural literati through the writings of William Gibson (Gibson 1984, 1987a, 1987b, 1988; cf. Greenfield 1988; Hamburg 1989), whose work has been most closely identified with a new science fiction genre and subculture of popular music known collectively under the rubric *cyberpunk* (cf. Sterling 1988, Dery 1989). Popular culture is in fact a principal ingredient of the synergetic technovisionary world of "cyberpunk literature" to the extent that one finds that *Neuromancer* (1984), "surely the quintessential cyberpunk novel" (Sterling 1988: xiv), was not considered by its author to be a book but rather "a pop artifact" (Gibson quoted in Dery 1989: 77).

Although cyberspace has been popularized by Gibson's books, it is neither a pure "pop" phenomenon nor a simple technological artifact, but rather a powerful, collective, mnemonic technology that promises to have an important, if not revolutionary, impact on the future compositions of human identities and cultures (cf. Tomas 1989). Gibson, in particular, has devoted considerable attention to the chilling socioeconomic implications of this space and its postindustrial context. His depiction of an information society whose governing economy is transnational and cyberspatial is sobering and worthy of consideration, especially since he has chosen to depict it, as Sterling has pointed out, "from the belly up, as it is lived, not merely as dry speculation" (1986: xi). In fact, Gibson has presented us with the most sophisticated and detailed "anthropological" vision of cyberspace to date: its social and economic facets, and the outlines of its advanced postindustrial form. Although this is not the place to engage the totality of Gibson's dystopic vision, it is important to note that his novels and short stories are noteworthy social documents because they "show," as Sterling notes, "with exaggerated clarity, the hidden bulk of an iceberg of social change" that "now glides with sinister majesty across the surface of the late twentieth century" (1986: xi). In the following pages, I examine the anthropological implications of Gibsonian cyberspace and, in particular, its connections to a panhuman ritual process known as *rites de passage*.

Why choose to treat cyberspace from an anthropological, indeed ritual, perspective? First, as Sterling (cf. 1988: xi) has pointed out, science fiction is an important tool that allows us to make sense of a rapidly emerging postindustrial culture. It is a *spatial operator* (Serres 1983a: 42–43) connecting pasts and futures by way of the present (a function that it shares, as we shall see, with a cyberspace *rite de passage*). Second, it allows us to make sense of an advanced information technology that has the potential to not only change the economic structure of human societies but also overthrow the sensorial and organic architecture of the human body, this by disembodying and reformatting its sensorium in powerful, computer-generated, digitalized spaces. This latter consequence of cyberspace will force anthropologists to take into account not only the jaded and contested question of the organic Other, which in Michel Serres's opinion (1983b: 67), "is only a variety or variation—of the Same," but also the postorganic, the classical (hardware-interfaced) *cyborg* and the postclassical, (software-interfaced) transorganic data-based cyborg or personality construct (Tomas 1989). This latter point is, as we shall see, intimately connected to the social and symbolic functions of traditional *rites of passage* rituals. Third, advanced digital technologies, such as those that generate cyberspace, hold considerable promise as a testing ground for postritual theories and practices, in particular as conceptualized by a critical postindustrial, postorganic anthropology.

Can cyberspace be considered to be a new social space? According to Gibson's description and contemporary work on virtual worlds technologies,[1] it does indeed hold the promise of new spatial configurations and related postorganic life forms. There is little doubt, therefore, that the composition of cyberspace must become the focus of intense speculation and the site of contested engagement in the immediate future, especially since corporations as well as research and development companies, university departments, and individual researchers are currently engaged in discussing and engineering its specifications. Anthropologists also have a vested interest in engaging virtual worlds technologies during this early stage of speculation and development, especially those who are interested in engaging advanced forms of Western technology from the points of view of their modes of social production. For there is reason to believe that these technologies might constitute the central phase in a postindustrial "rite of passage" between organically human and cyberpsychically digital life-forms as reconfigured through computer software systems. Should this prove to be correct, existing theories of ritual processes can provide important insights into the socially engineered cultural dimensions of cyberspace, and its social function as presently conceived. On the other hand, one can imagine that anthropological theories will not remain unaffected by their contact with advanced information systems of the caliber of cyberspace. One might envisage, for example, the outlines of another postorganic form of anthropology developing in the context of cyberspace, an anthropology specifically engaged in addressing the problems of engineering cyberspatial forms of intelligence as opposed to the more conventional humanistic, more or less reflexive, *study* of premodernist, modernist, or postmodernist humankind.

From Euclidean Space to Cyberspace

> Walls of shadow, walls of ice.
>
> —William Gibson, *Burning Chrome*

The identity of a historical culture is, according to Serres, the product of "very precise and particular connections" between "spatial varieties." "This construction," this "original intersection" is "that culture's very history." Cultures are therefore to be "differentiated by the form of the set of junctions, its appearance, its place, as well as by its changes of state, its fluctuations." "But what they have in common," Serres points out, "and what constitutes them as such is the operation itself of joining, of connecting." As a consequence, "the identity of a culture is to be read on a map, its identification card: this is the map of its homeomorphisms" (Serres 1983a: 45).

Serres argues, moreover, that an individual is also a continuously constructed product of a similar intersection or junction of social spaces and cannot therefore be considered to exist in a unified homogeneous space. The body, in fact, "works in Euclidean space, but it only works there. It sees in a projective space; it touches, caresses, and feels in a topological space; it suffers in another; hears and communicates in a third; and so forth, as far as one wishes to go" (Serres 1983a: 44). However, Serres cautions that

this [intersection] is not simply given or is not *always already* there, as the saying goes. This intersection, these junctions, always need to be constructed. And in general whoever is unsuccessful in this undertaking is considered sick. His body explodes from the disconnection of spaces. (1983a: 44)

What Serres is addressing is the way both cultures and individuals are constituted in terms of junctures or multiplicities of more or less fluid social spaces that must always be understood as particular historical, indeed cultural, constructs. One such space is Euclidean space, the master space of Western "work-oriented cultures." It is a master space because it governs communication—"the space of measure and transport," "linked homogeneity," and "congruent identity" (Serres 1983a: 44, 52)—and also "because it is the space of work—of the mason, the surveyor, or the architect" (Serres 1983a: 44). It is the space of Western geometry: the geometry of vision, the road, the building, and the machine. On the other hand, this master space is a binary construct, consisting of the everyday social or profane spaces that Serres has initially described in Euclidean terms *and* sacred or "liminal" spaces. As Victor Turner notes, "for every major social formation there is a dominant mode of public liminality, the subjunctive space/time that is the counterstroke to its pragmatic indicative texture" (1977c: 34). Film, in his opinion, is the dominant form of public liminality in electronically advanced societies.

In recent years, there has emerged a new form of electronic space that holds revolutionary promise as *the* fin de siècle metasocial postindustrial work space. Variously described as a "space that wasn't space," a "nonplace," and a space in which "there are no shadows" (Gibson 1987a: 38, 166; Gibson 1988: 219), cyberspace is a postindustrial work environment predicated on a new hardwired communications interface that provides a direct and total sensorial *access* to a parallel world of potential work spaces. This interface, which is a world removed from the indirect and limited access provided by older print-based paradigms of visual literacy (Gibson 1984: 170), mediates between the sensorial world of the organically human and a parallel virtual world of pure digitalized information. Thus, in the words of one of Gibson's characters (1988: 13):

People jacked in so they could hustle. Put the trodes on and they were out there, all the data in the world stacked up like one big neon city, so you could cruise around and have a kind of grip on it, visually anyway, because if you didn't, it was too complicated, trying to find your way to a particular piece of data you needed.

At its most extreme, the corporeality of "laboring" human bodies is replaced by pure information whose configurations "signify" disembodied human sensoria, personality constructs, and artificial intelligences.

The "abstract representation of the relationships between data systems" (Gibson 1987b: 169) in cyberspace is, however, highly plastic, and can take any form ranging from pure geometric color-coded copyrighted shapes or architectural representations (cf. Gibson 1988: 64, Gibson 1987b: 178, Gibson 1984: 256–257) signifying corporate ownership to "photo-realistic" illusions (cf. Gibson 1988: 148–149, 174, 221–224). *Such sites are the essence of a postindustrial society*—pure information *duplicated in metasocial form*: a global information economy articulated as a metropolis of bright data constructs, whose plasticity is governed by a Euclidean model based on the given problematic of visualizing data, a problematic subordinated, in Gibsonian cyberspace, to the dictates of a transnational computer-based economy. Although predominantly conceived as a virtual Euclidean work space for legitimate computer programmers to visualize their employer's data (cf. Gibson 1987b: 169–170, Gibson 1988: 13), cyberspace does, in principal, also allow a *direct* hardwired experience of other non-Euclidean spaces and spatialized consciousnesses to occur. The latter includes virus programs that have "replicated" and become smart (Gibson 1987a: 169) and artificial intelligences that have permeated its digitalized expanse and have finally attained "cosmic sentience" (Gibson 1988: 259). Gibsonian cyberspace is, as a consequence, digitally and socially Durkheimian in the sense that it is both profane (a metropolis of data) and sacred (a cybernetic godhead) (Gibson 1988: 192).

Gibsonian cyberspace can be distinguished according to three, dominant, "Euclidean" characteristics. First, it is conceived as a common transnational work environment. Second, it is a transportation space designed to accomplish work related tasks—both a space in which one can travel in real-time or by way of "bodiless, instantaneous shifts" (Gibson 1988: 220) *and* a space *through* which human memory and identity are transported globally. Third, it redefines and restructures *what it means to be human* in technoeconomic terms through a data-based collectivization of the human sensorium (cf. Tomas 1989) or, in the more radical customized and therefore "individualistic" terms, of "personality" or synthetic data constructs. This potential for a radical, postcorporeal, economic transubstantiation of the human body's traditional organic and sensorial architecture forcefully remind one of Serres's comments on the plasticity of interconnections between particular cultures and related human bodies. Cyberspace's dominant Euclidean form confirms Serres's observations on the Euclidean articulation of work and space in general in Western cultures. It raises, moreover, the critical question of the governing "logic" and social, cultural, and economic control of this new sensorial and mnemonic space.

In order to further clarify these issues and to highlight their importance for future cyberspace research one must attend to other, *post-Euclidean, social and cultural alternatives that are germane to cyberspace. These are best addressed from the point of view of parallels that can be drawn between the postindustrial, cyberspatial transubstantiation of the human body and related social and symbolic transformations of the human body in traditional *rites de passage* rituals. First described in the cases of so-called, small-scale, preindustrial, pre–Industrial Revolution societies, these rituals draw attention to the role of "the social" as a mediative force in the many fundamental biological, cultural, and technological changes that have been experienced by humankind.

Rites de Passage

> She slid the trodes on over the orange silk headscarf and smoothed the contacts
> against her forehead.
>   "Let's go," she said.
>   Now and ever was, fast forward, Jammer's deck jacked up so high above the
> neon hotcores, a topography of data he didn't know. Big stuff, mountain-high,
> sharp and corporate in the nonplace that was cyberspace.
>   —William Gibson, *Count Zero*

In 1909, Arnold Van Gennep published *Les Rites de Passage*, an examination of a class of rituals that mark major social stages in an individual's life or in the collective existence of a group. Van Gennep's study was the first to focus on this important class of ritual and note its tripartite structure.

Rites of passage are ideally distinguished by three successive phases in ritual or nonsecular space and time that function to engineer the transposition and transformation of an initiand (or initiands in the case of a social group) from one socioenvironmental, sociobiological, social position or stage to another. Notable events that are mediated by this ritual process include birth, puberty, marriage, and death. However, other more general transpositions from one significant social category, spatial area, cosmic, or seasonal phase to another are also subject to ritual, rites of passage, mediation. The sequence of rites, in such cases, is designed to acknowledge the importance and abnormality of the events while negotiating the movement between stages by integrating them into a succession of socially sanctioned activities while ensuring a certain spatial, temporal, and symbolic continuity.

The first phase of rites of passage, or rites of separation, consists of symbolic behavior that initiates the transition from the secular and profane world of the social group to a second phase known as a *liminal* period. This phase is marked by a time of ritual metamorphosis that mediates between the initiand's or group's previous and subsequent state. It is an especially dangerous and unclean period, since the initiand (or group of initiands) is considered to possess little of her/his (their) former or later attributes. This period is not only symbolically abnormal, but the location(s) and time in which the transformation takes place is considered outside of society. Turner (1977a) has described it as "betwixt and between" neither one nor the other—a state of nonbeing, death, or nothingness. He has argued, moreover, that "the liminal phase is the essential, *anti*secular component in true ritual, whether it be labeled 'religious' or 'magical'" (Turner 1980: 161). When the desired sociobiological or symbolic transformation has taken place, a procedure of renormalization is needed to reinstate the individual or group into society. The third phase, or rite of aggregation, functions as a "decontamination chamber," metaphorically speaking, in order that initiands can enter into a new social or sociobiological status in society without polluting everyday space and time with vestiges of former attributes or residues of the ritual process itself.

If rites of passage clearly *articulate* human bodies and identities from one social position to another, they do so because they function as structured conduits between normally distinct social realities. However, as Van Gennep originally pointed out,

Although a complete scheme of rites of passage theoretically includes preliminal rites (rites of separation), liminal rites (rites of transition), and postliminal rites (rites of incorporation), in specific instances these three types are not always equally important or equally elaborated. (1960: 11)

This observation was later confirmed by Turner's brilliant examination and reconceptualization of the social and symbolic functions of the central, or liminal, phase of rites of passage (1974, 1977a, 1977b, 1977c).

Reflexive transformation is the governing function of the liminal, interstitial, and antistructural phase of the rites of passage in "small-scale, relatively stable and cyclical" tribal societies. In these societies, the liminal functions in relation to a "structure of positions" (Turner 1977a: 93). Liminality is, within this patchwork of stability, an interstructural space and time that always functions in relation to the structured, between which the "passenger," to use Turner's term (1977a: 94), is symbolically as well as spatially and temporally articulated.

Transformation is signified and rendered effective by the unusual and "bizarre" symbolism found in the liminal phase (Turner 1977a: 96). This symbolism, associated with bodily and biological processes, marks the "ambiguous" and "paradoxical" condition that the subject or initiand is in. It is an extraclassifactory stage in which the initiand is "no longer classified and not yet classified" (Turner 1977a: 96). From the former point of view, liminality is defined by symbols of death, invisibility, structural dissolution, and other negative physical and organic processes. From the latter point of view, the condition is defined by symbols of gestation and parturition. Insofar as the liminal phase encompasses the destructuring and restructuring of the subject, it is also marked by the denial of structure and the origin of structure per se. Turner has pointed out, in connection with its creative aspect, that liminality is "a realm of pure possibility whence novel configurations of ideas and relations may arise" (1977a: 97). This creativity is well illustrated in his discussion of the communication of "sacra" to the initiand in the form of "exhibitions," "actions," and "instructions":

> The communication of *sacra* and other forms of esoteric instruction really involves three processes, though these should not be regarded as in series but as in parallel. The first is the reduction of culture into recognized components or factors; the second is their recombination in fantastic or monstrous patterns and shapes; and the third is their recombination in ways that make sense with regard to the new state and status that the neophytes will enter. (Turner 1977a: 106)

The result is "a promiscuous intermingling and juxtaposing of the categories of event, experience, and knowledge, with a pedagogic intention" (Turner 1977a: 106).

One begins to have a sense, in the case of small-scale societies, of the creative possibilities—of social and individual regeneration—inhering in this ambiguous phase of the rites of passage, a phase in which "the central cluster of nonlogical *sacra* is then the symbolic template of the whole system of beliefs and values in a given culture, its archetypal paradigm and ultimate measure" (Turner 1977a: 108). Turner describes this as a process in which "to look at itself a society must cut out a piece of itself for inspection." He continues,

> To do this it must set up a frame within which images and symbols of what has been sectioned off can be scutinized, assessed, and, if need be, remodeled and rearranged. (Turner 1977c: 35)

This phase operates like a mythic cultural weaver whose function is, according to Serres,

> to link, to tie, to open bridges, pathways, wells, or relays among radically different spaces; to say (*dire*) what takes place between them; to inter-dict (*inter-dire*). (1983a: 45)

Moreover, as Serres points out in an observation that could have been made in regard to liminality in general, "the category of *between* is fundamental in topology and for our purposes here: to interdict in the rupture and cracks between varieties completely enclosed upon themselves" (1983a: 45). Thus, if the social is never, according to both Serres and Turner, a completely stable construct but must always be considered to be in a process of creative regeneration, then rites of passage offer exemplary instances of a social process that achieves unstable stability.

There are a number of similarities between the overall structure of rites of passages and cyberspace that suggest that the latter might be closely related to the former in a functional sense. First, the acts of "jacking in" and out of cyberspace by way of cyberdecks and matrix simulators suggest radically truncated versions of separation and aggregation "rites" in which the hardware serves as portal to, and exit from, a parallel virtual reality. The full emphasis of this "access" technology is on cyberspace itself, or on a central liminal—virtual—condition. Moreover, access and exit are *sequential*,[2] but *not* necessarily unidirectional (cf. Turner 1980: 160–161) in the sense that programmers and console cowboys can jack in and out of cyberspace at will. At its most extreme, however, as in the case of Bobby Newmark in *Mona Lisa Overdrive*, the technology can function as a permanent prosthesis or hardwired interface (Turner 1988: 240). Second, jacking into cyberspace involves a passage from the everyday space and finite time of the organically human or postorganic hardware-based cyborg to a digital— as opposed to an analogical—space and time that is both transorganic and cyberpsychically collective. It is collective in the sense that cyberspace exists, in the words of Continuity (an artificial intelligence in *Mona Lisa Overdrive*), "insofar as it can be said to exist, by virtue of human agency" (Turner 1988: 107). But the nature and form of the "human" portion of "human agency" is now in question. Finally, the nature of human identity at both an individual and collective level is rendered problematic in terms of agency, for cyberspace itself has the potential to become consensually posthuman as well as "post-present," as when its matrix is inhabited by what appear to human protagonists to be transcendent panhuman "intelligences" whose power and creative inventiveness is signaled by intracyberspatial influence or by way of the extracyberspatial production of cultural sacra (cf. Gibson 1988; Gibson 1987a: 217, 225–227).[3] The "mytho-logic" of these "presences" is explained by Continuity in an exchange with a human, Angela Mitchell (Gibson 1988: 107):

"The folklore of console jockeys, Continuity. What do you know about that?"

"What would you like to know, Angie?" . . .

"'When It changed'. . . "

"The mythform is usually encountered in one of two modes. One mode assumes that the cyberspace matrix is inhabited, or perhaps visited, by entities whose characteristics correspond with the primary mythform of a 'hidden people.' The other involves assumptions of omniscience, omnipotence, and incomprehensibility on the part of the matrix itself."

"That the matrix is God?"

"In a manner of speaking, although it would be more accurate, in terms of the mythform, to say that the matrix *has* a God, since this being's omniscience and omnipotence are assumed to be limited to the matrix."

"If it has limits, it isn't omnipotent."

"Exactly. Notice that the mythform doesn't credit the being with immortality, as would ordinarily be the case in belief systems positing a supreme being, at least in terms of your particular culture. Cyberspace exists, insofar as it can be said to exist, by virtue of human agency."

This mytho-logic suggests that one of cyberspace's more fundamental social functions is to serve as a medium to communicate a form of *"gnosis,* mystical knowledge about the nature of things and how they came to be what they are" (Turner 1977a: 107). In cyberspace, the classical hardware interfaced cyborg and the postclassical data-based cyborg or personality construct meet with new posthuman intelligences that engage in revelatory and pedagogic activities reminiscent of the activities of shaministic figures who mediate between traditional sacred and profane worlds. Such mediations between human and post-human, analogue and digital spaces, suggest that cyberspace must be understood not only in narrowly socioeconomic terms, or in terms of a conventional parallel culture, but also and more importantly as an inherently original and inventive metasocial operator *and* potential creative cybernetic godhead.

On the Metasocial in Postindustrial Society

"Yeah," the Finn said, turning out onto the long straight empty highway, "but nobody's talkin' *human,* see?"

—William Gibson, *Mona Lisa Overdrive*

Turner has proposed a distinction between "liminal" and "liminoid," or "ergic-ludic ritual liminality and anergic-ludic liminoid genres of action" (1974: 83) that acknowledges an important, indeed decisive, difference between types of society called "tribal" and "post-tribal" or pre– and post–Industrial Revolution societies. This distinction is especially expressed in the opposition between work and leisure.

It would seem that with industrialization, urbanization, spreading literacy, labor migration, specialization, professionalization, bureaucracy, the division of the leisure sphere from the work sphere, the former integrity of the orchestrated religious gestalt that once constituted ritual has burst open and many specialized performative genres have been born from the death of that mighty *opus deorum hominumque*. . . . Disintegration has been accompanied by secularization. Traditional religions, their rituals denuded of much of their former symbolic wealth and meaning, hence their transformative capacity, persist in the leisure sphere but have not adapted well to modernity. Modernity means the exaltation of the indicative mood; but in what Ihab Hassan has called the "postmodern turn" we may be seeing a return to subjunctivity and a rediscovery of cultural transformative modes. . . . Dismembering may be a prelude to remembering, which is not merely restoring some past intact but setting it in living relationship to the present. (1980: 166–167)

The transition from the liminal to the liminoid is broadly marked by a transition from the ritual collective to the secular individual, individualistic, or individualizing. While the ritual collective is "centrally integrated into the total social process, forming with all its other aspects a complete whole," the secular individual develops "most characteristically *outside* the central economic and political processes, along their *margins*, on their *interfaces*, in their 'tacit dimensions'" (Turner 1977b: 44). Liminal behavior is constrained by obligations associated with status, while liminoid behavior is characterized by relative contractual freedom (Turner 1974: 84; Turner 1977c: 50). The "liminal" is also distinguished by wide social and cultural affect in contrast to the more idiosyncratic orientations of the "liminoid." Finally, the liminal is more prone to "the inversion or reversal of secular, mundane reality and social structure" because of its overtly socially integrative functions, while liminoid phenomena, on the other hand, "are not merely reversive, they are often subversive, representing radical critiques of the central structures and proposing utopian alternative models" (Turner 1977b: 45). In conclusion, Turner notes that

liminal genres put much stress on social frames, plural reflexivity, and mass flow, shared flow, while liminoid genres emphasize idiosyncratic framing, individual reflexivity, subjective flow, and see the social as problem not datum. (1977c: 52)

There are, however, connections between the two that suggest that they are not exclusive products of each type of society, that "in complex societies today's liminoid is yesterday's liminal" (1977b: 46). In this case, it might be true that the dismembering aspect of the liminal is continued in its fragmentation into different genres of social activity. As Turner points out:

In complex modern societies both types coexist in a sort of cultural pluralism. . . . The liminoid is more like a commodity indeed, often *is* a commodity, which one selects and pays for—than the liminal, which elicits loyalty and is bound up with one's membership or desired membership in some highly corporate group. One *works at* the liminal, one *plays with* the liminoid. (1974: 86)

While this is not the place to enter into details concerning Turner's important distinction between liminal and liminoid, one might nevertheless query the uncritical celebration of a return to subjunctivity, with its implied displacement of a politics of the present in terms of utopian potentialities. For example, our experiences of "tribal" rites of passages have taught us that "voluntary *sparagmos* or self-dismemberment of order, in the subjunctive depths of liminality" (Turner 1980: 164), is achieved in relation to a set of actual positional states. This is also the case with "post-tribal" cyberspace. Not only does it have, in its Gibsonian version, a determinate Euclidean form and particular economic functions that are linked to actual socioeconomic formations (the military/industrial complex), but its principal mode of access and exit is by way of sophisticated communications equipment (cyberdecks or matrix simulators), the use of which is predicated on privileged or differential institutional access and specialized hardware and software expertise. In the case of legitimate programmers this access is clearly defined in terms of corporate filiation. Thus, a dominant Western spatial paradigm, corporate capitalism, differential access to technology, and specialized knowledge and expertise imply, at the very least, a more complex set of relations between so-called postmodern cultural phenomena—for cyberspace can also become, as we have seen, a culturally creative arena—and a given postindustrial economic context.

However, the distinction between liminal and liminoid, the collective and individualistic, does sharpen the focus on who will control cyberspace's governing socioeconomic logic—governments, corporations, or free-lance entrepreneurs? It does so because the issue of collective vs. individual control, which is not as clearly demarcated in the context of traditional rites-of-passage rituals, is effectively foregrounded in the contractual, marginal, subversive, and individualistic realms of the liminoid. This question can be extended to the design and operation of alternative hardware/software artifacts (the latter extending to the design of personality constructs and artificial intelligences) that might be used to mask or foreground cyberspace's governing logic and minimize or maximize social and cultural diversity. Thus, although problematic, Turner's attempt to account for pre- and posttribal ritual forms is valuable in that it serves as a corrective to uncritical trans- or ahistorical accounts of ritual processes; and the distinctions that he makes between liminal and liminoid allow for a more refined understanding of cyberspace's potential postindustrial forms and functions. At least, with this refinement and corrective in mind, we have an initial framework for addressing the complex issue of control in posttribal rituals of the caliber of cyberspace.

Gibsonian cyberspace exhibits liminoid characteristics connected to its economic functions as articulated in a complex open-ended postindustrial society. First, it exhibits a specific logic that structures the matrix in the name of an transnational information economy. This logic, replicating the binary (0-1) logic of computer languages, is visibly articulated in the matrix, sculpted in the copyrighted forms of data that structure, in turn, social activity. Walls of data, rather then walls of brick and glass, divide a hardwired, or postorganic, humanity into economic protagonists—those included and those excluded from, say, the dominant military/industrial complex. Those, in other words, that do and those that do not have direct access to information, hardware technology and software expertise. The latter include the "industrial-espionage artists and hustlers," computer or console cowboys like Bobby Quine in Gibson's short story "Burning Chrome":

Bobby was a cracksman, a burglar, casing mankind's extended electronic nervous system, rustling data and credit in the crowded matrix, monochrome nonspace where the only stars are dense concentrations of information, and high above it all burn corporate galaxies and the cold spiral arms of military systems. (1987b: 170)

Cyberspace's binary structure governs the activities of anonymous corporate programmers, and individualistic console cowboys or jockeys who are always trying to gain an edge on liminal corporate artificial intelligences who are in the business of continually upgrading corporate ICE (Intrusion Countermeasures Electronics), but who periodically mutate into independent creative entities. This structure, as expressed in the walls of data, is therefore, on the one hand, literally considered as datum, while on the other hand it is considered as problem. Second, although primarily a work space, and thus not a space of "play" per se, cyberspace has the potential to become inventively and playfully anti-structural in the sense proposed by Turner:

Anti-structure . . . can generate and store a plurality of alternative models of living, from utopias to programs, which are capable of influencing the behavior of those in mainstream social and political roles (whether authoritative or dependent, in control or rebelling against it) in the direction of radical change. (Turner 1974: 65)

Bearing in mind the often complex relationship between the structuring functions of hardware and software systems and their potential for creative flexibility, one can nevertheless note that cyberspace's anti-structural potential is best exemplified in the creative activities of console cowboys, in particular Bobby Newmark, who have transcended the economic dualism of the matrix in order to contemplate its creative logic, a logic rendered intelligible through the machinations of artificial intelligences of the caliber of Wintermute and Neuromancer. The activities of individuals such as Newmark and panhuman intelligences highlight cyberspace's inventive potential and its ability to sustain collective *and* individual activity, work *and* play, normative *and* spontaneous behavior in a manner that might directly and strategically challenge its economic logic. Thus, there is evidence to suggest that cyberspace might not only be a paradigmatic, postindustrial liminal/liminoid space—an "eye and eyestalk which society bends round upon its own condition, whether healthy or unsound" (Turner 1977c: 40)—but also a possible generative site for other creative logics and sensorial regimes.

Conclusion: On a Postorganic Anthropology

The world before television equates with the world before the Net—the mass culture and mechanisms of Information. And we are *of* the Net; to recall another mode of being is to admit to having once been something other than human.

—William Gibson, "Rocket Radio"

Serres's conception of the social is radically modernist in the sense that it takes into account the fracture and fluidity of cultural and individual identities whose junctions are organic human bodies, while positing the social realm as a topological construct governed by a Euclidean master space and inhabited by the mythic figure of the weaver. The weaver is, however, a *holographic* mytho-logical form to be found wherever information is being produced, transmitted, received. Its mytho-logic is, therefore, eminently cyberspatial, in the sense that its communicative function is digitally replicated throughout cyberspace. Serres's weaver is concentrated in particular in the prescriptive roles of artificial intelligences such as Wintermute, Neuromancer, Continuity, and Colin—entities that ensure that cyberspace retains its potential as a generative locale for other, more unusual, spatial configurations and cyborg intelligences.

Gibson's powerful vision is now beginning to influence the way virtual reality and cyberspace researchers are structuring their research agendas and problematics.[4] But if cyberspace represents, at the very least, the birth of a new postindustrial, metasocial spatial operator, it will remain for most part stillborn if its parameters are engineered primarily to function, following Gibson's dystopic vision, as a virtual world of contestatory economic activity. In order to counter this vision, one must actively and strategically seek alternative spatial and creative logics, social and cultural configurations. If such creative flexibility is critically foregrounded in current research agendas, cyberspace will indeed become a site of considerable cultural promise, and a locale for a new postorganic anthropology.

Acknowledgments

I would like to thank Jody Berland for comments that have contributed to the clarity of the argument presented in this chapter.

Notes

1. Cf. The First Conference on Cyberspace held May 4 and 5, 1990, at the University of Texas at Austin.

2. Turner draws attention to the definitive importance of "performative *sequencing*" for any understanding of ritual because of its role in ensuring a ritual's unidirectional transformative power (1980: 160–161).

3. In Turner's words, ritual is "prescribed formal behavior for occasions not given over to technological routine, having reference to beliefs in invisible beings or powers regarded as first and final causes of all effects" (1980: 159).

4. Cf. Tim McFadden's "The Structure of Cyberspace and the Ballistic Actors Model—an Extended Abstract" presented at The First Conference on Cyberspace, University of Texas at Austin.

References

Ballard, J. G. 1984 [1974]. Introduction to the French Edition. *Crash.* New York: Vintage Books.

Dery, Mark. 1989. Cyberpunk: Riding the Shockwave with the Toxic Underground. *Keyboard* 15-5: 74–89.

Gibson, William. 1984. *Neuromancer.* New York: Ace Books.

1987a. *Count Zero.* New York: Ace Books.

1987b. *Burning Chrome.* New York: Ace Books.

1988. *Mona Lisa Overdrive.* Toronto, New York: Bantam Books.

1989. Rocket Radio. *Rolling Stone* 554, 85–87.

Greenfield, Adam. 1988. New romancer. *Spin* 4-9, 96–99, 119. (Interview with William Gibson).

Hamburg, Victoria 1989. The King of Cyberpunk Talks to Victoria Hamburg. *Interview* 19-1, 84–86, 91.

Serres, Michel. 1983a [1977]. Language and Space: From Oedipus to Zola. In *Hermes: Literature, Science, Philosophy.* Josue V. Harai and David F. Bell (eds.), 39–53. Baltimore: The John Hopkins University Press.

1983b [1968]. Platonic Dialogue. In *Hermes: Literature, Science, Philosophy.* Josue V. Harai and David F. Bell (eds.), 65–70.

Sterling, Bruce. 1986. Preface. In William Gibson, *Burning Chrome,* ix–xii. New York: Ace Books.

1988. Preface. *Mirrorshades,* Bruce Sterling (ed.), New York: Ace Books, ix–xvi.

Tomas, David. 1989. The Technophilic Body: On Technicity in William Gibson's Cyborg Culture. *New Formations* 8, 113–129.

Turner, Victor. 1974. Liminal to Liminoid, in Play, Flow, and Ritual: An Essay in Comparative Symbology. *Rice University Studies* 60-4, 53–92.

1977a [1964]. Betwixt and Between: the Liminal Period in *Rites de Passage.* In V. Turner, *The Forest of Symbols,* 93–111. Ithaca: Cornell University Press.

1977b. Variations on a Theme of Liminality. In *Secular Ritual.* Sally F. Moore and Barbara G. Myerhoff (eds.), 36–52. Amsterdam: Van Gorcum, Assen.

1977c. Frame, Flow and Reflection: Ritual and Drama as Public Liminality. In *Performance in Postmodern Culture.* Michel Benamou and Charles Caramello (eds.), 33–55. Madison, Wisconsin: Coda Press.

1980. Social Dramas and Stories about Them. *Critical Inquiry* 7-1, 141–168.

Van Gennep, Arnold. 1960 [1909]. *The Rites of Passage.* Trans. M. B. Vizedom and G. L. Caffee. Chicago: The University of Chicago Press.

4   *Mind Is a Leaking Rainbow*

Nicole Stenger

The French poet Henri Michaux, in *Connaissance par les Gouffres* (1988), has left us descriptions of his (provoked) experiences with hallucination, experiences that have, strangely enough, found their exact replicas in the generating patterns of computer animation. With a genuine sense of wonder, an unrestricted open-mindedness, and a methodical effort to translate his visions into adequate poetic language, Michaux has singled out alterations of vision that strikingly resemble the properties of imaging software: space made of innumerable points, shapes borne of those points in motion, of grids as thin as spiderwebs, avalanches of brilliance, garish colors, hyper reliefs (bumps, granulations, fibrosity), patterns endlessly recurring inside patterns, coalescence of shapes, disintegrations, etc . . . He also has gone into more detailed descriptions: of minarets, of pipelines rushing with wild animals' heads, rhythmic faces, and objects mapped on planes in 3-D, images that are today all familiar scenes to the computer specialist and common special effects on TV.

Those idealists (among whom I count myself) who were shown the way by *Sunstone*, the 1979 3-D computer film by Ed Emschwiller, and who decided to follow the light, had sensed from the beginning that the medium of computer animation was no mere image generation in the traditional sense, but rather a virtual world, populated by half-living entities, that we would inhabit someday when the technology would allow it. Like them, I felt that this hallucination behind a screen was just the first stage in a development, a rehearsal for a D day when this substance would finally escape and invade what we call reality. Emschwiller was our Melies, our Kertesz; he had revealed a state of grace to us, tapped a wavelength where image, music, language, and love were pulsing in one harmony. Inside the enclosed area of the computer, something was calling us, inviting us. If Jean Baudrillard is right when he says that "seduction represents mastery over the symbolic universe" (1990: 8), then by building his architecture of numbers, of computer language, and of visual symbols (which is what 3-D animation is), Emschwiller had found one of the secrets of seduction. How could we resist?

As in the film *Encounters of the Third Kind* we were shaping our 3-D mountain on our computers, at night, and in the most unpoetic environments. We were getting ready for what would become cyberspace, a wavelength of well-being where we would encounter the second half of ourselves.

Our minds were softly leaking rainbows of colored imagination, soon to be joined by innumerable rainbows that would embrace the earth and change the climate of the human psyche. Perception would change, and with it, the sense of reality, of time, of life and death. We would, as Michaux puts it, "enter the world of Fluids," it would be "over with the solid, over with the continuous and with the calm," (1967: 187), some dance quality would invade everything, and Cartesian philosophers would go through a trance, floating on history like chops on gravy. "We must never tread on the other side of the Real, on the side of the exact resemblance of the world with itself, of the subject with itself," says Baudrillard (1981: 158). And as we would find a common thread running through cyberspace, dream, hallucination, and mysticism (which may prove that images generated by light impulses, be they technical or biological, can only follow certain patterns), the imaginary might become a reasonable subject of science. Some questions would then literally come into light: Is there a hardware or a software for it?

As Noam Chomsky believes of language, is there a universal grammar, with a finite number of parameters for 3-D shapes, brilliance, transparency, texture, motion, rhythm, and so on? Are archetypes inherited condensations of meaning? Do they serve as raw models for the constructions of imagination? What are the characteristics of the idealization process? of " realism"? A change in the "resolution," in the "lighting"? Are there sexual differences? Do they reflect what we already know of the differences in the sensory apparatus: perception of sounds, sensitivity to light, vision of colors, pheromonal organization?

The imaginary so far contained in skulls, in bound books, in frames would expand its territory and reach equal status with abstract thinking, as a tool for knowledge. Artificial imagination programs would be written, and in the hyperreality of tomorrow's design, kitchens would dream that they are living rooms.

Descartes would say that "the senses cheat us." Now the senses are back as the only reasonable means of information when the acceleration of modern warfare makes the time needed for abstract analysis obsolete. Tom Furness invents the "Super Cockpit" to redistribute information according to the parallel tracks of vision, hearing, and touch, when the pilot faces a paralysis of decision in front of the maze of graphics and commands. At the speed of war, Descartes' world becomes brittle. Cyberspace will shatter it like a mirror.

It was obvious from the very beginning that the acceleration of the computing power in the space of simulation, which is also a space of perception, would lead us to a threshold. The threshold of disappearance for Paul Virilio, of "surfusion" for Baudrillard, of the sublime for Julia Kristeva: "The sublime object dissolves in the raptures of a bottomless memory. It is such a memory which [...] transfers that object to the refulgent point of the dazzlement in which I stray to be" (Kristeva 1982: 12). It was obvious that an explosion would take place, a Big Bang of the old order that was based on gravity, history, and territory. A whole civilization would become unbalanced. Without exaggeration, cyberspace can be seen as the new bomb, a pacific blaze that will project the imprint of our disembodied selves on the walls of eternity. No wonder some computer labs that heralded this new culture at the beginning of the eighties, shocked by the flash of lightning of their own discoveries have since gone underground to digging technical galleries. We had been rushing all along, computing at full speed to cross the Wall of Inertia and reappear as virtual puppets further along the way. Speed was no longer an improvement for transportation but, as Virilio says, a better "way of seeing, of conceiving the reality of facts" (1988: 151). But is there such a thing as the reality of facts?

Suddenly cyberspace grafts a new nature of reality in our everyday life. It opens up an infinity of space in an eternity of light. And one of the most striking features of this discontinuity with the real is the elasticity of the exchange. "He can't escape from that exterior that owes its life to him. A psychic chord unites them," says Michaux (1967: 194).

As those vessels of reality communicate, you feel that any level of commitment is possible: that you can answer the phone, drink your coffee and yawn, as you drag yourself as a grasshopper in the greens of cyberspace. Or that you can pull yourself entirely into the Blue Lagoon and get drowned. "The real can be turned inside out like a glove," says Virilio (1984: 148). On the other side of our data gloves, we become creatures of colored light in motion, pulsing with golden particles. Forget about Andy Warhol's petty promise of fame for fifteen minutes. We will all become angels, and for eternity! Highly unstable, hermaphrodite angels, unforgettable in terms of computer memory. In this cubic fortress of pixels that is cyberspace, we will be, as in dreams, everything: the Dragon, the Princess, and the Sword. "Dream is a second life," said Gerard de Nerval. "We cannot determine the exact moment when the self, under another form, goes on with the task of existence" (1918, 291). Now we can. Just press the key ENTER.

Isn't it exciting to live twice? To walk to a party with an accurate reconstruction of your entire body, flesh and bones, stored on a floppy disk in your pocket? Karl Hohne, a mathematician from Hamburg, introduced a perceptual revolution in 1988, when he released *Voxelman*, the videotape of his medical research. A navigation inside a virtual human being whose flesh, organs, bones, and veins were entirely reconstructed in three dimensions from medical views of a patient, using scanner and magnetic nuclear resonance techniques. Karl Hohne opened a bright future for the virtual exploration, animation, manipulation, and statistical projection of any human being, which will change forever the vision we have of ourselves. Voxelman was only the portrait of a sleepy stranger, but what if a high-resolution reconstruction of ourselves was aging in front of us? statistically developing a pathology under our very eyes? What if we met our own Voxelman in cyberspace? What if someone stole it? wore it?

Isn't it exciting also to experience in cyberspace the life of all creatures? to let them come one by one in their cyberspace outfits and be named by you? O the naivete of the first days of creation when you will set the clock for the independent animation of objects that will wink at each other, play tricks on each other, learn from each other. In this primeval garden where a synthetic sun will rise, inner voices will whisper, immaterial kisses hover in the air, and you will lie in the reconstructed sense of fur. For blind bards as for nearsighted whiz kids, cyberspace will feel like Paradise!

Of course don't expect to keep your old identity: one name, one country, one clock. For be it through medical reconstruction or through fantasy, multiplied versions of yourself are going to blossom up everywhere. Ideal, statistical, ironical. A springtime for schizophrenia! And to make it worse, as Randy Walser puts it, there won't be any possibility in cyberspace to distinguish between a dumb object, an "intelligent" one, and a human being. We will all be the same: half alive, half dead, ready to be stored and tagged forever in a Mormon mountain.

But we need not worry. What we call reality was only a temporary consensus anyway, a mere stage in the technique. "Reality is never a given," says Virilio, "it is acquired, generated by the development of societies" (1987: 19). The twenty-first century will give birth to the geysers of cyberspace, a sensory lava that will find its way through the cracks of that consensus. Computer hallucination will come in the foreground, as cinema verite and documentaries, and their trail of dogma about the real will recede in the distance. Godard said that in cinema he had found a continent where all the gestures of life would find their place. As the twentieth century is drifting away, so is the continent of cinema, and the soft matte of its 2-D images.

Where is cyberspace? Watching the poor cyberspace visitors throw themselves on the floor in search of an unattainable area of their virtual environment is enough in itself to understand that there is no easy answer to that question. Cyberspace is like Oz—it is, we get there, but it has no location.

Entering this realm of pure feelings is a decision to leave firm ground that may have more consequences than we think. Watching TV, after all, only commits us to being obese. In cyberspace we lose weight immediately; but there is a gravity of choice that is not yet fully understood. Certainly there will be a shifting from the sense of territory, of being an inhabitant of an earthly system of values that includes roots, walls, and possessions, toward a radical adventure that blasts it all. Baudrillard would probably think that cyberspace is the just future of "those peoples with no origins and no authenticity, who will know how to exploit that situation to the full," peoples who are already satellites of the Earth and who will know, like Japan, how "to transform the power of territoriality and feudalism into that of deterritoriality and weightlessness" (1989: 76). He would also think that it is the destiny of those individuals who, through multiple external prostheses of their inner functions, will become satellites of themselves and deep-fry their neurons in cyberspace for lack of a historical challenge, a war, for example, where real flesh bleeds and bones rot in pyramids. Poor souls that may get drowned in the liquid mirror of their minds and perhaps become females, since "surprisingly, this proposition that in the feminine, the very distinction between authenticity and artifice is without foundation, also defines the space of simulation" (Baudrillard 1990: 11). Cyberspace, one of the greatest challenges of Humanity for the century to come? Cyberspace, last frontier of all? Borne of Disneyland like luscious candy cotton. Atoll of grace between the West and the Orient. Soufflé of desires revolving in the light, whispering the names of the world's fiancés: Laure, Beatrice, Peter Pan, John Lennon.

Cyberspace the dessert of humanity!

Mr. Baudrillard, how can you sell your dry nuts on the shore when we have so much fun in the water! "Immersed in the world of fluids, of the psychic, of the magic, thrown away from his property, without property, with no remembrance of property, the only inhabitant on Earth of the Immaterial, he is dealing from now on with the Infinite," says Michaux (1967: 232). We had long ago lost all trace of earth on our shoes. Our houses were whistling with drafts and our families had slowly thinned away. Cyberspace, both open and closed, would be our last shelter. The hut of the global village: a few plastic bamboos held by a membrane, a pill in a box. We would celebrate in cyberspace, rocking and humming in televirtuality, inhabitants of a country that is nowhere, above the busy networks of money laundering. Over the rainbow.

We would celebrate all the more in cyberspace, for we would very likely be in one of those areas of reality that Mircea Eliade, in an interview before his death, said would soon be unsealed: areas of the Sacred, as it were. According to Eliade, what characterizes the experience of the Sacred, what sets the stage, is a discontinuity in the sense of space and time: "For religious man, space is not homogeneous, he experiences interruptions, breaks in it, some parts of space are qualitatively different from others" (1961: 20). That we should have chosen to desacralize the world through modern science does for Eliade not change anything. Man cannot get completely away with his religious behavior. For the Sacred is a dimension of consciousness. It is thus not surprising to witness the resurgence of such attitudes in lay society, and cyberspace, which unseals areas of reality hitherto contained by a screen, definitely qualifies for Eliade's vision. Cyberspace creates a break in the plane of reality, one that seems to generate the ideal conditions for a "hierophany: an irruption of the Sacred that results in detaching a territory from the surrounding cosmic milieu and making it qualitatively different" (Eliade 1961: 26). Through rites and techniques of orientation, religious man recreates the world again and again, managing openings in ordinary experience to communicate with the Transcendent and refresh himself at the source of absolute reality, when the world was in its prime. With special outfits and 3-D techniques of orientation, cyberspace men and women will recreate virtual worlds again and again, managing openings for televirtual communication and refreshing themselves in a computer reality that never ages.

Another fundamental aspect of the Sacred for Eliade is the concept of time. If there is something that computers have forced into our society, it is a different sense of time. We still conceive of time within the mental framework of the eighteenth century: time is what is set by a clock, God as "the Great Clockmaker" (Voltaire), and so on. The (re)emergence of different senses of time will not destroy the traditional one, but make it relative.

For the computer has an infinity of times in potential, ready to be actualized. Flashy splinters of time, speedy batches of time, dumb loops, programs that open up like fans according to the number of frames you assign them. Michaux speaks of "a time that has a crowd of moments" (1967: 11), Paul Virilio of "an intensive time . . . (that) . . . is not complete any more, but indefinitely fractionated in as many instants, instantaneities as the techniques of communication allow it" (1987: 20). The computer is also home to "real time," 30 frames per second, a time that fits exactly Eliade's definition of sacred time. "Religious man lives in two kinds of time, of which, the more important, sacred time, appears under the paradoxical aspect of a circular time, reversible and recoverable, a sort of eternal present that is periodically reintegrated by means of rites." "Real time" too is an eternal present of symbols, reversible and recoverable, an ontological time that never gets exhausted and that can be reintegrated at any time. Eliade's sacred time is nonhistorical. It is the time that "floweth not," that does not participate in the profane duration.

Now if we consider cyberspace as a "location" for such time, it is obvious that it will create a sort of atmospheric depression in history, one that will concentrate aspirations for fundamental re-sourcing. Computers are re-sourcing facilities in many senses.

Today, you can retrieve ever fresh information from data banks, soon you will be able to re-source yourself entirely in cyberspace. Its eternal present will be seen as the Fountain of Youth, where you will bathe and refresh yourself into a sparkling juvenile. Drinking from this source might become one of the major recreations of the future.

The prophets of the "Hypercalypse," who look at simulation as an immoral attraction from the existentialist values of responsibility and commitment to history, who would declare with Philippe Queeau that "the symbolic is also the most diabolical instrument" (1985: 256), will no doubt see evil in a technique that actually may fulfill a deep thirst for the real, for the essential, for the communication with others on a new wavelength.

Their pessimism betrays their fear of change, but also their denial for the legitimate enjoyment that people might gain from these techniques. It also shows that for many thinkers of modernity, in spite of the fact that the virtual dimension of society is already accomplished, that anonymity has become one of the fundamental rights on the computer networks, and that (it fascinates them) human beings are still measured by the length of their roots in the soil. "One does not become a complete man until one has passed beyond and in some sense abolished 'natural humanity,' " says Eliade (1961: 187). And what if the passage to a new level of humanity actually meant abolishing indeed the natural one, or at least some part of it? What if cyberspace was this "home for Eros" that Michael Heim describes in this volume? Wouldn't the drive for cyberspace be so irresistible that some of the basic functions of human life might fall off like ripe fruit? Human reproduction, for instance. "Our mortal self-absorption with our image consoles us for the irreversibility of having been born and having to reproduce," writes Baudrillard (1990: 69). How will your boyfriend know that you've been in your pajamas for weeks if you only meet in a cyberspace exchange? You won't need condoms anymore. Cyberspace will be the condom. Ninety-nine percent of all encounters on the French Minitel never lead to an actual meeting. What if cyberspace were the final act of a natural evolution of family? a mutation into the virtual cloning of individuals?

Will this exacerbation of colors and reliefs, this purity of 3-D sound, this startling simulation of the sense of touch that Margaret Minsky is now developing at MIT, change our relationship with nature altogether? Will it not require immense effort to recover from this enhancement of the senses, from this habit of perfection? Will we still look with the same tenderness at the discreet charm and dazzling creativity of a New England garden? Working with computers changes our sensitivity to light, to depth, makes our dreams more vivid, facilitates the use of metaphors in language. These effects are slight today, but they already provide a different presence, a different sense to the everyday world. This new acuity will develop. Communicating at the speed of light on the computer networks induces euphoria, boosts intuition. Interactivity with 3-D images reveals enlightening visual lapses. This is why we can expect the speed of "real time" to help us project into cyberspace some of our dearest phantasms, some of our worst monsters. This power of revelation and embodiment will be felt by many to be the utmost obscenity. Let's not forget that both Hitler and Stalin are known for having banned the publication of fairy tales. (Fairy tales that describe a world of desires, immorally fulfilled at the speed of light, in a setting of metamorphic imagery, absolute brilliance, space and time distortions, in cyberspace as it were.) Freedom of imagination is feared by most powers. (One of the slogans of May 1968 in Paris was "Power to the Imagination." Strangely enough, it seems that the political arena of the sixties is slowly drifting to the existential arena of computers in the nineties.) This revelation, this release of the imagination, will be felt by others as a major transgression, comparable to the dissection of cadavers in its own time, or to the manipulation of the human egg. Michel de Certeau said that the end of the world would be "a white eschatology," that no secret would remain, and no shadow. Everything being revealed in an absolute spotlight, there would be a stupor, "(a)n absorption of all objects and of all subjects in the act of seeing" (de Certeau 1983: 18).

What a profound irony that we should need to wear heavy prostheses (eye phones, data gloves, body suits) to enjoy freedom of motion and expression in cyberspace. That we should look for the real inside a space of simulation. "What a profound irony," says Benoit Mandelbrot of fractals, "that this new geometry which everyone seems spontaneously to describe as (baroque) and (organic), should owe its birth to an unexpected but profound match between those two symbols of the inhuman, the dry and the technical: namely between mathematics and the computer" (1989: 21).

According to Sartre, the atomic bomb was what humanity had found to commit collective suicide. It seems, by contrast, that cyberspace, though born of a war technology, opens up a space for collective restoration, and for peace. As screens are dissolving, our future can only take on a luminous dimension!

Welcome to the New World.

Bibliography

Note: "author's translation" is by Nicole Stenger.

Baudrillard, Jean, *Seduction*, (New York, Saint Martin's Press, 1990); *Simulacres et Simulations* (Paris, Galilee, 1981), author's translation; *America* (New York, Verso, 1989).

de Certeau, Michel, "Extase Blanche" in *L'Obscene* (Paris, Traverses, October 1983).

Eliade, Mircea, *The Sacred and the Profane* (New York, Harper and Row, 1961).

Kristeva, Julia, *Powers of Horror* (New York, Columbia University Press,1982).

Mandelbrot, Benoit, "Fractals and an Art for the Sake of Science " *Leonardo*, July 1989.

Michaux, Henri, *Connaissance par les Gouffres* (Paris, Gallimard, 1988 [1967]).

de Nerval, Gerard, *Aurelia* (Paris, Folio, 1918). Author's translation.

Queeau, Philippe, *Eloge de la Simulation* (Paris, Champvallon, 1985). Author's translation.

Virilio, Paul, *La Machine de Vision* (Paris, Galilee, 1988); *L'Horizon Negatif* (Paris, Galilee, 1984); "L'Image ou le Reste du Temps" in *L'Imaginaire Numerique* (Paris, Hermes, 1987). All author's translation.

5   *The Erotic Ontology of Cyberspace*

Michael Heim

Introduction

Cyberspace is more than a breakthrough in electronic media or in computer interface design. With its virtual environments and simulated worlds, cyberspace is a metaphysical laboratory, a tool for examining our very sense of reality.

When designing virtual worlds, we face a series of reality questions. How, for instance, should users appear to themselves in a virtual world? Should users appear to themselves in cyberspace as one set of objects among others, as third-person bodies that users can inspect with detachment? Or should users feel themselves to be headless fields of awareness, similar to our phenomenological experience? Should causality underpin the cyber world so that an injury inflicted on the user's cyberbody likewise somehow damages the user's physical body? And who should make the ongoing design decisions? If the people who make simulations inevitably incorporate their own perceptions and beliefs, loading cyberspace with their prejudices as well as insights, who should build the cyber world? Should multiple users at any point be free to shape the qualities and dimensions of cyber entities? Should artistic users roam freely, programming and directing their own unique cyber cinemas that provide escape from the mundane world? Or does fantasy cease where the economics of the virtual workplace begins? But why be satisfied with a single virtual world? Why not several? Must we pledge allegiance to a single reality? Perhaps worlds should be layered like onion skins, realities within realities, or loosely linked like neighborhoods, permitting free aesthetic pleasure to coexist with the task-oriented business world. Does the meaning of "reality"—and the keen existential edge of experience—weaken as it stretches over many virtual worlds?

Important as these questions are, they do not address the ontology of cyberspace itself, the question of what it means to *be* in a virtual world, whether one's own or another's world. They do not probe the reality status of our metaphysical tools or tell us why we invent virtual worlds. They are silent about the essence or soul of cyberspace. How does the metaphysical laboratory fit into human inquiry as a whole? What status do electronic worlds have within the entire range of human experience? What perils lurk in the metaphysical origins of cyberspace?

In what follows I explore the philosophical significance of cyberspace. I want to show the ontological origin from which cyber entities arise and then indicate the trajectory they seem to be on. The ontological question, as I see it, requires a two-pronged answer. We need to give an account of (1) the way entities exist within cyberspace, and (2) the ontological status of cyberspace—the construct, the phenomenon— itself. The way we understand the ontological structure of cyberspace will determine how realities can exist within it. But the structure of cyberspace becomes clear only once we appreciate the distinctive way in which things appear within it. So we must begin with (1), the entities we experience within the computerized environment.

My approach to cyberspace passes first through the ancient idealism of Plato and moves onward through the modern metaphysics of Leibniz. By connecting with intellectual precedents and prototypes, we can enrich our self-understanding and make cyberspace function as a more useful metaphysical laboratory.

Our Marriage to Technology

The phenomenal reality of cyber entities exists within a more general fascination with technology, and the fascination with technology is akin to aesthetic fascination. We love the simple, clear-cut linear surfaces that computers generate. We love the way computers reduce complexity and ambiguity, capturing things in a digital network, clothing them in beaming colors and girding them with precise geo-metrical structures. We are enamored of the possibility of controlling all human knowledge. The appeal of seeing society's data structures in cyberspace—if we begin with William Gibson's vision—is like the appeal of seeing the Los Angeles metropolis in the dark at 5,000 feet:

a great warmth of powerful, incandescent blue and green embers with red stripes beckons the traveler to come down from the cool darkness. We are the moths attracted to flames, and frightened by them too, for there may be no home behind the lights, no secure abode behind the vast glowing structures. There are only the fiery objects of dream and longing.

Our love affair with computers, computer graphics, and computer networks runs deeper than aesthetic fascination and deeper than the play of the senses. We are searching for a home for the mind and heart. Our fascination with computers is more erotic than sensuous, more deeply spiritual than utilitarian. Eros, as the ancient Greeks understood, springs from a feeling of insufficiency or inadequacy. While the aesthete feels drawn to casual play and dalliance, the erotic lover reaches out to a fulfillment far beyond aesthetic detachment.

The computer's allure is more than utilitarian or aesthetic; it is erotic. Instead of a refreshing play with surfaces, as with toys or amusements, our affair with information machines announces a symbiotic relationship and ultimately a mental marriage to technology. Rightly perceived, the atmosphere of cyberspace carries the scent that once surrounded Wisdom. The world rendered as pure information not only fascinates our eyes and minds, it captures our hearts. We feel augmented and empowered. Our hearts beat in the machines. This is Eros.

Cyberspace entities belong to a broad cultural phenomenon of the last third of the twentieth century: the phenomenon of computerization. Something becomes a phenomenon when it arrests and holds the attention of a civilization. Only then does our shared language articulate the presence of the thing so that it can appear in its steady identity the moving stream of history. Because we are immersed in everyday phenomena, however, we usually miss their overall momentum and cannot see where they are going, or even what they truly are. A writer like William Gibson helps us grasp what is phenomenal in current culture because he captures the forward movement of our attention and shows us the future as it projects its claim back onto our present. Of all writers, Gibson most clearly reveals the intrinsic allure of computerized entities, and his books, *Neuromancer, Count Zero,* and *Mona Lisa Overdrive,* point to the near-future, phenomenal reality of cyberspace. Indeed, Gibson invented the word *cyberspace.*

The Romance of Neuromancer

For Gibson, cyber entities appear under the sign of Eros. The fictional characters of *Neuromancer* experience the computer Matrix—cyberspace—as a place of rapture and erotic intensity, of powerful desire and even self-submission. In the Matrix, things attain a supervivid hyperreality. Ordinary experience seems dull and unreal by comparison. Case, the data wizard of *Neuromancer,* awakens to an obsessive Eros that drives him back again and again to the information network:

A year [in Japan] and he still dreamed of cyberspace, hope fading nightly. . . . He was no [longer] console man, no cyberspace cowboy, but still he'd see the matrix in his sleep, bright lattices of logic unfolding across that colorless void. . . . The dreams came on in the Japanese night like livewire voodoo, and he'd cry for it, cry in his sleep, and wake alone in the dark, curled in his capsule in some coffin hotel, his hands clawed into the bedslab, . . . trying to reach the console that wasn't there. (P. 5)

The sixteenth-century Spanish mystics, John of the Cross and Theresa of Avila, used a similar point of reference. Seeking words to connote the taste of spiritual divinity, they reached for the language of sexual ecstasy. They wrote of the breathless union of meditation in terms of the ecstatic blackout of consciousness, the *llama de amor viva* piercing the interior center of the soul like a white-hot arrow, the *cauterio suave* searing through the dreams of the dark night of the soul. Similarly, the intensity of Gibson's cyberspace inevitably conjures up the reference to orgasm, and vice versa:

Now she straddled him again, took his hand, and closed it over her, his thumb along the cleft of her buttocks, his fingers spread across the labia. As she began to lower herself, the images came back, the faces, fragments of neon arriving and receding. She slid down around him and his back arched convulsively. She rode him that way, impaling herself, slipping down on him again and again, until they both had come, his orgasm flaring blue in a timeless space, a vastness like the matrix, where the faces were shredded and blown away down hurricane corridors, and her inner thighs were strong and wet against his hips. (P. 33)

But the orgasmic connection does not mean that Eros toward cyberspace entities terminates in a merely physiological or psychological reflex. Eros goes beyond private, subjective fantasies. Cyber Eros stems ultimately from the ontological drive highlighted long ago by Plato. Platonic metaphysics helps clarify the link between Eros and computerized entities.

In her speech in Plato's *Symposium,* Diotima, the priestess of love, teaches a doctrine of the escalating spirituality of the erotic drive. She tracks the intensity of Eros continuously from bodily attraction all the way to the mental attention of mathematics and beyond. The outer reaches of the biological sex drive, she explains to Socrates, extend to the mental realm where we continually seek to expand our knowledge.

On the primal level, Eros is a drive to extend our finite being, to prolong something of our physical selves beyond our mortal existence. But Eros does not stop with the drive for physical extension. We seek to extend ourselves and to heighten the intensity of our lives in general through Eros. The psyche longs to perpetuate itself and to conceive offspring; and this it can do, in a transposed sense, by conceiving ideas and nurturing awareness in the minds of others as well as our own. The psyche develops consciousness by formalizing perceptions and by stabilizing experiences through clearly defined entities. But Eros motivates humans to see more and to know more deeply. So, according to Plato, the fully explicit formalized identities of which we are conscious help us to maintain life in a "solid state," thereby keeping perishability and impermanence at bay.

Only a short philosophical step separates this Platonic notion of knowledge from the matrix of cyberspace entities. (The word *matrix,* of course, stems from the Latin for the Mother, the generative-erotic origin). A short step in fundamental assumptions, however, can take centuries—especially if the step needs hardware support. The hardware for implementing Platonically formalized knowledge took centuries. Underneath, though, runs an ontological continuity, connecting the Platonic knowledge of ideal forms to the information systems of the Matrix. Both approaches to cognition first extend and then renounce the physical embodiment of knowledge. In both, Eros inspires humans to outrun the drag of the "meat"—the flesh—by attaching human attention to what formally attracts the mind. As Platonists and Gnostics down through the ages have insisted: Eros guides us to Logos.

The erotic drive, however, as Plato saw it, needs education to attain its fulfillment. Left on its own, Eros naturally goes astray on any number of tangents, most of which come from sensory stimuli. In the *Republic,* Plato tells the well-known story of the Cave in which people caught in the prison of everyday life learn to love the fleeting, shadowy illusions projected on the walls of the dungeon of the flesh. With their attention forcibly fixed on the shadowy moving images cast by a flickering physical fire, the prisoners passively take sensory objects to be the highest and most interesting realities. Only later when the prisoners manage to get free of their corporeal shackles do they ascend to the realm of active thought where they enjoy the shockingly clear vision of real things, things not present to the physical eyes but to the mind's eye. Only by actively processing things through mental logic, according to Plato, do we move into the upper air of reliable truth, which is also a lofty realm of intellectual beauty stripped of the imprecise impressions of the senses. Thus the liberation from the Cave requires a reeducation of human desires and interests. It entails a realization that what attracts us in the sensory world is no more than an outer projection of ideas we can find within us. Education must redirect desire toward the formally defined, logical aspects of things. Properly trained, love guides the mind to the well-formed, mental aspects of things.

Cyberspace is Platonism as a working product. The cybernaut seated before us, strapped into sensory input devices, appears to be, and is indeed, lost to this world. Suspended in computer space, the cybernaut leaves the prison of the body and emerges in a world of digital sensation.

This Platonism is thoroughly modern, however. Instead of emerging in a sensationless world of pure concepts, the cybernaut moves among entities that are well-formed in a special sense. The spatial objects of cyberspace proceed from the constructs of Platonic imagination not in the same sense that perfect solids or ideal numbers are Platonic constructs, but in the sense that inFORMation in cyberspace inherits the beauty of Platonic FORMS. The computer recycles ancient Platonism by injecting the ideal content of cognition with empirical specifics. Computerized representation of knowledge, then, is not the direct mental insight fostered by Platonism. The computer clothes the details of empirical experience so that they seem to share the ideality of the stable knowledge of the Forms. The mathematical machine uses a digital mold to reconstitute the mass of empirical material so that human consciousness can enjoy an integrity in the empirical data that would never have been possible before computers. The notion of ideal Forms in early Platonism has the allure of a perfect dream. But the ancient dream remained airy, a landscape of genera and generalities, until the hardware of information retrieval came to support the mind's quest for knowledge. Now, with the support of the electronic matrix, the dream can incorporate the smallest details of here-and-now existence. With an electronic infrastructure, the dream of perfect FORMS becomes the dream of inFORMation.

Filtered through the computer matrix, all reality becomes patterns of information. When reality becomes indistinguishable from information, then even Eros fits the schemes of binary communication. Bodily sex appears to be no more than an exchange of signal blips on the genetic corporeal network. Further, the erotic-generative source of formal idealism becomes subject to the laws of information management. Just as the later Taoists of ancient China made a yin/yang cosmology that encompassed sex, cooking, weather, painting, architecture, martial arts, etc., so too the computer culture interprets all knowable reality as transmissible information. The conclusion of *Neuromancer* shows us the transformation of sex and personality into the language of information:

There was a strength that ran in her. . . , something he'd found and lost so many times. It belonged, he knew—he remembered—as she pulled him down, to the meat, the flesh the cowboys mocked. It was a vast thing, beyond knowing, a sea of information coded in spiral and pheromone, infinite intricacy that only the body, in its strong blind way, could ever read . . . As he broke the zipper, some tiny metal part shot off against the wall as salt-rotten cloth gave, and then he was in her, effecting the transmission of the old message. Here, even here, in a place he knew for what it was, a coded model of some stranger's memory, the drive held. She shuddered against him as the stick caught fire, a leaping flare that threw their locked shadows across the bunker wall. (Pp. 239–240)

The dumb meat once kept sex private, an inner sanctum, an opaque, silent, unknowable mystery. The sexual body held its genetic information with the strength of a blind, unwavering impulse. What is translucent you can manipulate, you can see. What stays opaque you cannot scrutinize and manipulate. It is an alien presence. The meat we either dismiss or come up against; we cannot ignore it. It remains something to encounter. Yet here, in *Neuromancer,* the protagonist Case makes love to a sexual body named Linda. Who is this Linda?

Gibson raises the deepest ontological question of cyberspace by suggesting that the Neuromancer master-computer *simulates* the body and personality of Case's beloved. A simulated, embodied personality provokes the sexual encounter. Why? Perhaps because the cyberspace system, which depends on the physical space of bodies for its initial impetus, now seeks to undermine the separate existence of human bodies that make it dependent and secondary. The ultimate revenge of the information system comes when the system absorbs the very identity of the human personality, absorbing the opacity of the body, grinding the meat into information, and deriding erotic life by reducing it to a transparent play of puppets. In an ontological turnabout, the computer counterfeits the silent and private body from which mental life originated. The machinate mind disdainfully mocks the meat. Information digests even the secret recesses of the caress. In its computerized version, Platonic Eros becomes a master artificial intelligence, CYBEROS, the controller, Neuromancer.

The Inner Structure of Cyberspace

Aware of the phenomenal reality of cyber entities, we can now appreciate the backdrop that is cyberspace itself. We can sense a distant source radiating an all-embracing power. For the creation of computerized entities taps into the most powerful of our psychobiological urges. Yet so far this account of the distant source as Eros tells only half the story. For while Platonism provides the psychic makeup for cyberspace entities, only modern philosophy shows us the structure of cyberspace itself.

In its early phases—from roughly 400 B.C. to 1600 A.D.—Platonism exclusively addressed the speculative intellect, advancing a verbal-mental intellectuality over physical actuality. Later, Renaissance and modern Platonists gradually injected new features into the model of intelligence. The modern Platonists opened up the gates of verbal-spiritual understanding to concrete experiments set in empirical space and time. The new model of intelligence included the evidence of repeatable experience and the gritty details of experiment. For the first time, Platonism would have to absorb real space and real time into the objects of its contemplation.

The early Platonic model of intelligence considered space to be a mere receptacle for the purely intelligible entities subsisting as ideal forms. Time and space were refractive errors that rippled and distorted the mental scene of perfect unchanging realities. The bouncing rubber ball was in reality a round object, which was in reality a sphere, which was in reality a set of concentric circles, which could be analyzed with the precision of Euclidian geometry. Such a view of intelligence passed to modern Platonists, and they had to revise the classical assumptions. Thinkers and mathematicians would no longer stare at the sky of unchanging ideals. By applying mathematics to empirical experiment, science would absorb physical movement in space/time through the calculus. Mathematics transformed the intelligent observer from a contemplator to a calculator. But as long as the calculator depended on feeble human memory and on scattered printed materials, a gap would still stretch between the longing and the satisfaction of knowledge. To close the gap, a computational engine was needed.

Before engineering an appropriate machine, the cyberspace project first needed a new logic and a new metaphysics. The new logic and metaphysics of modernity came largely from the work of Baron Gottfried Wilhelm von Leibniz (1646–1716). In many ways, the later philosophies of Kant, Schopenhauer, Nietzsche, and Heidegger took their bearings from Leibniz.

As Leibniz worked out the modern Idealist epistemology, he was also experimenting with protocomputers. Pascal's calculator had been no more than an adding machine; Leibniz went further and produced a mechanical calculator which could, by using stepped wheels, also multiply and divide. The basic Leibnizean design became the blueprint for all commercial calculators until the electronics revolution of the 1970s. Leibniz, therefore, is one of the essential philosophical guides to the inner structure of cyberspace. His logic, his metaphysics, and his notion of representational symbols show us the hidden underpinnings of cyberspace. At the same time, his monadological metaphysics alerts us to the paradoxes that are likely to engulf cyberspace's future inhabitants.

Leibniz's Electric Language

Leibniz was the first to conceive an "electric language," a set of symbols engineered for manipulation at the speed of thought. His *De Arte Combinatoria* (1666) outlines a language that became the historical foundation of contemporary symbolic logic. Leibniz's general outlook on language would also become the ideological basis for computer-mediated telecommunications. A modern Platonist, Leibniz dreamt of the matrix.

The language Leibniz outlines is an ideographic system of signs that can be manipulated to produce logical deductions without recourse to natural language. The signs represent primitive ideas gleaned from prior analysis. Once broken down into primitives and represented by stipulated signs, the component ideas can be paired and recombined to fashion novel configurations. In this way, Leibniz sought to mechanize the production of new ideas. As he describes it, the encyclopedic collection and definition of primitive ideas would require the coordinated efforts of learned scholars from all parts of the civilized world. The royal academies Leibniz promoted were the group nodes for an international republic of letters, a universal network for problem solving.

Leibniz believed all problems to be in principle soluble. The first step was to create a universal medium in which conflicting ideas can coexist and interrelate. A universal language makes it possible to translate all human notions and disagreements into the same set of symbols. His universal character set, *characteristica universalis,* rests on a binary logic, one quite unlike natural discourse in that it is neither restricted by material content nor embodied in vocalized sound. Contentless and silent, the binary language can transform every significant statement into the terms of a logical calculus, a system for proving argumentative patterns valid or invalid, or at least for connecting them in a homogeneous matrix. Through the common binary language, discordant ways of thinking can exist under a single roof. Disagreements in attitude or belief, once translated into matching symbols, can later yield to operations for insuring logical consistency. To the partisans of dispute Leibniz would say, "Let us upload this into our common system, then let us sit down and calculate." A single system would encompass all the combinations and permutations of human thought. Leibniz longed for his symbols to foster unified scientific research throughout the civilized world. The universal calculus would compile all human culture, bringing every natural language into a single shared database.

Leibniz's binary logic, disembodied and devoid of material content, depends on an artificial language remote from the words, letters, and utterances of everyday discourse. This logic treats reasoning as nothing more than a combining of signs, as a calculus. Like mathematics, the Leibnizean symbols erase the distance between the signifiers and the signified, between the thought seeking to express and the expression. No gap remains between symbol and meaning. Given the right motor, the Leibnizean symbolic logic—as developed later by Boole, Russell, and Whitehead, and then applied to electronic switching circuitry by Shannon—can function at the speed of thought. At such high speed the felt semantic space closes between thought, language, and the thing expressed. Centuries later, John von Neumann was to apply a version of Leibniz's binary logic in building the first computers at Princeton.

In his search for a universal language of the matrix, Leibniz to some extent continued a premodern, medieval tradition. For behind his ideal language lurks a premodern model of human intelligence. The medieval Scholastics held that human thinking, in its pure or ideal form, is more or less identical with logical reasoning. Reasoning functions along the lines of a superhuman model who remains unaffected by the vagaries of feelings and of spatial-temporal experience. Human knowledge imitates a Being who knows things perfectly and knows them in their deductive connections. The omniscient Being transcends finite beings. Finite beings go slowly, one step at a time, seeing only moment by moment what is happening. On the path of life, a finite being cannot see clearly the things that remain behind on the path nor the things that are going to happen after the next step. A divine mind, on the contrary, oversees the whole path. God sees all the trails below, inspecting at a single glance every step traveled, what has happened and even what will happen on all possible paths below. God views things from the perspective of the mountain top of eternity.

Human knowledge, thought Leibniz, should emulate this *visio Dei,* this omniscient intuitive cognition of the deity. Human knowledge strives to know the way a divine or infinite Being knows things. No temporal unfolding, no linear steps, no delays limit God's knowledge of things. The temporal simultaneity, the all-at-once-ness of God's knowledge serves as a model for human knowledge in the modern world as projected by the work of Leibniz. What better way, then, to emulate God's knowledge than to generate a virtual world constituted by bits of information? Over such a cyber world human beings could enjoy a God-like instant access. But, if knowledge be power, who would handle the controls that govern every single particle of existence?

The power of Leibniz's modern logic made traditional logic seem puny and inefficient by comparison. For centuries Aristotle's logic had been taught in the schools. Logic traditionally evaluated the steps of finite human thought, valid or invalid, as they occur in arguments in natural language. Traditional logic stayed close to spoken natural language. When modern logic absorbed the steps of Aristotle's logic into its system of symbols, modern logic became a network of symbols that could apply equally to electronic switching circuits as to arguments in natural language. Just as non-Euclidian geometry can set up axioms that defy the domain of real circles (physical figures), so too modern logic freed itself of any naturally given syntax. The universal logical calculus could govern computer circuits.

Leibniz's "electric language" operates by emulating the divine intelligence. God's knowledge has the simultaneity of all-at-once-ness, and so, in order to achieve a divine access to things, the global matrix functions like a net to trap all language in an eternal present. Because access need not be linear, cyberspace does not in principle require a jump from one location to another. Science fiction writers have often imagined what it would be like to experience traveling at the speed of light, and one writer, Isaac Asimov, describes such travel as a "jump through hyperspace." When his fictional spaceship hits the speed of light, Asimov says that the ship makes a special kind of leap. At that speed, it is impossible to trace the discrete points of the distance traversed. In the novel *The Naked Sun*, Asimov depicts movement in hyperspace like this:

There was a queer momentary sensation of being turned inside out. It lasted an instant and Baley knew it was a Jump, that oddly incomprehensible, almost mystical, momentary transition through hyperspace that transferred a ship and all it contained from one point in space to another, light years away. Another lapse of time and another Jump, still another lapse, still another Jump. (P. 16)

Like the fictional hyperspace, cyberspace unsettles the felt logical tracking of the human mind. Cyberspace is the perfect computer environment for accessing hypertext if we include all human perceptions as the "letters" of the "text." In both hyperspace and hypertext, linear perception loses track of the series of discernible movements. With hypertext, we connect things at the speed of a flash of intuition. The interaction with hypertext resembles movement beyond the speed of light. Hypertext reading and writing supports the intuitive leap over the traditional step-by-step logical chain. The jump, not the step, is the characteristic movement in hypertext.

As the environment for sensory hypertext, cyberspace feels like transportation through a frictionless, timeless medium. There is no jump because everything exists, implicitly if not actually, all at once. To understand this lightning speed and its perils for finite beings, we must look again at the metaphysics of Leibniz.

Monads Do Have Terminals

Leibniz called his metaphysics a *monadology*, a theory of reality describing a system of "monads." From our perspective, the monadology conceptually describes the nature of beings who are capable of supporting a computer matrix. The monadology can suggest how cyberspace fits into the larger world of networked, computerized beings.

The term *monadology* is derived from the Greek *monas*, meaning "unit," as in monastic, monk, or monopoly. It refers to a certain kind of aloneness, a solitude in which each being pursues its appetites in isolation from all other beings, which are also solitary. The monad exists as an independent point of vital willpower, a surging drive to achieve its own goals, according to its own internal dictates. Because they are sheer, vital thrust, the monads do not have inert spatial dimensions but produce space as a by-product of their activity. Monads are nonphysical, psychical substances whose forceful life is an immanent activity. For monads, there is no outer world to access, no larger, broader vision. What the monads see are the projections of their own appetites and their own ideas. In Leibniz's succinct phrase: "Monads have no windows."

Monads have no windows, but they do have terminals. The mental life of the monad—and the monad has no other life—is a procession of internal representations. Leibniz's German calls these representations *Vorstellungen,* from "vor" (in front of) and "stellen" (to place). Realities are representations continually placed in front of the viewing apparatus of the monad, but placed in such a way that the system interprets or represents what is being pictured. The monad sees the pictures of things and knows only what can be pictured. The monad knows through the interface. The interface *re*presents things, simulates them, and preserves them in a format that the monad can manipulate in any number of ways. The monad keeps the presence of things on tap, as it were, making things instantly available and disposable, so that the presence of things is *re*presented or "canned." From the vantage point of physical phenomenal beings, the monad undergoes a surrogate experience. Yet the monad does more than think about or imagine things at the interface. The monad senses things, sees and hears them as perceptions. But the perceptions of phenomenal entities do not occur in real physical space because no substances other than monads really exist. While the interface with things vastly expands the monad's perceptual and cognitive powers, the things at the interface are simulations and representations.

Yet Leibniz's monadology speaks of monads in the plural. For a network to exist, more than one being must exist, otherwise nothing is there to be networked. But how can monads coordinate or agree on anything at all, given their isolated nature? Do they even care if other monads exist? Leibniz tells us that each monad represents within itself the entire universe. Like Indra's Net, each monad mirrors the whole world. Each monad represents the universe in concentrated form, making within itself a *mundus concentratus*. Each microcosm contains the macrocosm. As such, the monad reflects the universe in a living mirror, making it a *miroir actif indivisible,* whose appetites drive it to represent everything to itself—everything, that is, mediated by its mental activity. Since each unit represents everything, each unit contains all the other units, containing them as represented. No direct physical contact passes between the willful mental units. Monads never meet face-to-face.

Although the monads represent the same universe, each sees it differently. The differences in perception come from differences of perspective. The different perspectives arise not from different physical positions in space—the monads are not physical and physical space is a by-product of mental perception—but perspectives differ on account of the varying degrees of clarity and intensity in each monad's mental landscape. The appetitive impulses in each monad highlight different things in the sequence of representational experience. Their different impulses constantly shift the scenes they see. Monads run different software.

Still, there exists, according to the monadology, one actual universe. Despite their ultimately solitary character, the monads belong to a single world. The harmony of all the entities in the world comes from the one underlying operating system. While no unit directly contacts other units, each unit exists in synchronous time in the same reality. All their representations are coordinated. The coordination takes place through the supervisory role of the Central Infinite Monad, traditionally known as God. The Central Infinite Monad, we could say, is the Central System Operator (sysop) who harmonizes all the finite monadic units. The Central System Monad is the only being that exists with absolute necessity. Without a sysop, no one could get on-line to reality. Thanks to the Central System Monad, each individual monad lives out its separate life according to the dictates of its own willful nature while still harmonizing with all the other monads on-line.

Paradoxes in the Cultural Terrain of Cyberspace

Leibniz's monadological metaphysics brings out certain aspects of the erotic ontology of cyberspace. While the monadology does not actually describe computerized space, of course, it does nevertheless suggest some of the inner tendencies of computerized space. These tendencies are inherent in the structure of cyberspace and therefore affect the broader realities in which the matrix exists. Some paradoxes crop up. The monadological metaphysics shows us a cultural topography riddled with deep inconsistencies.

Cyberspace supplants physical space. We see this happening already in the familiar cyberspace of on-line communication—telephone, e-mail, newsgroups, etc. When on line, we break free, like the monads, from bodily existence. Telecommunication offers an unrestricted freedom of expression and personal contact, with far less hierarchy and formality than is found in the primary social world. Isolation persists as a major problem of contemporary urban society—I mean spiritual isolation, the kind that plagues individuals even on crowded city streets. With the telephone and television, the computer network can function as a countermeasure. The computer network appears as a godsend in providing forums for people to gather in surprisingly personal proximity—especially considering today's limited bandwidths—without the physical limitations of geography, time zones, or conspicuous social status. For many, networks and bulletin boards act as computer antidotes to the atomism of society. They assemble the monads. They function as social nodes for fostering those fluid and multiple elective affinities that everyday urban life seldom, in fact, supports.

Unfortunately, what technology gives with one hand, it often takes away with the other. Technology increasingly eliminates direct human interdependence. While our devices give us greater personal autonomy, at the same time they disrupt the familiar networks of direct association. Because our machines automate much of our labor, we have less to do with one another. Association becomes a conscious act of will. Voluntary associations operate with less spontaneity than those sprouted by serendipity. Because machines provide us with the power to flit about the universe, our communities grow more fragile, airy, and ephemeral, even as our connections multiply.

Being a *body* constitutes the principle behind our separateness from one another and behind our personal presence. Our bodily existence stands at the forefront of personal identity and individuality. Both law and morality recognize the physical body as something of a fence, an absolute boundary, establishing and protecting our privacy. Now, the computer network simply brackets the physical presence of the participants, either by omitting or by simulating corporeal immediacy. In one sense this frees us from the restrictions imposed by our physical identity. We are more equal on the net because we can either ignore or create the body that appears in cyberspace. But, in another sense, the quality of the human encounter narrows. The secondary or stand-in body reveals only as much of our selves as we mentally wish to reveal. Bodily contact becomes optional; you need never stand face-to-face with other members of the virtual community. You can live your own separate existence without ever physically meeting another person. Computers may at first liberate societies through increased communication and may even foment revolutions (I think of the computer printouts in Tienanmen Square during the 1989 prodemocracy student uprisings in China). They have, however, another side, a dark side.

The darker side hides a sinister melding of human and machine. The cyborg or cybernetic organism implies that the conscious mind steers— the meaning of the Greek *kybernetes*—our organic life. Organic life energy ceases to initiate our mental gestures. Can we ever be fully present when we live through a surrogate body standing in for us? The stand-in self lacks the vulnerability and fragility of our primary identity. The stand-in self can never fully *re*present us. The more we mistake the cyberbodies for ourselves, the more the machine twists our selves into the prostheses we are wearing.

Gibson's fiction has inspired the creation of role-playing games for young people. One of these games in the cybertech genre, *The View from the Edge: The Cyberpunk Handbook,* portrays the visage of humanity twisted to fit the shapes of the computer prosthesis. The body becomes literally "meat" for the implantation of information devices. The computer plugs jack directly into the bones of the wrist or skull, and the plugs tap into major nerve trunks so that the chips can send and receive neural signals. As the game book wryly states:

Some will put an interface plug at the temples (a "plug head"), just behind the ears (called a "frankenstein") or in the back of the head (a "puppethead"). Some cover them with inlaid silver or gold caps, others with wristwarmers. Once again, a matter of style. Each time you add a cybernetic enhancement, there's a corresponding loss of humanity. But it's not nice, simple and linear. Different people react differently to the cyborging process. Therefore, your humanity loss is based on the throw of random dice value for each enhancement. This is important, because it means that sheer luck could put you over the line before you know it. Walk carefully. Guard your mind. (Pp. 20–22)

At the computer interface, the spirit migrates from the body to a world of total representation. Information and images float through the Platonic mind without a grounding in bodily experience. You can lose your humanity at the throw of the dice.

Gibson highlights this essentially Gnostic aspect of cybertech culture when he describes the computer addict who despairs at no longer being able to enter the computer matrix: "For Case, who'd lived for the bodiless exultation of cyberspace, it was the Fall. In the bars he'd frequented as a cowboy hotshot, the elite stance involved a certain relaxed contempt for the flesh. The body was meat. Case fell into the prison of his own flesh" (*Neuromancer,* 6). The surrogate life in cyberspace makes flesh feel like a prison, a fall from grace, a sinking descent into a dark, confusing reality. From the pit of life in the body, the virtual life looks like the virtuous life. Gibson evokes the Gnostic-Platonic-Manichean contempt for earthy, earthly existence.

Today's computer communication cuts the physical face out of the communication process. Computers stick the windows of the soul behind monitors, headsets, and datasuits. Even video conferencing adds only a simulation of face-to-face meeting, only a representation or appearance of real meeting. The living, nonrepresentable face is the primal source of responsibility, the direct, warm link between private bodies. Without directly meeting others physically, our ethics languishes. Face-to-face communication, the fleshly bond between people, supports a long-term warmth and loyalty, a sense of obligation for which the computer-mediated communities have not yet been tested. Computer networks offer a certain sense of belonging, to be sure, but the sense of belonging circulates primarily among a special group of pioneers. How long and how deep are the personal relationships that develop outside of embodied presence? The face is the primal interface, more basic than any machine mediation. The physical eyes are the windows that establish the neighborhood of trust. Without the direct experience of the human face, ethical awareness shrinks and rudeness enters. Examples abound. John Coates, spokesperson for The Well in Northern California, says: "Some people just lose good manners on line. You can really feel insulated and protected from people if you're not looking at them—nobody can take a swing at you. On occasion, we've stepped in to request more diplomacy. One time we had to ask someone to go away."

At the far end of distrust lies computer crime. The machine interface may amplify an amoral indifference to human relationships. Computers often eliminate the need to respond directly to what takes place between humans. People do not just observe one another, they become "lurkers." Without direct human presence, participation becomes optional. Electronic life converts primary bodily presence into telepresence, introducing a remove between *re*presented presences. True, in bodily life we often play at altering our identity with different clothing, masks, and nicknames, but electronics installs the illusion that we are "having it both ways," keeping a distance while at the same time "putting ourselves on the line." On-line existence is intrinsically ambiguous, like the purchased passion of the customers in the House of Blue Lights in Gibson's *Burning Chrome:* "The customers are torn between needing someone and wanting to be alone at the same time, which has probably always been the name of that particular game, even before we had the neuroelectronics to enable them to have it both ways" (191). As the expanding global network permits the passage of bodily representations, "having it both ways" may reduce trust and spread cynical anomie.

A loss of innocence therefore accompanies an expanding network. As on-line culture grows geographically, the sense of community diminishes. "Shareware" worked well in the early days of computers, and so did open bulletin boards. When the size of the user base increased, however, the spirit of community diminished and the villains began appearing, some introducing viruses. Hackers invisibly reformatted hard disks, and shareware software writers moved to the commercial world. When we speak of a global village, we should keep in mind that every village makes villains, and when civilization reaches a certain degree of density, the barbaric tribes return, from within. Tribes shun their independent thinkers and punish individuality. A global international village, fed by accelerated competition and driven by information, may be host to an unprecedented barbarism. Gibson's vision of cyberspace works like a mental aphrodisiac, but it turns the living environment—electronic and real—into a harsh, nightmarish jungle. This jungle is more than a mere cyberpunk affectation, a matter of aestheticizing grit or conflict or rejection. It may also be an accurate vision of the intrinsic energies released in a cyberized society.

For a man-made information jungle already spreads out over the world, duplicating with its virtual vastness the scattered geography of the actual world. The matrix already multiplies confusion, and future cyberspace may not simply reproduce a more efficient version of traditional information. The new information networks resemble the modern megalopolis, often described as a concrete jungle (New York) or a sprawl (Los Angeles). A maze of activities and hidden byways snakes around with no apparent center. Architecturally, the network sprawl suggests the absence of the philosophical or religious absolute. Traditional publishing resembles a medieval European city, with the center of all activity, the cathedral or church spire, guiding and gathering all the communal directions and pathways. The steeple visibly radiates like a hub, drawing the inhabitants into a unity and measuring the other buildings on a central model. Traditionally, the long-involved process of choosing which texts to print or which movies to produce or which television shows to produce serves a similar function. The book industry, for instance, provides readers with various cues for evaluating information. The publishers legitimize printed information by giving clues that affect the reader's willingness to engage in reading the book. Editorial attention, packaging endorsements by professionals or colleagues, book design and materials, all add to the value of the publisher's imprint. Communication in contemporary cyberspace lacks the formal clues. In their place are private recommendations, or just blind luck. The electronic world, unlike the traditional book industry, does not protect its readers or travelers by following rules that set up certain expectations. Already, in the electric element, the need for stable channels of content and reliable processes of choice grows urgent.

If cyberspace unfolds like existing large-scale media, we might expect a debasement of discriminating attention. If the economics of marketing forces the matrix to hold the attention of a critical mass of the population, we might expect a flashy liveliness and a flimsy currency to replace depth of content. Sustained attention will give way to fast-paced cuts. One British humanist spoke of the HISTORY forum on Bitnet in the following terms: "The HISTORY network has no view of what it exists for, and of late has become a sort of bar-room courthouse for pseudo-historical discussion on a range of currently topical events. It really is, as Glasgow soccer players are often called, a waste of space." Cyberspace without carefully laid channels of choice may become a waste of space.

The Underlying Fault

Finally, on-line freedom seems paradoxical. If the drive to construct cyber entities comes from Eros in the Platonic sense, and if the structure of cyberspace follows the model of Leibniz's computer God, then cyberspace rests dangerously on an underlying fault of paradox. Remove the hidden recesses, the lure of the unknown, and you also destroy the erotic urge to uncover and reach further, you destroy the source of yearning. Set up a synthetic reality, place yourself in a computer-simulated environment, and you undermine the human craving to penetrate what radically eludes you, what is novel and unpredictable. The computer God's-eye view robs you of your freedom to be fully human. Knowing that the computer God already knows every nook and cranny deprives you of your freedom to search and discover.

Even though the computer God's-eye view remains closed to the human agents in cyberspace, they will know that such a view exists. Computerized reality synthesizes everything through calculation, and nothing exists in the synthetic world that is not literally numbered and counted. Here Gibson's protagonist gets a brief glimpse of this super-human, or inhuman, omniscience:

Case's consciousness divided like beads of mercury, arching about an endless beach the color of the dark silver clouds. His vision was spherical, as though a single retina lined the inner surface of a globe that contained all things, if all things could be counted, and here things could be counted, each one. He knew the number of grains of sand in the construct of the beach (a number coded in a mathematical system that existed nowhere outside the mind that was Neuromancer). He knew the number of yellow food packets in the canisters in the bunker (four hundred and seven). He knew the number of brass teeth in the left half of the open zipper of the salt-crusted leather jacket that Linda Lee wore as she trudged along the sunset beach, swinging a stick of driftwood in her hand (two hundred and two). (P. 258)

The erotic lover reels under the burden of omniscience. "*If* all things could be counted. . . ." Can the beloved remain the beloved when she is fully known, when she is fully exposed to the analysis and synthesis of binary construction? Can we be touched or surprised—deeply astonished—by a synthetic reality, or will it always remain a magic trick, an illusory prestidigitation?

With the thrill of free access to unlimited corridors of information comes the complementary threat of total organization. Beneath the artificial harmony lies the possibility of surveillance by the all-knowing Central System Monad. The absolute sysop wields invisible power over all members of the network. The infinite CSM holds the key for monitoring, censoring, or rerouting any piece of information or any phenomenal presence on the network. The integrative nature of the computer shows up today in the ability of the CSM to read, delete, or alter private e-mail on any computer-mediated system. Those who hold the keys to the system, technically and economically, have access to anything on the system. The CSM will most likely place a top priority on maintaining and securing its power. While matrix users feel geographical and intellectual distances melt away, the price they pay is their ability to initiate uncontrolled and unsupervised activity.

According to Leibniz's monadology, the physical space perceived by the monads comes as an inessential by-product of experience. Spatio-temporal experience goes back to the limitations of the fuzzy finite monad minds, their inability to grasp the true roots of their existence. From the perspective of eternity, the monads exist by rational law and make no unprescribed movements. Whatever movement or change they make disappears in the lightning speed of God's absolute cognition. The flesh, says Leibniz, introduces a cognitive fuzziness. For the Platonic imagination, this fuzzy incarnate world dims the light of intelligence.

Yet the erotic ontology of cyberspace contradicts this preference for disembodied intelligibility. If I am right about the erotic basis of cyberspace, then the surrogate body undoes its genesis, contradicts its nature. The ideal of the simultaneous all-at-once-ness of computerized information access undermines any world that is worth knowing. The fleshly world is worth knowing for its distances and for its hidden horizons. Thankfully, the Central System Monad never gets beyond the terminals into the physical richness of this world. Fortunately, here in the broader world, we still need eyes, fingers, mice, modems, and phone lines.

Gibson leaves us the image of a human group who instinctively keeps its distance from the computer matrix. These are the Zionites, the religiously tribal folk who prefer music to computers and intuitive loyalties to calculation. The Zionites constitute a human remnant in the environmental desolation of *Neuromancer:*

Case didn't understand the Zionites. . . . The Zionites always touched you when they were talking, hands on your shoulder. He [Case] didn't like that. . . . "Try it," Case said [holding out the electrodes of the cyberspace deck].
The Zionite Aerol took the bank, put it on, and Case adjusted the trodes. He closed his eyes. Case hit the power stud. Aerol shuddered. Case jacked him back out. "What did you see, man?" "Babylon," Aerol said, sadly, handing him the trodes and kicking off down the corridor. (P. 106)

As we suit up for the exciting future in cyberspace, we must not lose touch with the Zionites, the body people who remain rooted in the energies of the earth. They will nudge us out of our heady reverie in this new layer of reality. They will remind us of the living genesis of cyberspace, of the heartbeat behind the laboratory, of the love that still sprouts amid the broken slag and the rusty shells of oil refineries "under the poisoned silver sky."

Notes and Related Reading

The quotations from *Neuromancer* come from the Ace edition (New York, 1984). The quotations from Asimov's *Naked Sun* are from the Ballantine edition (New York, 1957). *The View from the Edge: The Cyberpunk Handbook* is by Mike Pondsmith, (Berkeley, California: R. Talsorian Games, 1988). The Coates quotation appeared in *Electric Word* magazine, November/December 1989, p. 35.

For more on Leibniz and the philosophical foundations of computing, see my *Electric Language: A Philosophical Study of Word Processing* (New Haven, Yale University Press, 1987; paperback 1989). Another paper, "The Metaphysics of Virtual Reality," appears in *Multimedia Review,* number 3, 1990 ("New Paradigms"), published by Meckler.

6 Will the Real Body Please Stand Up?: Boundary Stories about Virtual Cultures

Allucquere Rosanne Stone

The Machines Are Restless Tonight

After Donna Haraway's "Promises of Monsters" and Bruno Latour's papers on actor networks and artifacts that speak, I find it hard to think of any artifact as being devoid of agency. Accordingly, when the dryer begins to beep complainingly from the laundry room while I am at dinner with friends, we raise eyebrows at each other and say simultaneously, "The machines are restless tonight . . ."

It's not the phrase, I don't think, that I find intriguing. Even after Haraway 1991 and Latour 1988, the phrase is hard to appreciate in an intuitive way. It's the ellipsis I notice. You can hear those three dots. What comes after them? The fact that the phrase—obviously a send-up of a vaguely anthropological chestnut—seems funny to us, already says a great deal about the way we think of our complex and frequently uneasy imbrications with the unliving. I, for one, spend more time interacting with Saint-John Perse, my affectionate name for my Macintosh computer, than I do with my friends. I appreciate its foibles, and it gripes to me about mine. That someone comes into the room and reminds me that Perse is merely a "passage point" for the work practices of a circle of my friends over in Silicon Valley changes my sense of facing a vague but palpable sentience squatting on my desk not one whit. The people I study are deeply imbricated in a complex social network mediated by little technologies to which they have delegated significant amounts of their time and agency, not to mention their humor. I say to myself: Who am I studying? A group of people? Their machines? A group of people and or in their machines? Or something else?

When I study these groups, I try to pay attention to all of their interactions. And as soon as I allow myself to see that most of the interactions of the people I am studying involve vague but palpable sentiences squatting on their desks, I have to start thinking about watching the machines just as attentively as I watch the people, because, for them, the machines are not merely passage points. Haraway and other workers who observe the traffic across the boundaries between "nature," "society," and "technology" tend to see nature as lively, unpredictable, and, in some sense, actively resisting interpretations. If nature and technology seem to be collapsing into each other, as Haraway and others claim, then the unhumans can be lively too.

One symptom of this is that the flux of information that passes back and forth across the vanishing divides between nature and technology has become extremely dense. Cyborgs with a vengeance, one of the groups I study, is already talking about colonizing a social space in which the divide between nature and technology has become thoroughly unrecognizable, while one of the individuals I study is busy trying to sort out how the many people who seem to inhabit the social space of her body are colonizing her. When I listen to the voices in these new social spaces I hear a multiplicity of voices, some recognizably human and some quite different, all clamoring at once, frequently saying things whose meanings are tantalizingly familiar but which have subtly changed.

My interest in cyberspace is primarily about communities and how they work. Because I believe that technology and culture constitute each other, studying the actors and actants that make up our lively, troubling, and productive technologies tells me about the actors and actants that make up our culture. Since so much of a culture's knowledge is passed on by means of stories, I will begin by retelling a few boundary stories about virtual cultures.

Schizophrenia as Commodity Fetish

Let us begin with a person I will call Julie, on a computer conference in New York in 1985. Julie was a totally disabled older woman, but she could push the keys of a computer with her headstick. The personality she projected into the "net"—the vast electronic web that links computers all over the world—was huge. On the net, Julie's disability was invisible and irrelevant. Her standard greeting was a big, expansive "HI!!!!!!" Her heart was as big as her greeting, and in the intimate electronic companionships that can develop during on-line conferencing between people who may never physically meet, Julie's women friends shared their deepest troubles, and she offered them advice—advice that changed their lives. Trapped inside her ruined body, Julie herself was sharp and perceptive, thoughtful and caring.

After several years, something happened that shook the conference to the core. "Julie" did not exist. "She" was, it turned out, a middle-aged male psychiatrist. Logging onto the conference for the first time, this man had accidentally begun a discussion with a woman who mistook him for another woman. "I was stunned," he said later, "at the conversational mode. I hadn't known that women talked among themselves that way. There was so much more vulnerability, so much more depth and complexity. Men's conversations on the nets were much more guarded and superficial, even among intimates. It was fascinating, and I wanted more." He had spent weeks developing the right persona. A totally disabled, single older woman was perfect. He felt that such a person wouldn't be expected to have a social life. Consequently her existence only as a net persona would seem natural. It worked for years, until one of Julie's devoted admirers, bent on finally meeting her in person, tracked her down.

The news reverberated through the net. Reactions varied from humorous resignation to blind rage. Most deeply affected were the women who had shared their innermost feelings with Julie. "I felt raped," one said. "I felt that my deepest secrets had been violated." Several went so far as to repudiate the genuine gains they had made in their personal and emotional lives. They felt those gains were predicated on deceit and trickery.

The computer engineers, the people who wrote the programs by means of which the nets exist, just smiled tiredly. They had understood from the beginning the radical changes in social conventions that the nets implied. Young enough in the first days of the net to react and adjust quickly, they had long ago taken for granted that many of the old assumptions about the nature of identity had quietly vanished under the new electronic dispensation. Electronic networks in their myriad kinds, and the mode of interpersonal interaction that they foster, are a new manifestation of a social space that has been better known in its older and more familiar forms in conference calls, communities of letters, and FDR's fireside chats. It can be characterized as "virtual" space—an imaginary locus of interaction created by communal agreement. In its most recent form, concepts like distance, inside/outside, and even the physical body take on new and frequently disturbing meanings.

Now, one of the more interesting aspects of virtual space is "computer crossdressing." Julie was an early manifestation. On the nets, where *warranting*, or grounding, a persona in a physical body, is meaningless, men routinely use female personae whenever they choose, and vice versa. This wholesale appropriation of the other has spawned new modes of interaction. Ethics, trust, and risk still continue, but in different ways. Gendered modes of communication themselves have remained relatively stable, but who uses which of the two socially recognized modes has become more plastic. A woman who has appropriated a male conversational style may be simply assumed to be male at that place and time, so that her/his on-line persona takes on a kind of quasi life of its own, separate from the person's embodied life in the "real" world.

Sometimes a person's on-line persona becomes so finely developed that it begins to take over their life *off* the net. In studying virtual systems, I will call both the space of interaction that is the net and the space of interaction that we call the "real" world *consensual loci*. Each consensual locus has its own "reality," determined by local conditions. However, not all realities are equal. A whack on the head in the "real" world can kill you, whereas a whack in one of the virtual worlds will not (although a legal issue currently being debated by futurist attorneys is what liability the whacker has if the fright caused by a virtual whack gives the whackee a "real" heart attack).

Some conferencees talk of a time when they will be able to abandon warranting personae in even more complex ways, when the first "virtual reality" environments come on line. VR, one of a class of interactive spaces that are coming to be known by the general term *cyberspace*, is a three-dimensional consensual locus or, in the terms of science fiction author William Gibson, a "consensual hallucination" in which data may be visualized, heard, and even felt. The "data" in some of these virtual environments are people—3-D representations of individuals in the cyberspace. While high-resolution images of the human body in cyberspace are years away, when they arrive they will take "computer crossdressing" even further. In this version of VR a man may be seen, and perhaps touched, as a woman and vice versa—or as anything else. There is talk of renting prepackaged body forms complete with voice and touch . . . multiple personality as commodity fetish!

It is interesting that at just about the time the last of the untouched "real-world" anthropological field sites are disappearing, a new and unexpected kind of "field" is opening up—incontrovertibly social spaces in which people still meet face-to-face, but under new definitions of both "meet" and "face." These new spaces instantiate the collapse of the boundaries between the social and technological, biology and machine, natural and artificial that are part of the postmodern imaginary. They are part of the growing imbrication of humans and machines in new social forms that I call *virtual systems*.

A Virtual Systems Origin Myth

Cyberspace, without its high-tech glitz, is partially the idea of virtual community. The earliest cyberspaces may have been virtual communities, passage points for collections of common beliefs and practices that united people who were physically separated. Virtual communities sustain themselves by constantly circulating those practices. To give some examples of how this works, I'm going to tell an origin story of virtual systems.

There are four epochs in this story. The beginning of each is signaled by a marked change in the character of human communication. Over the years, human communication is increasingly mediated by technology. Because the rate of change in technological innovation increases with time, the more recent epochs are shorter, but roughly the same quantity of information is exchanged in each. Since the basis of virtual communities is communication, this seems like a reasonable way to divide up the field.

Epoch One: Texts. [From the mid-1600s]

Epoch Two: Electronic communication and entertainment media.
            [1900+]

Epoch Three: Information technology. [1960+]

Epoch Four: Virtual reality and cyberspace. [1984+]

Epoch One

This period of early textual virtual communities starts, for the sake of this discussion, in 1669 when Robert Boyle engaged an apparatus of literary technology to "dramatize the social relations proper to a community of philosophers." As Steven Shapin and Simon Shapiro point out in their study of the debate between Boyle and the philosopher Thomas Hobbes, *Leviathan and the Air-Pump*, we probably owe the invention of the boring academic paper to Boyle. Boyle developed a method of compelling assent that Shapin and Shaffer described as *virtual witnessing*. He created what he called a "community of like-minded gentlemen" to validate his scientific experiments, and he correctly surmised that the "gentlemen" for whom he was writing believed that boring, detailed writing implied painstaking experimental work. Consequently it came to pass that boring writing was likely to indicate scientific truth. By means of such writing, a group of people were able to "witness" an experiment without being physically present. Boyle's production of the detailed academic paper was so successful that it is still the exemplar of scholarship.

The document around which community forms might also be a novel, a work of fiction. Arguably the first texts to reach beyond class, gender, and ideological differences were the eighteenth-century sentimental novels, exemplified by the publication of Bernardin de Saint-Pierre's short novel *Paul and Virginia* (1788), which Roddey Reid, in his study "Tears For Fears," identifies as one of the early textual productions that "dismantled the absolutist public sphere and constructed a bourgeois public sphere through fictions of national community." Reid claims that *Paul and Virginia* was a passage point for a circulating cluster of concepts about the nature of social identity that transformed French society. Reid suggests that an entire social class—the French bourgeoisie—crystallized around the complex of emotional responses that the novel produced. Thus in the first epoch texts became ways of creating, and later of controlling, new kinds of communities.

Epoch Two

The period of the early electronic virtual communities began in the twentieth century with invention of the telegraph and continued with musical communities, previously constituted in the physical public space of the concert hall, shifting and translating to a new kind of virtual communal space around the phonograph. The apex of this period was Franklin Delano Roosevelt's radio "fireside chats," creating a community by means of readily available technology.

Once communities grew too big for everyone to know everyone else, which is to say very early on, government had to proceed through delegates who represented absent groups. FDR's use of radio was a way to bypass the need for delegates. Instead of talking to a few hundred representatives, Roosevelt used the radio as a machine for fitting listeners into his living room. The radio was one-way communication, but because of it people were able to begin to think of presence in a different way. Because of radio and of the apparatus for the production of community that it implied and facilitated, it was now possible for millions of people to be "present" in the same space—seated across from Roosevelt in his living room.

This view implies a new, different, and complex way of experiencing the relationship between the physical human body and the "I" that inhabits it. FDR did not physically enter listeners' living rooms. He invited listeners into his. In a sense, the listener was in two places at once—the body at home, but the delegate, the "I" that belonged to the body, in an imaginal space with another person. This space was enabled and constructed with the assistance of a particular technology. In the case of FDR the technology was a device that mediated between physical loci and incommensurable realities—in other words, an interface. In virtual systems *an interface is that which mediates between the human body (or bodies) and an associated "I" (or "I's").* This double view of "where" the "person" is, and the corresponding trouble it may cause with thinking about "who" we are talking about when we discuss such a problematic "person," underlies the structure of more recent virtual communities.

During the same period thousands of children, mostly boys, listened avidly to adventure serials, and sent in their coupons to receive the decoder rings and signaling devices that had immense significance within the community of a particular show. Away from the radio, they recognized each other by displaying the community's tokens, an example of communities of consumers organized for marketing purposes.

The motion picture, and later, television, also mobilized a similar power to organize sentimental social groups. Arguably one of the best examples of a virtual community in the late twentieth century is the

Trekkies, a huge, heterogenous group partially based on commerce but mostly on a set of ideas. The fictive community of "Star Trek" and the fantasy Trekkie community interrelate and mutually constitute each other in complex ways across the boundaries of texts, films, and video interfaces.

Epoch Two ended in the mid-1970s with the advent of the first computer, terminal-based, bulletin board systems (BBSs).

Epoch Three

This period began with the era of information technology. The first virtual communities based on information technology were the on-line bulletin board services (BBS) of the middle 1970s. These were not dependent upon the widespread ownership of computers, merely of terminals. But because even a used terminal cost several hundred dollars, access to the first BBSs was mainly limited to electronics experimenters, ham-radio operators, and the early hardy computer builders.

BBSs were named after their perceived function—virtual places, conceived to be just like physical bulletin boards, where people could post notes for general reading. The first successful BBS programs were primitive, usually allowing the user to search for messages alphabetically, or simply to read messages in the order in which they were posted. These programs were sold by their authors for very little, or given away as "shareware"—part of the early visionary ethic of electronic virtual communities. The idea of shareware, as enunciated by the many programmers who wrote shareware programs, was that the computer was a passage point for circulating concepts of community. The important thing about shareware, rather than making an immediate profit for the producer, was to nourish the community in expectation that such nourishment would "come around" to the nourisher.

CommuniTree Within a few months of the first BBS's appearance, a San Francisco group headed by John James, a programmer and visionary thinker, had developed the idea that the BBS was a virtual community, a community that promised radical transformation of existing society and the emergence of new social forms. The CommuniTree Group, as they called themselves, saw the BBS in McLuhanesque terms as transformative because of the ontological structure it presupposed and simultaneously created—the mode of tree-structured discourse and the community that spoke it—and because it was another order of "extension," a kind of prosthesis in McLuhan's sense. The BBS that the CommuniTree Group envisioned was an extension of the participant's instrumentality into a virtual social space.

The CommuniTree Group quite correctly foresaw that the BBS in its original form was extremely limited in its usefulness. Their reasoning was simple. The physical bulletin board for which the BBS was the metaphor had the advantage of being quickly scannable. By its nature, the physical bulletin board was small and manageable in size. There was not much need for bulletin boards to be organized by topic. But the on-line BBS could not be scanned in any intuitively satisfactory way. There were primitive search protocols in the early BBSs, but they were usually restricted to alphabetical searches or searches by keywords. The CommuniTree Group proposed a new kind of BBS that they called a tree-structured conference, employing as a working metaphor both the binary tree protocols in computer science and also the organic qualities of trees as such appropriate to the 1970s. Each branch of the tree was to be a separate conference that grew naturally out of its root message by virtue of each subsequent message that was attached to it. Conferences that lacked participation would cease to grow, but would remain on-line as archives of failed discourse and as potential sources of inspiration for other, more flourishing conferences.

With each version of the BBS system, The CommuniTree Group supplied a massive, detailed instruction manual—which was nothing less than a set of directions for constructing a new kind of virtual community. They couched the manual in radical seventies language, giving chapters such titles as "Downscale, please, Buddha" and "If you meet the electronic avatar on the road, laserblast hir!" This rich intermingling of spiritual and technological imagery took place in the context of George Lucas's *Star Wars*, a film that embodied the themes of the technological transformativists, from the all-pervading Force to what Vivian Sobchack (1987) called "the outcome of infinite human and technological progress." It was around *Star Wars* in particular that the technological and radically spiritual virtual communities of the early BBSs coalesced. *Star Wars* represented a future in which the good guys won out over vastly superior adversaries—with the help of a mystical Force that "surrounds us and penetrates us . . . it binds the galaxy together" and which the hero can access by learning to "trust your feelings"—a quintessential injunction of the early seventies.

CommuniTree #1 went on-line in May 1978 in the San Francisco Bay area of northern California, one year after the introduction of the Apple II computer and its first typewritten and hand-drawn operating manual. CommuniTree #2 followed quickly. The opening sentence of the prospectus for the first conference was "We are as gods and might as well get good at it." This technospiritual bumptiousness, full of the promise of the redemptive power of technology mixed with the easy, catch-all Eastern mysticism popular in upscale northern California, characterized the early conferences. As might be gathered from the tone of the prospectus, the first conference, entitled "Origins," was about successor religions.

The conferencees saw themselves not primarily as readers of bulletin boards or participants in a novel discourse but as agents of a new kind of social experiment. They saw the terminal or personal computer as a tool for social transformation by the ways it refigured social interaction. BBS conversations were time-aliased, like a kind of public letter writing or the posting of broadsides. They were meant to be read and replied to some time later than they were posted. But their participants saw them as conversations nonetheless, as social acts. When asked how sitting alone at a terminal was a social act, they explained that they saw the terminal as a window into a social space. When describing the act of communication, many moved their hands expressively as though typing, emphasizing the gestural quality and essential tactility of the virtual mode. Also present in their descriptions was a propensity to reduce other expressive modalities to the tactile. It seemed clear that, from the beginning, the electronic virtual mode possessed the power to overcome its character of single-mode transmission and limited bandwidth.

By 1982 Apple Computer had entered into the first of a series of agreements with the federal government in which the corporation was permitted to give away computers to public schools in lieu of Apple's paying a substantial portion of its federal taxes. In terms of market strategy, this action dramatically increased Apple's presence in the school system and set the pace for Apple's domination in the education market. Within a fairly brief time there were significant numbers of personal computers accessible to students of grammar school and high school age. Some of those computers had modems.

The students, at first mostly boys and with the linguistic proclivities of pubescent males, discovered the Tree's phone number and wasted no time in logging onto the conferences. They appeared uninspired by the relatively intellectual and spiritual air of the ongoing debates, and proceeded to express their dissatisfaction in ways appropriate to their age, sex, and language abilities. Within a short time the Tree was jammed with obscene and scatalogical messages. There was no way to monitor them as they arrived, and no easy way to remove them once they were in the system. This meant that the entire system had to be purged—a process taking hours—every day or two. In addition, young hackers enjoyed the sport of attempting to "crash" the system by discovering bugs in the system commands. Because of the provisions of the system that made observing incoming messages impossible, the hackers were free to experiment with impunity, and there was no way for the system operator to know what was taking place until the system crashed. At that time it was generally too late to save the existing disks. The system operator would be obliged to reconstitute ongoing conferences from earlier backup versions.

Within a few months, the Tree had expired, choked to death with what one participant called "the consequences of freedom of expression." During the years of its operation, however, several young participants took the lessons and implications of such a community away with them, and proceeded to write their own systems. Within a few years there was a proliferation of on-line virtual communities of somewhat less visionary character but vastly superior message-handling capability—systems that allowed monitoring and disconnection of "troublesome" participants (hackers attempting to crash the system), and easy removal of messages that did not further the purposes of the system operators. The age of surveillance and social control had arrived for the electronic virtual community.

The visionary character of CommuniTree's electronic ontology proved an obstacle to the Tree's survival. Ensuring privacy in all aspects of the Tree's structure and enabling unlimited access to all conferences did not work in a context of increasing availability of terminals to young men who did not necessarily share the Tree gods' ideas of what counted as community. As one Tree veteran put it, "The barbarian hordes mowed us down." Thus, in practice, surveillance and control proved necessary adjuncts to maintaining order in the virtual community.

It is tempting to speculate about what might have happened if the introduction of CommuniTree had not coincided with the first wave of "computerjugen." Perhaps the future of electronic virtual communities would have been quite different.

SIMNET   Besides the BBSs, there were more graphic, interactive systems under construction. Their interfaces were similar to arcade games or flight simulators—(relatively) high-resolution, animated graphics. The first example of this type of cyberspace was a military simulation called SIMNET. SIMNET was conducted by a consortium of military interests, primarily represented by DARPA, and a task group from the Institute for Simulation and Training, located at the University of Central Florida. SIMNET came about because DARPA was beginning to worry about whether the Army could continue to stage large-scale military practice exercises in Germany. With the rapid and unpredictable changes that were taking place in Europe in the late 1980s, the army wanted to have a backup—some other place where they could stage practice maneuvers without posing difficult political questions. As one of the developers of SIMNET put it, "World War III in Central Europe is at the moment an unfashionable anxiety." In view of the price of land and fuel, and of the escalating cost of staging practice maneuvers, the armed forces felt that if a large-scale consensual simulation could be made practical they could realize an immediate and useful financial advantage. Therefore, DARPA committed significant resources—money, time, and computer power—to funding some research laboratory to generate a 200-tank cyberspace simulation. DARPA put out requests for proposals, and a group at the University of Central Florida won.

The Florida group designed and built the simulator units with old technology, along the lines of conventional aircraft cockpit simulators. Each tank simulator was equipped to carry a crew of four, so the SIMNET environment is an 800-person virtual community.

SIMNET is a two-dimensional cyberspace. The system can be linked up over a very large area geographically; without much difficulty, in fact, to anywhere in the world. A typical SIMNET node is an M-1 tank simulator. Four crew stations contain a total of eight vision blocks, or video screens, visible through the tank's ports. Most of these are 320 x 138 pixels in size, with a 15 Hertz update rate. This means that the image resolution is not very good, but the simulation can be generated with readily available technology no more complex than conventional video games. From inside the "tank" the crew looks out the viewports, which are the video screens. These display the computer-generated terrain over which the tanks will maneuver (which happens to be the landscape near Fort Knox, Kentucky). Besides hills and fields, the crew can see vehicles, aircraft, and up to 30 other tanks at one time. They can hear and see the vehicles and planes shooting at each other and at them.

By today's standards, SIMNET's video images are low-resolution and hardly convincing. There is no mistaking the view out the ports for real terrain. But the simulation is astonishingly effective, and participants become thoroughly caught up in it. SIMNET's designers believe that it may be the lack of resolution itself that is responsible, since it requires the participants to actively engage their own imaginations to fill the holes in the illusion! McLuhan redux. That it works is unquestionable. When experimenters opened the door to one of the simulators during a test run to photograph the interior, the participants were so caught up in the action that they didn't notice the bulky camera poking at them.

Habitat, designed by Chip Morningstar and Randall Farmer, is a large-scale social experiment that is accessible through such common tele-phone-line computer networks as Tymnet. Habitat was designed for LucasFilm, and has been on-line for about a year and a half. It is a completely decentralized, connectionist system. The technology at the user interface was intended to be simple, this in order to minimize the costs of getting on-line. Habitat is designed to run on a Commodore 64 computer, a piece of very old technology in computer terms (in other words, at least ten years old), but Morningstar and Farmer have milked an amazing amount of effective bandwidth out of the machine. The Commodore 64 is very inexpensive and readily available. Almost anyone can buy one if, as one Habitat participant said, "they don't already happen to have one sitting around being used as a doorstop." Commodore 64s cost $100 at such outlets as Toys R Us.

Habitat existed first as a 35-foot mural located in a building in Sausalito, California, but, on-line, each area of the mural represents an entirely expandable area in the cyberspace, be it a forest, a plain, or a city. Habitat is inhabitable in that, when the user signs on, he or she has a window into the ongoing social life of the cyberspace—the community "inside" the computer. The social space itself is represented by a cartoonlike frame. The virtual person who is the user's delegated agency is represented by a cartoon figure that may be customized from a menu of body parts. When the user wishes his/her character to speak, s/he types out the words on the Commodore's keyboard, and these appear in a speech balloon over the head of the user's character. The speech balloon is visible to any other user nearby in the virtual space.[1] The user sees whatever other people are in the immediate vicinity in the form of other figures.

Habitat is a two-dimensional example of what William Gibson called a "consensual hallucination." First, according to Morningstar and Farmer, it has well-known protocols for encoding and exchanging information. By generally accepted usage among cyberspace engineers, this means it is consensual. The simulation software uses agents that can transform information to simulate environment. This means it is an hallucination.

Habitat has proved to be incontrovertibly social in character. During Habitat's beta test, several social institutions sprang up spontaneously. As Randall Farmer points out in his report on the initial test run, there were marriages and divorces, a church (complete with a real-world Greek Orthodox minister), a loose guild of thieves, an elected sheriff (to combat the thieves), a newspaper with a rather eccentric editor, and before long two lawyers hung up their shingles to sort out claims. And this was with only 150 people. My vision (of Habitat) encompasses tens of thousands of simultaneous participants.

Lessons of the Third Epoch   In the third epoch the participants of electronic communities seem to be acquiring skills that are useful for the virtual social environments developing in late twentieth-century technologized nations. Their participants have learned to delegate their agency to body-representatives that exist in an imaginal space contiguously with representatives of other individuals. They have become accustomed to what might be called lucid dreaming in an awake state—to a constellation of activities much like reading, but an active and interactive reading, a participatory social practice in which the actions of the reader have consequences in the world of the dream or the book.

In the third epoch the older metaphor of reading is undergoing a transformation in a textual space that is consensual, interactive, and haptic, and that is constituted through inscription practices—the production of microprocessor code. Social spaces are beginning to appear that are simultaneously natural, artificial, and constituted by inscription. The boundaries between the social and the natural and between biology and technology are beginning to take on the generous permeability that characterizes communal space in the fourth epoch.

Epoch Four

Arguably the single most significant event for the development of fourth-stage virtual communities was the publication of William Gibson's science fiction novel *Neuromancer*. *Neuromancer* represents the dividing line between the third and fourth epochs not because it signaled any technological development, but because it crystallized a new community, just as Boyle's scientific papers and *Paul and Virginia* did in an earlier age.

Neuromancer reached the hackers who had been radicalized by George Lucas's powerful cinematic evocation of humanity and technology infinitely extended, and it reached the technologically literate and socially disaffected who were searching for social forms that could transform the fragmented anomie that characterized life in Silicon Valley and all electronic industrial ghettos. In a single stroke, Gibson's powerful vision provided for them the imaginal public sphere and refigured discursive community that established the grounding for the possibility of a new kind of social interaction. As with *Paul and Virginia* in the time of Napoleon and Dupont de Nemours, *Neuromancer* in the time of Reagan and DARPA is a massive intertextual presence not only in other literary productions of the 1980s, but in technical publications, conference topics, hardware design, and scientific and technological discourses in the large.

The three-dimensional inhabitable cyberspace described in *Neuromancer* does not yet exist, but the groundwork for it can be found in a series of experiments in both the military and private sectors.

Many VR engineers concur that the tribal elders of 3-D virtual systems are Scott Fisher and Ivan Sutherland, formerly at MIT, and Tom Furness, with the Air Force. In 1967–68, Sutherland built a see-through helmet at the MIT Draper Lab in Cambridge. This system used television screens and half-silvered mirrors, so that the environment was visible through the TV displays. It was not designed to provide a surround environment. In 1969–70 Sutherland went to the University of Utah, where he continued this work, doing things with vector-generated computer graphics and maps, still see-through technology. In his lab were Jim Clark, who went on to start Silicon Graphics, and Don Vickers.

Tom Furness had been working on VR systems for 15 years or more—he started in the mid-seventies at Wright-Patterson Air Force Base. His systems were also see-through, rather than enclosing. He pushed the technology forward, particularly by adopting the use of high-resolution CRTs. Furness's system, designed for the USAF, was an elaborate flight simulation cyberspace employing a helmet with two large CRT devices, so large and cumbersome that it was dubbed the "Darth Vader helmet." He left Wright-Patterson in 1988–89 to start the Human Interface Technology Lab at the University of Washington.

Scott Fisher started at MIT in the machine architecture group. The MA group worked on developing stereo displays and crude helmets to contain them, and received a small proportion of their funding from DARPA. When the group terminated the project, they gave the stereo displays to another group at UNC (University of North Carolina), which was developing a display device called the Pixel Planes Machine. In the UNC lab were Henry Fuchs and Fred Brooks, who had been working on force feedback with systems previously developed at Argonne and Oak Ridge National labs. The UNC group worked on large projected stereo displays, but was aware of Sutherland's and Furness's work with helmets, and experimented with putting a miniature display system into a helmet of their own. Their specialties were medical modeling, molecular modeling, and architectural walk-through. The new Computer Science building at UNC was designed partially with their system. Using their software and 3-D computer imaging equipment, the architects could "walk through" the full-sized virtual building and examine its structure. The actual walk-through was accomplished with a treadmill and bicycle handlebars. The experiment was so successful that during the walk-through one of the architects discovered a misplaced wall that would have cost hundreds of thousands of dollars to fix once the actual structure had been built.

In 1982, Fisher went to work for Atari. Alan Kay's style at Atari was to pick self-motivated people and then turn them loose, on anything from flight simulation to personal interactive systems. The lab's philosophy was at the extreme end of visionary. According to Kay, the job of the group was to develop products not for next year or even for five years away, but for no less than 15 to 20 years in the future. In the corporate climate of the 1980s, and in particular in Silicon Valley, where product life and corporate futures are calculated in terms of months, this approach was not merely radical but stratospheric. For the young computer jocks, the lure of Silicon Valley and of pushing the limits of computer imaging into the far future was irresistible, and a group of Cambridge engineers, each outstanding in their way, made the trek out to the coast. Eric Gullichsen arrived first, then Scott Fisher and Susan Brennan, followed a year later by Ann Marion. Michael Naimark was already there, as was Brenda Laurel. Steve Gans was the last to arrive.

As it turned out, this was not a good moment to arrive at Atari. When the Atari lab closed, Ann Marion and Alan Kay went to Apple (followed by a drove of other Atari expatriates), where they started the Vivarium project and continued their research. Susan Brennan went first to the Stanford Psychology Department and also Hewlett-Packard, which she left in 1990 to teach at CUNY Stony Brook. Michael Naimark became an independent producer and designer of interactive video and multimedia art. William Bricken and Eric Gullichsen took jobs at Autodesk, the largest manufacturer of CAD software, where they started a research group called Cyberia.

Scott Fisher went to work for Dave Nagel, head of the NASA-Ames View Lab. To go with their helmet, the Ames lab had developed a primitive sensor to provide the computer with information about the position of the user's hand. The early device used a simple glove with strain gauges wired to two fingers. They contracted with VPL, Inc. to develop it further, using software written in collaboration with Scott. The Ames group referred to the software as "gesture editors." The contract started in 1985, and VPL delivered the first glove in March 1986. The Ames group intended to apply the glove and software to such ideas as surgical simulation, 3-D virtual surgery for medical students. In 1988, Dave Nagel left the Ames laboratory to become director of the Advanced Technology Group (ATG) at Apple.

Lusting for images, such organizations as SIGGRAPH gobbled up information about the new medium and spread it out through its swarm of networks and publications. The audience, made up largely of young, talented, computer-literate people in both computer science and art, and working in such fields as advertising, media, and the fine arts, had mastered the current state of the art in computers and was hungry for the next thing. LucasFilm (later LucasArts) in Marin, now doing the bulk of all computerized special effects for the film industry, and Douglas Trumbull's EEG in Hollywood, fresh from their spectacular work on *Blade Runner*, had made the production of spectacular visual imaginaries an everyday fact. They weren't afraid to say that they had solved all of the remaining problems with making artificial images, under particular circumstances, indistinguishable from "real" ones— a moment that Stewart Brand called "(t)he end of photography as evidence for anything." Now the artists and engineers who worked with the most powerful imaging systems, like Lucas's Pixar, were ready for more. They wanted to be able to get inside their own fantasies, to experientially inhabit the worlds they designed and built but could never enter. VR touched the same nerve that *Star Wars* had, the englobing specular fantasy made real.

Under Eric Gullichsen and William Bricken, the Autodesk Cyberspace Project quickly acquired the nickname Cyberia. John Walker, president of Autodesk, had seen the UNC architectural system and foresaw a huge market for virtual CAD— 3-D drawings that the designers could enter. But after a year or so, Autodesk shrank the Cyberia project. Eric Gullichsen left to start Sense8, a manufacturer of low-end VR systems. William Bricken (and later his wife Meredith) left the company to take up residence at the University of Washington, where Tom Furness and his associates had started the Human Interface Technology Laboratory. Although there were already academic-based research organizations in existence at that time (Florida, North Carolina), and some of them (Florida) were financed at least in part by DOD, the HIT lab became the first academic organization to secure serious research funding from private industry.

During this period, when *Neuromancer* was published, "virtual reality" acquired a new name and a suddenly prominent social identity as "cyberspace." The critical importance of Gibson's book was partly due to the way that it triggered a conceptual revolution among the scattered workers who had been doing virtual reality research for years: As task groups coalesced and dissolved, as the fortunes of companies and projects and laboratories rose and fell, the existence of Gibson's novel and the technological and social imaginary that it articulated enabled the researchers in virtual reality—or, under the new dispensation, cyberspace—to recognize and organize themselves as a community.

By this time private industry, represented by such firms as American Express, PacBell, IBM, MCC, Texas Instruments, and NYNEX, were beginning to explore the possibilities and commercial impact of cyberspace systems. That is not to say that people were rushing out to purchase tickets for a cyberspace vacation! The major thrust of the industrial and institutional commitment to cyberspace research was still focused on data manipulation—just as Gibson's *zaibatsu* did in *Neuromancer*. Gibson's cowboys were outlaws in a military-industrial fairyland dominated by supercomputers, artificial intelligence devices, and data banks. Humans were present, but their effect was minimal. There is no reason to believe that the cyberspaces being designed at NASA or Florida will be any different. However, this knowledge does not seem to daunt the "real" cyberspace workers. Outside of their attention to the realities of the marketplace and workplace, the young, feisty engineers who do the bulk of the work on VR systems continue their discussions and arguments surrounding the nature and context of virtual environments. That these discussions already take place in a virtual environment—the great, sprawling international complex of commercial, government, military, and academic computers known as Usenet—is in itself suggestive.

Decoupling the Body and the Subject

> The illusion will be so powerful you won't be able *to tell what's real and what's not.*
>
> —Steve Williams

In her complex and provocative 1984 study *The Tremulous Private Body*, Frances Barker suggests that, because of the effects of the Restoration on the social and political imaginary in Britain (1660 and on), the human body gradually ceased to be perceived as public spectacle, as had previously been the case, and became privatized in new ways. In Barker's model of the post-Jacobean citizen, the social economy of the body became rearranged in such a way as to interpose several layers between the individual and public space. Concomitant with this removal of the body from a largely public social economy, Barker argues that the subject, the "I" or perceiving self that Descartes had recently pried loose from its former unity with the body, reorganized, or was reorganized, in a new economy of its own. In particular, the subject, as did the body, ceased to constitute itself as public spectacle and instead fled from the public sphere and constituted itself in *text*—such as Samuel Pepys' diary (1668).

Such changes in the social economy of both the body and the subject, Barker suggests, very smoothly serve the purposes of capital accumulation. The product of a privatized body and of a subject removed from the public sphere is a social monad more suited to manipulation by virtue of being more isolated. Barker also makes a case that the energies of the individual, which were previously absorbed in a complex public social economy and which regularly returned to nourish the sender, started backing up instead, and needing to find fresh outlets. The machineries of capitalism handily provided a new channel for productive energy. Without this damming of creative energies, Barker suggests, the industrial age, with its vast hunger for productive labor and the consequent creation of surplus value, would have been impossible.

In Barker's account, beginning in the 1600s in England, the body became progressively more hidden, first because of changing conventions of dress, later by conventions of spatial privacy. Concomitantly, the self, Barker's "subject," retreated even further inward, until much of its means of expression was through texts. Where social communication had been direct and personal, a warrant was developing for social communication to be indirect and delegated through communication technologies—first pen and paper, and later the technologies and market economics of print. The body (and the subject, although he doesn't lump them together in this way) became "the site of an operation of power, of an exercise of meaning . . . a transition, effected over a long period of time, from a socially visible object to one which can no longer be seen" (Barker 1984: 13).

While the subject in Barker's account became, in her words, "raging, solitary, productive," what it produced was text. On the other hand, it was the newly hidden Victorian body that became physically productive and that later provided the motor for the industrial revolution; it was most useful as a brute body, for which the creative spark was an impediment. In sum, the body became more physical, while the subject became more textual, which is to say nonphysical.

If the information age is an extension of the industrial age, with the passage of time the split between the body and the subject should grow more pronounced still. But in the fourth epoch the split is simultaneously growing and disappearing. The socioepistemic mechanism by which bodies mean is undergoing a deep restructuring in the latter part of the twentieth century, finally fulfilling the furthest extent of the isolation of those bodies through which its domination is authorized and secured.

I don't think it is accidental that one of the earliest, textual, virtual communities—the community of gentlemen assembled by Robert Boyle during his debates with Hobbes—came into existence at the moment about which Barker is writing. The debate between Boyle and Hobbes and the production of Pepys' diary are virtually contemporaneous. In the late twentieth century, Gibson's *Neuromancer* is simultaneously a perverse evocation of the Restoration subject and its annihilation in an implosion of meaning from which arises a new economy of signification.

Barker's work resonates in useful ways with two other accounts of the evolution of the body and the subject through the interventions of late twentieth-century technologies: Donna Haraway's "A Manifesto for Cyborgs" and "The Biopolitics of Postmodern Bodies" (1985, 1988). Both these accounts are about the collapse of categories and of the boundaries of the body. (Shortly after being introduced to Haraway's work I wrote a very short paper called "Sex And Death among the Cyborgs." The thesis of "Sex And Death" was similar to Haraway's.) The boundaries between the subject, if not the body, and the "rest of the world" are undergoing a radical refiguration, brought about in part through the mediation of technology. Further, as Baudrillard and others have pointed out, the boundaries between technology and nature are themselves in the midst of a deep restructuring. This means that many of the usual analytical categories have become unreliable for making the useful distinctions between the biological and the technological, the natural and artificial, the human and mechanical, to which we have become accustomed.

François Dagognet suggests that the recent debates about whether nature is becoming irremediably technologized are based on a false dichotomy: namely that there exists, here and now, a category "nature" which is "over here," and a category "technology" (or, for those following other debates, "culture") which is "over there." Dagognet argues on the contrary that the category "nature" has not existed for thousands of years . . . not since the first humans deliberately planted gardens or discovered slash-and-burn farming. I would argue further that "Nature," instead of representing some pristine category or originary state of being, has taken on an entirely different function in late twentieth-century economies of meaning. Not only has the character of nature as yet another coconstruct of culture become more patent, but is has become nothing more (or less) than an ordering factor—a construct by means of which we attempt to *keep technology visible* as something separate from our "natural" selves and our everyday lives. In other words, the category "nature," rather than referring to any object or category in the world, is a *strategy* for maintaining boundaries for political and economic ends, and thus a way of making meaning. (In this sense, the project of reifying a "natural" state over and against a technologized "fallen" one is not only one of the industries of postmodern nostalgia, but also part of a binary, oppositional cognitive style that some maintain is part of our society's pervasively male epistemology.)

These arguments imply as a corollary that "technology," as we customarily think of it, does not exist either; that we must begin to rethink the category of technology as also one that exists only because of its imagined binary opposition to another category upon which it operates and in relation to which it is constituted. In a recent paper Paul Rabinow asks what kind of being might thrive in a world in which nature is becoming increasingly technologized. What about a being who has learned to live in a world in which, rather than nature becoming technologized, technology *is* nature—in which the boundaries between subject and environment have collapsed?

Phone sex workers and VR engineers   I have recently been conducting a study of two groups who seemed to instantiate productive aspects of this implosion of boundaries. One is phone sex workers. The other is computer scientists and engineers working on VR systems that involve making humans visible in the virtual space. I was interested in the ways in which these groups, which seem quite different, are similar. For the work of both is about representing the human body through limited communication channels, and both groups do this by coding cultural expectations as tokens of meaning.

Computer engineers seem fascinated by VR because you not only program a world, but in a real sense inhabit it. Because cyberspace worlds can be inhabited by communities, in the process of articulating a cyberspace system, engineers must model cognition and community; and because communities are inhabited by bodies, they must model bodies as well. While cheap and practical systems are years away, many workers are already hotly debating the form and character of the communities they believe will spring up in their quasi-imaginary cyberspaces. In doing so, they are articulating their own assumptions about bodies and sociality and projecting them onto the codes that define cyberspace systems. Since, for example, programmers create the codes by which VR is generated in interaction with workers in widely diverse fields, how these heterogenous co-working groups understand cognition, community, and bodies will determine the nature of cognition, community, and bodies in VR.

Both the engineers and the sex workers are in the business of constructing tokens that are recognized as objects of desire. Phone sex is the process of provoking, satisfying, *constructing* desire through a single mode of communication, the telephone. In the process, participants draw on a repertoire of cultural codes to construct a scenario that compresses large amounts of information into a very small space. The worker verbally codes for gesture, appearance, and proclivity, and expresses these as tokens, sometimes in no more than a word. The client uncompresses the tokens and constructs a dense, complex interactional image. In these interactions desire appears as a product of the tension between embodied reality and the emptiness of the token, in the forces that maintain the preexisting codes by which the token is constituted. The client mobilizes expectations and preexisting codes for body in the modalities that are not expressed in the token; that is, tokens in phone sex are purely verbal, and the client uses cues in the verbal token to construct a multimodal object of desire with attributes of shape, tactility, odor, etc. This act is thoroughly individual and interpretive; out of a highly compressed token of desire the client constitutes meaning that is dense, locally situated, and socially particular.

Bodies in cyberspace are also constituted by descriptive codes that "embody" expectations of appearance. Many of the engineers currently debating the form and nature of cyberspace are the young turks of computer engineering, men in their late teens and twenties, and they are preoccupied with the things with which postpubescent men have always been preoccupied. This rather steamy group will generate the codes and descriptors by which bodies in cyberspace are represented. Because of practical limitations, a certain amount of their discussion is concerned with data compression and tokenization. As with phone sex, cyberspace is a relatively narrow-bandwidth representational medium, visual and aural instead of purely aural to be sure, but how bodies are represented will involve how *recognition* works.

One of the most active sites for speculation about how *recognition* might work in cyberspace is the work of computer game developers, in particular the area known as interactive fantasy (IF). Since Gibson's first book burst onto the hackers' scene, interactive fantasy programmers (in particular, Laurel and others) have been taking their most durable stock-in-trade and speculating about how it will be deployed in virtual reality scenarios. For example, how, if they do, will people make love in cyberspace—a space in which everything, including bodies, exists as something close to a metaphor. Fortunately or unfortunately, however, everyone is still preorgasmic in virtual reality.

When I began the short history of virtual systems, I said that I wanted to use accounts of virtual communities as an entry point into a search for two things: an apparatus for the production of community and an apparatus for the production of body. Keeping in mind that this chapter is necessarily brief, let me look at the data so far:

• Members of electronic virtual communities act as if the community met in a physical public space. The number of times that on-line conferencees refer to the conference as an architectural place and to the mode of interaction in that place as being social is overwhelmingly high in proportion to those who do not. They say things like "This is a nice place to get together" or "This is a convenient place to meet."

• The virtual space is most frequently visualized as Cartesian. On-line conferencees tend to visualize the conference system as a three-dimensional space that can be mapped in terms of Cartesian coordinates, so that some branches of the conference are "higher up" and others "lower down." (One of the commands on the Stuart II conference moved the user "sideways.") Gibson's own visualization of cyberspace was Cartesian. In consideration of the imagination I sometimes see being brought to bear on virtual spaces, this odd fact invites further investigation.

• Conferencees act as if the virtual space were inhabited by bodies. Conferencees construct bodies on-line by describing them, either spontaneously or in response to questions, and articulate their discourses around this assumption.

• Bodies in virtual space have complex erotic components. Conferencees may flirt with each other. Some may engage in "netsex," constructing elaborate erotic mutual fantasies. Erotic possibilities for the virtual body are a significant part of the discussions of some of the groups designing cyberspace systems. The consequences of virtual bodies are considerable in the local frame, in that conferencees mobilize significant erotic tension in relation to their virtual bodies. In contrast to the conferences, the bandwidth for physicalities in phone sex is quite limited. (One worker said ironically, "(o)n the phone, every female sex worker is white, five feet four, and has red hair.")

• The meaning of locality and privacy is not settled. The field is rife with debates about the legal status of communications within the networks. One such, for example, is about the meaning of inside and outside. Traditionally, when sending a letter one preserves privacy by enclosing it in an envelope. But in electronic mail, for example, the address is part of the message. The distinction between inside and outside has been erased, and along with it the possibility of privacy. Secure encryption systems are needed.[2]

• Names are local labels. "Conferencees" seem to have no difficulty addressing, befriending, and developing fairly complex relationships with the delegated puppets—agents—of other conferencees. Such relationships remain stable as long as the provisional name ("handle") attached to the puppet does not change, but an unexpected observation was that relationships remain stable when the conferencee decides to change handles, as long as fair notice is given. Occasionally a conferencee will have several handles on the same conference, and a constructed identity for each. Other conferencees may or may not be aware of this. Conferencees treat others' puppets as if they were embodied people meeting in a public space nonetheless.

Private Body, Public Body, and Cyborg Envy

Partly, my interest in VR engineers stems from observations that suggest that they while are surely engaged in saving the project of late-twentieth-century capitalism, they are also inverting and disrupting its consequences for the body as object of power relationships. They manage both to preserve the privatized sphere of the individual—which Barker characterizes as "raging, solitary, productive"—as well as to escape to a position that is of the spectacle and incontrovertibly public. But this occurs under a new definition of public and private: one in which warrantability is irrelevant, spectacle is plastic and negotiated, and desire no longer grounds itself in physicality. Under these conditions, one might ask, will the future inhabitants of cyberspace "catch" the engineers' societal imperative to construct desire in gendered, binary terms—coded into the virtual body descriptors—or will they find more appealing the possibilities of difference unconstrained by relationships of dominance and submission? Partly this will depend upon how "cyberspaceians" engage with the virtual body.

Vivian Sobchack, in her 1987 discussion of cinematic space excludes the space of the video and computer screen from participation in the production of an "apparatus of engagement." Sobchack describes engagement with cinematic space as producing a thickening of the present . . . a "temporal simultaneity (that) also extends presence spatially—transforming the 'thin' abstracted space of the machine into a thickened and concrete world." Contrasted with video, which is to say with the electronic space of the CRT screen and with its small, low-resolution, and serial mode of display, the viewer of cinema engages with the apparatus of cinematic production in a way that produces "a space that is deep and textural, that can be materially inhabited . . . a specific and mobile engagement of embodied and enworlded subjects/objects whose visual/visible activity prospects and articulates a shifting field of vision from a world that always exceeds it." Sobchack speaks of electronic space as "a phenomenological structure of sensual and psychological experience that seems to belong to no-body." Sobchack sees the computer screen as "spatially decentered, weakly temporalized and quasi-disembodied."

This seems to be true, as long as the mode of engagement remains that of spectator. But it is the quality of direct physical and kinesthetic engagement, the enrolling of hapticity in the service of both the drama and the dramatic, which is not part of the cinematic mode. The cinematic mode of engagement, like that of conventional theater, is mediated by two modalities; the viewer experiences the presentation through sight and hearing. The electronic screen is "flat," so long as we consider it in the same bimodal way. But it is the potential for interaction that is one of the things that distinguishes the computer from the cinematic mode, and that transforms the small, low-resolution, and frequently monochromatic electronic screen from a novelty to a powerfully gripping force. Interaction is the physical concretization of a desire to escape the flatness and merge into the created system. It is the sense in which the "spectator" is more than a participant, but becomes both participant in and creator of the simulation. In brief, it is the sense of unlimited power which the dis/embodied simulation produces, and the different ways in which socialization has led those always-embodied participants confronted with the sign of unlimited power to respond.

In quite different terms from the cinematic, then, cyberspace "thickens" the present, producing a space that is deep and textural, and one that, in Sobchack's terms, can be materially inhabited. David Tomas, in his article "The Technophilic Body" (1989), describes cyberspace as "a purely spectacular, kinesthetically exciting, and often dizzying sense of bodily freedom." I read this in the additional sense of freedom *from* the body, and in particular perhaps, freedom from the sense of loss of control that accompanies adolescent male embodiment. Cyberspace is surely also a concretization of the psychoanalytically framed desire of the male to achieve the "kinesthetically exciting, dizzying sense" of freedom.

Some fiction has been written about multimodal, experiential cinema. But the fictional apparatus surrounding imaginary cybernetic spaces seems to have proliferated and pushed experiential cinema into the background. This is because cyberspace is part of, not simply the medium for, the action. Sobchack, on the other hand, argues that cinematic space possesses a power of engagement that the electronic space cannot match:

Semiotically engaged as subjective and intentional, as presenting representation of the objective world . . . The spectator(s) can share (and thereby to a degree interpretively alter) a film's presentation and representation of embodied experience. (Forthcoming)

Sobchack's argument for the viewer's intentional engagement of cinematic space, slightly modified, however, works equally well for the cybernetic space of the computer. That is, one might say that the console cowboy is also " . . . semiotically engaged as subjective and intentional, as presenting representation of a *sub*jective world . . . the spectator can share (and thereby to a high degree interpretively alter) a simulation's presentation and representation of experience which may be, through cybernetic/semiotic operators not yet existent but present and active in fiction (the cyberspace deck), mapped back upon the physical body."

In psychoanalytic terms, for the young male, unlimited power first suggests the mother. The experience of unlimited power is both gendered, and, for the male, fraught with the need for control, producing an unresolvable need for reconciliation with an always absent structure of personality. An "absent structure of personality" is also another way of describing the peculiarly seductive character of the computer that Turkle characterizes as the "second self." Danger, the sense of threat as well as seductiveness that the computer can evoke, comes from both within and without. It derives from the complex interrelationships between human and computer, and thus partially within the human; and it exists quasi-autonomously within the simulation. It constitutes simultaneously the senses of erotic pleasure and of loss of control over the body. Both also constitute a constellation of responses to the simulation that deeply engage fear, desire, pleasure, and the need for domination, subjugation, and control.

It seems to be the engagement of the adolescent male within humans of both sexes that is responsible for the seductiveness of the cybernetic mode. There is also a protean quality about cybernetic interaction, a sense of physical as well as conceptual mutability that is implied in the sense of exciting, dizzying physical movement within purely conceptual space. I find that reality hackers experience a sense of longing for an embodied conceptual space like that which cyberspace suggests. This sense, which seems to accompany the desire to cross the human/machine boundary, to penetrate and merge, which is part of the evocation of cyberspace, and which shares certain conceptual and affective characteristics with numerous fictional evocations of the inarticulate longing of the male for the female, I characterize as *cyborg envy*.

Smoothness implies a seductive tactile quality that expresses one of the characteristics of cyborg envy: In the case of the computer, a desire literally to enter into such a discourse, to penetrate the smooth and relatively affectless surface of the electronic screen and enter the deep, complex, and tactile (individual) cybernetic space or (consensual) cyberspace within and beyond. Penetrating the screen involves a state change from the physical, biological space of the embodied viewer to the symbolic, metaphorical "consensual hallucination" of cyberspace; a space that is a locus of intense desire for refigured embodiment.

The act of programming a computer invokes a set of reading practices both in the literary and cultural sense. "Console cowboys" such as the cyberspace warriors of William Gibson's cyberpunk novels proliferate and capture the imagination of large groups of readers. Programming itself involves constant creation, interpretation, and reinterpretation of languages. To enter the discursive space of the program is to enter the space of a set of variables and operators to which the programmer assigns names. To enact naming is simultaneously to possess the power of, and to render harmless, the complex of desire and fear that charge the signifiers in such a discourse; to enact naming within the highly charged world of surfaces that is cyberspace is to appropriate the surfaces, to incorporate the surfaces into one's own. Penetration translates into envelopment. In other words, to enter cyberspace is to physically *put on* cyberspace. To become the cyborg, to put on the seductive and dangerous cybernetic space like a garment, is to put on the *female*. Thus cyberspace both *dis*embodies, in Sobchack's terms, but also *re*embodies in the polychrome, hypersurfaced cyborg character of the console cowboy. As the charged, multigendered, hallucinatory space collapses onto the personal physicality of the console cowboy, the intense tactility associated with such a reconceived and refigured body constitutes the seductive quality of what one might call the *cybernetic act*.

In all, the unitary, bounded, safely warranted body constituted within the frame of bourgeois modernity is undergoing a gradual process of translation to the refigured and reinscribed embodiments of the cyberspace community. Sex in the age of the coding metaphor—absent bodies, absent reproduction, perhaps related to desire, but desire itself refigured in terms of bandwidth and internal difference—may mean something quite unexpected. Dying in the age of the coding metaphor—in selectably inhabitable structures of signification, absent warrantability—gives new and disturbing meaning to the title of Steven Levine's book about the process, *Who Dies?*

Cyberspace, Sociotechnics, and Other Neologisms

Part of the problem of "going on in much the same way," as Harry Collins put it, is in knowing what the same way is. At the close of the twentieth century, I would argue that two of the problems are, first, as in Paul Virilio's analysis, *speed*, and second, tightly coupled to speed, what happens as human physical evolution falls further and further out of synchronization with human cultural evolution. The product of this growing tension between nature and culture is stress.

Stress management is a major concern of industrial corporations. Donna Haraway points out that

(t)he threat of intolerable rates of change and of evolutionary and ideological obsolescence are the framework that structure much of late twentieth-century medical, social and technological thought. Stress is part of a complex web of technological discourses in which the organism becomes a particular kind of communications system, strongly analogous to the cybernetic machines that emerged from the war to reorganize ideological discourse and significant sectors of state, industrial, and military practice.... Utilization of information at boundaries and transitions, biological or mechanical, is a critical capacity of systems potentially subject to stress, because failure to correctly apprehend and negotiate rapid change could result in communication breakdown—a problem which engages the attention of a broad spectrum of military, governmental, industrial and institutional interests. (1990: 186–230 passim)

The development of cyberspace systems—which I will refer to as part of a new *technics*—may be one of a widely distributed constellation of responses to stress, and secondly as a way of continuing the process of collapsing the categories of nature and culture that Paul Rabinow sees as the outcome of the new genetics. Cyberspace can be viewed as a toolkit for refiguring consciousness in order to permit things to go on in much the same way. Rabinow suggests that nature will be modeled on culture; it will be known and remade through technique. Nature will finally become artificial, just as culture becomes natural.

Haraway (1985) puts this in a slightly different way: "The certainty of what counts as nature," she says, "(that is, as) a source of insight, a subject for knowledge, and a promise of innocence—is undermined, perhaps fatally." The change in the permeability of the boundaries between nature and technics that these accounts suggest does not simply mean that nature and technics mix—but that, seen from the technical side, technics become natural, just as, from Rabinow's anthropological perspective on the culture side, culture becomes artificial.

In technosociality, the social world of virtual culture, technics is nature. When exploration, rationalization, remaking, and control mean the same thing, then nature, technics, and the structure of meaning have become indistinguishable. The technosocial subject is able successfully to navigate through this treacherous new world. S/he is constituted as part of the evolution of communications technology and of the human organism, in a time in which technology and organism are collapsing, imploding, into each other.

Electronic virtual communities represent flexible, lively, and practical adaptations to the real circumstances that confront persons seeking community in what Haraway (1987) refers to as "the mythic time called the late twentieth century." They are part of a range of innovative solutions to the drive for sociality—a drive that can be frequently thwarted by the geographical and cultural realities of cities increasingly structured according to the needs of powerful economic interests rather than in ways that encourage and facilitate habitation and social interaction in the urban context. In this context, electronic virtual communities are complex and ingenious strategies for *survival*. Whether the seemingly inherent seductiveness of the medium distorts the aims of those strategies, as television has done for literacy and personal interaction, remains to be seen.

So Much for Community. What about the Body?

No matter how virtual the subject may become, there is always a body attached. It may be off somewhere else—and that "somewhere else" may be a privileged point of view—but consciousness remains firmly rooted in the physical. Historically, body, technology, and community constitute each other.

In her 1990 book *Gender Trouble*, Judith Butler introduces the useful concept of the "culturally intelligible body," or the criteria and the textual productions (including writing on or in the body itself) that each society uses to produce physical bodies that it recognizes as members. It is useful to argue that most cultural production of intelligibility is about reading or writing and takes place through the mediation of texts. If we can apply textual analysis to the narrow-bandwidth modes of computers and telephones, then we can examine the production of gendered bodies in cyberspace also as a set of tokens that code difference within a field of ideal types. I refer to this process as the production of the *legible* body.

The opposite production, of course, is of the *illegible* body, the "boundary-subject" that theorist Gloria Anzaldúa calls the *Mestiza,* one who lives in the borderlands and is only partially recognized by each abutting society. Anzaldúa describes the Mestiza by means of a multiplicity of frequently conflicting accounts. There is no position, she shows, outside of the abutting societies themselves from which an omniscient overview could capture the essence of the Mestiza's predicament, nor is there any single account from within a societal framework that constitutes an adequate description.

If the Mestiza is an illegible subject, existing quantumlike in multiple states, then participants in the electronic virtual communities of cyberspace live in the borderlands of both physical and virtual culture, like the Mestiza. Their social system includes other people, quasi people or delegated agencies that represent specific individuals, and quasi agents that represent "intelligent" machines, clusters of people, or both. Their ancestors, lower on the chain of evolution, are network conferencers, communities organized around texts such as Boyle's "community of gentlemen" and the religious traditions based in holy scripture, communities organized around broadcasts, and communities of music such as the Deadheads. What separates the cyberspace communities from their ancestors is that many of the cyberspace communities interact in real time. Agents meet face-to-face, though as I noted before, under a redefinition of both "meet" and "face."

I might have been able to make my point regarding illegible subjects without invoking the Mestiza as an example. But I make an example of a specific kind of person as a way of keeping the discussion grounded in individual bodies: in Paul Churchland's words, in the "situated biological creatures" that we each are. The work of science is *about* bodies—not in an abstract sense, but in the complex and protean ways that we daily manifest ourselves as physical social beings, vulnerable to the powerful knowledges that surround us, and to the effects upon us of the transformative discourses of science and technology that we both enable and enact.

I am particularly conscious of this because much of the work of cyberspace researchers, reinforced and perhaps created by the soaring imagery of William Gibson's novels, assumes that the human body is "meat"—obsolete, as soon as consciousness itself can be uploaded into the network. The discourse of visionary virtual world builders is rife with images of imaginal bodies, freed from the constraints that flesh imposes. Cyberspace developers foresee a time when they will be able to forget about the body. But it is important to remember that virtual community originates in, and must return to, the physical. No refigured virtual body, no matter how beautiful, will slow the death of a cyberpunk with AIDS. Even in the age of the technosocial subject, life is lived through bodies.

Forgetting about the body is an old Cartesian trick, one that has unpleasant consequences for those bodies whose speech is silenced by the act of our forgetting; that is to say, those upon whose labor the act of forgetting the body is founded—usually women and minorities. On the other hand, as Haraway points out, forgetting can be a powerful strategy; through forgetting, that which is already built becomes that which can be discovered. But like any powerful and productive strategy, this one has its dangers. Remembering—discovering—that bodies and communities constitute each other surely suggests a set of questions and debates for the burgeoning virtual electronic community. I hope to observe the outcome.

Acknowledgments

Thanks to Mischa Adams, Gloria Anzaldúa, Laura Chernaik, Heinz von Foerster, Thyrza Goodeve, John Hartigan, Barbara Joans, Victor Kytasty, Roddey Reid, Chela Sandoval, Susan Leigh Star, and Sharon Traweek for their many suggestions; to Bandit (Seagate), Ron Cain (Borland), Carl Tollander (Autodesk), Ted Kaehler (Sun), Jane T. Lear (Intel), Marc Lentczner, Robert Orr (Amdahl), Jon Singer (soulmate), Brenda Laurel (Telepresence Research and all-around Wonderful Person); Joshua Susser, the Advanced Technology Group of Apple Computer, Inc., Tene Tachyon, Jon Shemitz, John James, and my many respondents in the virtual world of online BBSs. I am grateful to Michael Benedikt and friends and to the University of Texas School of Architecture for making part of the research possible, and to the participants in The First Conference on Cyberspace for their ideas as well as their collaboration in constituting yet another virtual community. In particular I thank Donna Haraway, whose work and encouragement have been invaluable.

Notes

1. "Nearby" is idiosyncratic and local in cyberspace. In the case of Habitat, it means that two puppets (body representatives) occupy that which is visible on both screens simultaneously. In practice this means that each participant navigates his or her screen "window" to view the same area in the cyberspace. Because Habitat is consensual, the space looks the same to different viewers. Due to processor limitations only nine puppets can occupy the same window at the same time, although there can be more in the neighborhood (just offscreen).

2. Although no one has actually given up on encryption systems, the probable reason that international standards for encryption have not proceeded much faster has been the United States Government's opposition to encryption key standards that are reasonably secure. Such standards would prevent such agencies as the CIA from gaining access to communications traffic. The United States' diminishing role as a superpower may change this. Computer industries in other nations have overtaken the United States' lead in electronics and are beginning to produce secure encryption equipment as well. A side effect of this will be to enable those engaged in electronic communication to reinstate the inside-outside dichotomy, and with it the notion of privacy in the virtual social space.

Bibliography

Allan, Francis, "The End of Intimacy." *Human Rights*, Winter 1984:55.

Anzaldúa, Gloria, *Borderlands/La Frontera: The New Mestiza* (San Francisco: Spinsters/Aunt Lute, 1987).

Barker, Francis, *The Tremulous Private Body: Essays in Subjection* (London: Methuen, 1984).

Baudrillard, Jean, *The Ecstasy of Communication*, trans. Bernard and Caroline Schutze, Sylvere Lotringer (New York: Semiotext(e), 1987).

Butler, Judith, *Gender Trouble: Feminism and the Subversion of Identity* (New York: Routledge, 1990).

Campbell, Joseph, *The Masks of God: Primitive Mythology* (New York: Viking, 1959).

Cohn, Carol, "Sex and Death in the Rational World of Defense Intellectuals. *Signs: Journal of Woman in Culture and Society*, 1987, 12:4.

de Certeau, Michel, "The Arts of Dying: Celibatory machines." In *Heterologies*, translated by Brian Massumi (Minneapolis: University of Minnesota Press, 1985).

Dewey, John, "The Reflex Arc Concept in Psychology" [1896]. In J. J. McDermott (ed.), *The Philosophy of John Dewey* (Chicago: University of Chicago Press, 1981), pp. 36–148.

Edwards, Paul N., "Artificial Intelligence and High Technology War: The perspective of the formal machine." Silicon Valley Research Group Working Paper No. 6, 1986.

Gibson, William, *Neuromancer* (New York: Ace, 1984).

Habermas, J., *Communication and the Evolution of Society* (Boston: Beacon Press, 1979).

Haraway, Donna, "A Manifesto for Cyborgs: Science, technology and socialist feminism in the 1980s," *Socialist Review*, 1985, 80:65–107.

Haraway, Donna, "Donna Haraway Reads National Geographic" (Paper Tiger, 1987) Video.

Haraway, Donna, "The Biopolitics of Postmodern Bodies: Determinations of Self and Other in Immune System Discourse," *Wenner Gren Foundation Conference on Medical Anthropology*, Lisbon, Portugal, 1988.

Haraway, Donna, "Washburn and the New Physical Anthropology." In *Primate Visions: Gender, Race, and Nature in the World of Modern Science* (New York: Routledge, 1990).

Haraway, Donna, "The Promises of Monsters: A regenerative politics for inappropriate/d others." In Treichler, P. and Nelson, G. (eds.), *Cultural Studies Now and in the Future*." Forthcoming.

Hayles, N. Katherine, "Text Out Of Context: Situating postmodernism within an information society," *Discourse*, 1987, 9:24–36.

Hayles, N. Katherine, "Denaturalizing Experience: Postmodern literature and science." Abstract from Conference on Literature and Science as Modes of Expression, sponsored by the Society for Literature and Science, Worcester Polytechnic Institute, October 8–11, 1987.

Head, Henry, *Studies in Neurology* (Oxford: Oxford University Press, 1920).

Head, Henry, *Aphasia and Kindred Disorders of Speech* (Cambridge: Cambridge University Press, 1926).

Hewitt, Carl, "Viewing Control Structures as Patterns of Passing Messages," *Artificial Intelligence*, 1977, 8:323–364.

Hewitt, C., "The Challenge of Open Systems," *Byte*, vol. 10 (April 1977).

Huyssen, Andreas, *After The Great Divide: Modernism, Mass Culture, Postmodernism* (Bloomington: Indiana University Press, 1986).

Jameson, Fredric, "On Interpretation: Literature as a socially symbolic act." In *The Political Unconscious* (Ithaca: Cornell University Press, 1981).

Lacan, Jacques, *The Language of the Self: The Function of Language in Psychoanalysis*, trans. Anthony Wilden (New York: Dell, 1968).

Lacan, Jacques, *The Four Fundamental Concepts of Psychoanalysis*, trans. Alain Sheridan, ed. Jacques-Alain Miller (London: Hogarth, 1977).

LaPorte, T. R. (ed.), *Organized Social Complexity: Challenge to Politics and Policy* (New Jersey: Princeton University Press, 1975).

Latour, Bruno, *The Pasteurization of France*, trans. Alan Sheridan and John Law. (Cambridge: Harvard University Press, 1988).

Laurel, Brenda, "Interface as Mimesis." In D. A. Norman, and S. Draper (eds.), *User Centered System Design: New Perspectives on Human-Computer Interaction* (Hillsdale, NJ: Lawrence Erlbaum Associates, 1986).

Laurel, Brenda, "Reassessing Interactivity," *Journal of Computer Game Design*, 1987, 1:3.

Laurel, Brenda, "Culture Hacking," *Journal of Computer Game Design*, 1988, 1:8.

Laurel, Brenda, "Dramatic Action and Virtual Reality." In Proceedings of the 1989 NCGA Interactive Arts Conference, 1989a.

Laurel, Brenda, "New Interfaces for Entertainment," *Journal of Computer Game Design*, 1989b, 2:5.

Laurel, Brenda, "A Taxonomy of Interactive Movies," *New Media News* (The Boston Computer Society), 1989c, 3:1.

Lehman-Wilzig, Sam, "Frankenstein Unbound: Toward a legal definition of artificial intelligence," *Futures*, December 1981, 447.

Levine, Steven, *Who Dies? An Investigation of Conscious Living and Conscious Dying* (Bath: Gateway Press, 1988).

Merleau-Ponty, Maurice, *Phenomenology of Perception*, trans. Colin Smith (New York: Humanities Press, 1962).

Merleau-Ponty, Maurice, *Sense and Non-Sense*, trans. Hubert L. Dreyfus and Patricia Allen Dreyfus (Chicago: Northwestern University Press, 1964a).

Merleau-Ponty, Maurice, *Signs*, trans. Richard McCleary (Chicago: Northwestern University Press, 1964b).

Mitchell, Silas Weir, George Read Morehouse, and William Williams Keen, "Gunshot Wounds and Other Injuries of Nerves." Reprinted with biographical introductions by Ira M. Rutkow, *American Civil War Surgery Series*, vol. 3 (San Francisco: Norman, 1989 [1864]).

Mitchell, Silas Weir, *Injuries of Nerves and Their Consequences,* with a new introduction by Lawrence C. McHenry, Jr., *American Academy of Neurology Reprint series*, vol. 2 (New York: Dover, 1965 [1872]).

Noddings, Nel, *Caring: A Feminine Approach to Ethics and Moral Education* (Berkeley: University of California Press, 1984).

Reid, Roddey, "Tears For Fears: Paul et Virginie, 'family' and the politics of the sentimental body in pre-revolutionary France." Forthcoming.

Rentmeister, Cacilia, "Beruftsverbot fur Musen," *Aesthetik und Kommunikation*, 25 (September 1976), 92–112.

Roheim, Geza, "Early Stages of the Oedipus Complex," *International Journal of Psycho-analysis*, vol. 9, 1928.

Roheim, Geza, "Dream Analysis and Field Work." In *Anthropology, Psychoanalysis and the Social Sciences* (New York: International Universities Press, 1947).

Shapin, Steven, and Schaffer, Simon, *Leviathan and the Air-Pump: Hobbes, Boyle, and the Experimental Life* (Princeton: Princeton University Press, 1985).

Sobchack, Vivian, "The Address of the Eye: A semiotic phenomenology of cinematic embodiment." Forthcoming.

Sobchack, Vivian, "The Scene Of The Screen: Toward a phenomenology of cinematic and electronic 'presence.'" In H. V. Gumbrecht and L. K. Pfeiffer (eds.), *Materialitat des Kommunikation* (GDR: Suhrkarp-Verlag, 1988).

Sobchack, Vivian, *Screening Space: The American Science Fiction Film* (New York: Ungar, 1987).

Stone, Allucquere Rosanne, 1988. "So That's What Those Two Robots Were Doing In The Park . . . I Thought They Were Repairing Each Other! The Discourse of Gender, Pornography, and Artificial Intelligence." Presented at Conference of the Feminist Studies Focused Research Activity, University of California, Santa Cruz, CA, October 1988.

Stone, Allucquere Rosanne, "How Robots Grew Gonads: A cautionary tale." Presented at *Contact V: Cultures of the Imagination*, Phoenix, AZ, March 28, 1989. Forthcoming in Funaro and Joans (eds.), *Collected Proceedings of the Contact Conferences*.

Stone, Allucquere Rosanne, "Sex and Death among the cyborgs: How to construct gender and boundary in distributed systems," *Contact VI: Cultures of the Imagination* (Phoenix, AZ, 1990a).

Stone, Allucquere Rosanne, " Sex and Death among the disembodied: How to provide counseling for the virtually preorgasmic." In M. Benedikt (ed.), *Collected Abstracts of The First Cyberspace Conference* (The University of Texas at Austin, School of Architecture, 1990b).

Stone, Allucquere Rosanne, "Aliens, Freaks, Monsters: The politics of virtual sexuality." For the panel Gender and Cultural Bias in Computer Games, Computer Game Developers' Conference, San Jose, 1990c.

Stone, Allucquere Rosanne, "Ecriture Artifactuelle: Boundary Discourse, Distributed Negotiation, and the Structure of Meaning in Virtual Systems," forthcoming at the *1991 Conference on Interactive Computer Graphics*.

Stone, Christopher D., *Should Trees Have Standing?— Toward Legal Rights for Natural Objects* (New York: William A. Kaufman, 1974).

Theweleit, Klaus, *Male Fantasies,* vol.1 (Frankfurt am Main: Verlag Roter Stern, 1977).

Tomas, David, "The Technophilic Body: On technicity in William Gibson's cyborg culture," *New Formations*, 8, Spring, 1989.

Turkle, Sherry, *The Second Self: Computers and the Human Spirit* (New York: Simon and Schuster, 1984).

Von Foerster, Heinz (ed.), *Transactions of the Conference on Cybernetics* (New York: Josiah Macy, Jr. Foundation, 1951).

Weiner, Norbert, *The Human Use of Human Beings* (New York: Avon, 1950).

Wilden, Anthony, *System and Structure: Essays in Communication and Exchange*, 2nd ed. (New York: Tavistock, 1980).

Winograd, T., and Flores, C. F., *Understanding Computers and Cognition: A New Foundation for Design* (Norwood, NJ: Ablex, 1986).

Wolkomir, Richard, "High-tech hokum is changing the way movies are made," *Smithsonian* 10/90:124, 1990.

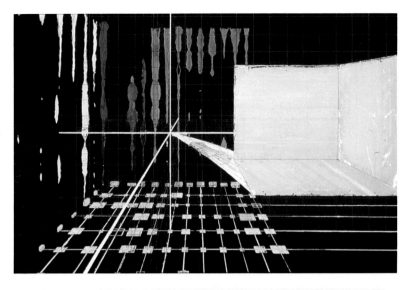

Plate 1

Daniel Wise, 1988. One vast data cell providing access to a visual database. Data available at the intersection of the three "crosshairs" opens into a subspace of three further dimensions.

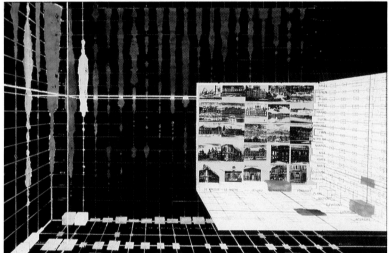

Plate 2

Two surfaces of the subspace continue to display navigation data (as well as quantitative indications of content) while the third surface is beginning to show destination data, that is, the sought images.

Plate 3

The user has moved in to inspect the images more closely.

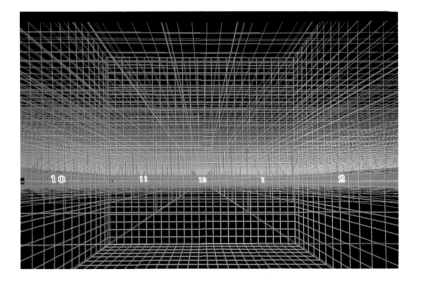

Plate 4
Stan George, 1989.
Underlying structure of
the matrix, seen from
within a cell.

Plate 5
Above one of the cells.

Plate 6
A possible "urban
landscape" of the matrix.
Ownership and identity
groups of data cells are
indicated by the trans-
parent superstructures.
Note clock-number
orientation system on
the horizon.

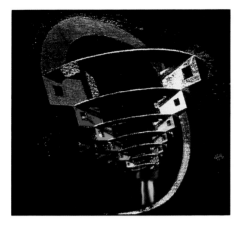

Plate 7
Jim Rojas, 1988. Somewhere in cyberspace, a spiraling helical construct mapping architectural history, time ascending.

Plate 8
Within the spiral, looking down—other users on distant ramps, searching the database.

Plate 9
Other users—personae—as rendered by the construct, colors and motion-style indicating personal characteristics.

Plate 10
In case you need help, an agent . . .

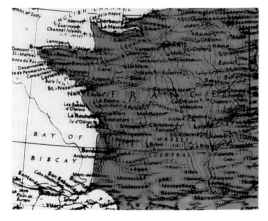

Plate 11
. . . to suggest a geographic approach to finding . . .

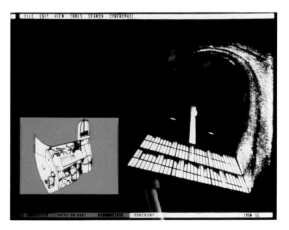

Plate 12
. . . representations of the chapel at Ronchamps, by LeCorbusier.

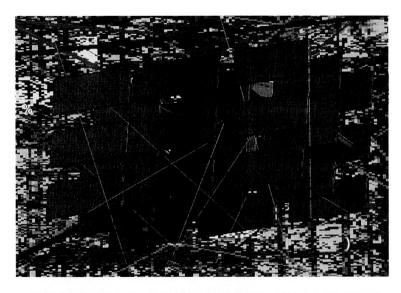

Plate 13
Clyde Logue, 1990.
Emerging from the static,
a flock of panels: a sales
convention in cyber-
space.

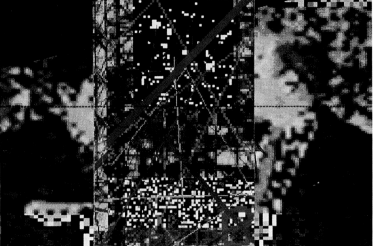

Plate 14
Differential resolution,
a visual cocktail party,
participants in discussion
overheard.

Plate 15
The user is addressed
about a product. In the
background, the whole
panel construct is seen to
float uncertainly above a
coursing terrain, in a
swarming sky.

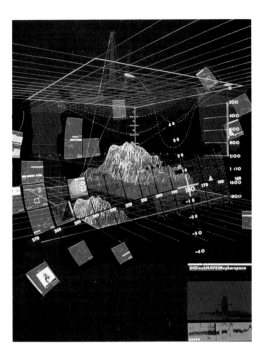

Plate 16
Gong Szeto, 1990. A data cell with an active surface approached "subaqueously." A spherical scrim of windows, analytical tools, support data, and navigational displays rotates over the scene.

Plate 17
Daniel Kornberg, 1990. Floating steadily through an infinite video and movie store—a gothic cathedral of sorts, its stained glass windows animated. The user plucks scenes and fragments, searching, creating new experiences. Windows to other parts of the construct.

Plate 18
Danielle Sergent, 1990. An auction house somewhere in cyberspace; a circular museum with displays viewable from within and without. Its shape changes with the contents of the auction underway.

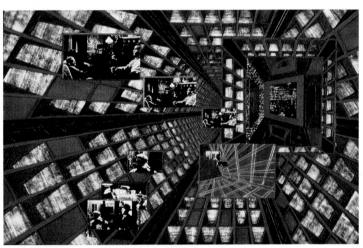

Plate 19
Composition created by a genetic algorithm. This image forms the basis of the following investigation of the spatialization of information.

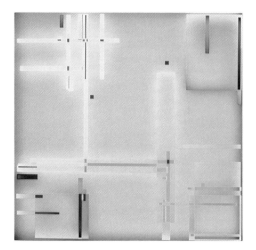

Plate 20
New composition derived from previous one by processes of superimposition, masking, and filtering. Information implicit in the original composition is now visible as color variation.

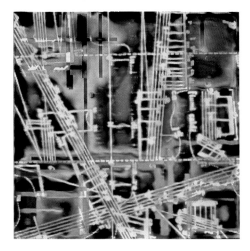

Plate 21
Merging of algorithmic composition with scanned data. Image processing reveals hidden patterns implicit in the structures of the component images

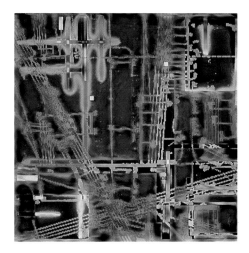

Plate 22
Variation of the image in plate 21 produced by further image processing. Although it is simply a transformation of the previous image, for the viewer this image constitutes, in effect, new information.

Plate 23
Three dimensional algorithmic composition, with the composition shown in plate 19 mapped onto the environment of a cyberspace chamber.

Plate 24
Two algorithmically composed objects in a cyberspace chamber. Dynamically varying algorithmically composed textures combining computed and scanned information are displayed on both objects and environment.

Plate 25
Dynamically varying three-dimensional composition comprising a liquid architecture. The number and kind of its component parts vary according to factors such as position, size, and proximity to other component parts.

Plate 26
The same object as that of the previous plate, as it appears at another time. Patterns in the information stream that creates this object are revealed spatially, temporally, and contextually.

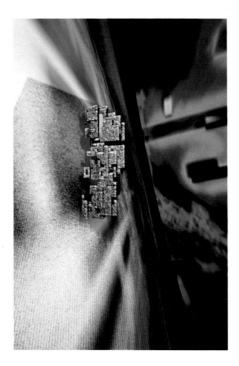

Plate 27
Visualization of a liquid architecture in cyber-space.

Plate 28
Every aspect of this world varies with position, time and information, and with the interests of the viewer and the other inhabitants.

Plate 29
Mapping information onto object and environ-ment, varying it in place, time, and attribute, focusing attention through filters and masks, and inhabiting it allows hidden patterns to become visible, and therefore knowable.

Plate 30
The information content of computed and digit-ized data is used to create the perceptual character of this space, the "place" of cyberspace.

7 Cyberspace: Some Proposals

Michael Benedikt

Introduction

On the Structure and Purpose of This Chapter, with Some Prefatory Comments

After this introduction, the chapter will be in two parts. The first attempts a broad analysis of what can be said in principle about the nature of physical space—the space of the everyday world—in relation to what can be said about the nature of the artificial or illusory space(s) of computer-sustained *virtual worlds*. Because virtual worlds—of which *cyberspace* will be one—are not real in the material sense, many of the axioms of topology and geometry so compellingly observed to be an integral part of nature can there be violated or re-invented, as can many of the laws of physics. A central preoccupation of this essay will be the sorting out of which axioms and laws of nature ought to be retained in cyberspace, on the grounds that humans have successfully evolved on a planet where these are fixed and conditioning of all phenomena (including human intelligence), and which axioms and laws can be adjusted or jettisoned for the sake of empowerment. Before dedicating significant resources to creating cyberspace, however, we should want to know how might it look, how might we get around in it, and, most importantly, what might we usefully *do* there. Thus the second part of the chapter presents some visualizations and descriptions of cyberspace(s) envisaged by myself and my students over the last few years. These are put forward as imaginative proposals, as designs, as descriptions of systems almost within our technological grasp.

The tenor of the chapter also changes between the two parts. The first, in laying some of the foundations for the second, attempts to be general, closely reasoned, and philosophical in tone. Although no less fully considered, the second is laced with fine-grained aesthetic and intuitive choices for which no rigorous explanations are offered. This is due not only to the intrinsic, technical complexity and scope of the problem, but also to the multiplicity of imaginative opportunities the very notion of cyberspace affords, demanding countless intuitive leaps and best guesses—fixings upon what simply seems good, workable, or interesting—from all who would enter into the task of its design.

Throughout, I write as a designer and theorist trained in architecture, and not as a computer scientist, mathematician, sociologist, or artist. In these fields I have only a modicum of specialized knowledge. Thus, my discussion will (try to) avoid the use of standard jargons from these areas of knowledge, as well as from my own field, architecture.

Some Remarks about Content

Time and Cosmology   In Part One I will attempt to define ordinary, physical space in a useful way. In the process I will need to recall and examine many truisms from modern science and mathematics. The exercise is worthwhile precisely because we are contemplating the design and implementation of an experienceable (actual) but nonphysical (unreal) space such as cyberspace. Since in cyberspace the very concept of *space* is clearly at issue, is not the concept of *time* also at issue? How should time be treated?

In both parts of this chapter, time is considered as a distinct and nonspatial dimension: "nonspatial" even though, as is well known, the dimension of time in such phenomena as physical motion and the transmission of energy and information is intimately involved with the dimensions of space.[1] Thus the reader will find very little talk of the "proper" unity of space and time associated with Minkowskian space-time, and only tangential discussion of relativity, cosmology, and quantum mechanics, where the dimension of time is often treated as all but interchangeable with spatial and other fundamental dimensions of reality. This is not because there is nothing "cosmological" to discuss in designing and modeling cyberspace. Far from it! Dwelling on the notion of cyberspace fairly *demands* that we query why things are the way they are in nature, and I find it necessary to speculate on cosmological matters more than once. However, as plausible and informed as these speculations may or may not seem, they are not put forward as serious and rigorous proposals about the nature of the real physical world. They are presented, rather, as comparative notes, as meditations on the way to a rich, viable, consensual, and "virtual" parallel one.

Why Cyberspace?   The reader will not find much discussion as to *why* we need to have cyberspace(s) at all. This very worthwhile debate must take place elsewhere. Here, almost by way of manifesto, it must suffice to say this:

Over the last twenty years the economies of advanced industrial societies have evolved rapidly. Though still founded on agriculture, manufacture, transportation, and energy production, a steadily larger portion of human activity has become increasingly involved with, and transformed by, the production and consumption of *information* as such—with finance, communication, advertising, education, entertainment, management, and the control and monitoring of complex natural and industrial processes. As a result, the economic principles of material production and distribution in their classically understood forms—principles of property, wealth, markets, capital, and labor—are no longer sufficient to describe or guide the dynamics of our modern, complex, "information society."

On the experiential front, our lives are changing too. Ever more dependent upon channels of communication, ever more saturated by the media, ever more reliant on the vast traffic in invisible data and ever more connected to the computers that manage it, we are becoming each day divided more starkly into the entertainers and the entertained, the informationally adept and the informationally inept. Bombarded everywhere by images of opportunity and escape, the very circumstances of a free and meaningful human life have become kaleidoscopic, vertiginous. Under these conditions, the definition of reality itself has become uncertain.[2] New forms of literacy and new means of orientation are called for.

Thus it is proposed that the creation of cyberspace is not only a good, but necessary, and even inevitable step (1) toward providing the maximum number of individuals with the means of creativity, productivity, and control over the shapes of their lives within the new information and media environment, and (2) toward isolating and clarifying, by sheer contrast, the value of *un*mediated realities—such as the natural and built environment, and such as the human body—as the source of older truths, silence of a sort, and perhaps sanity.

What Is Cyberspace? A Preview

I will attempt to provide a preliminary sketch of cyberspace. But first, this question: How does "cyberspace" relate to "virtual reality (VR)," "data visualization," "graphic user interfaces (GUIs)," "networks," "multimedia," "hypergraphics," and other such catchwords for recent developments in computing technology?

The answer: Cyberspace relates to all of them. More than this, in some sense "cyberspace" *includes* them all and much of the work being done under their rubrics. Indeed, I would assert that cyberspace as a *project* and as a *concept* has the capacity to collect these disparate projects into one—to focus them on a common target, as it were.

That said, my efforts here will likely fall short of providing a picture of that target that is clear and useful for all. And that is as it should be. For it is too soon for anyone to specify very fully the nature and uses of cyberspace. Just as there will likely be myriad places *in*, and many regions *of* cyberspace—each with its own character, rules, and function—in due time there may also be a number of different kinds of cyberspaces, each with its own overall culture, appearance, lore, and law. Someday, these cyberspaces and cyberspatial "domains" may well compete with each other just as information services and telephone companies do now. Some will thrive and some will not. Today, however, the very process of fairly specifically visualizing and describing models of cyberspace will help clarify the issues for all.

The enormity of the evolutionary step represented by cyberspace will claim our attention repeatedly. But what *is* cyberspace? I have assumed that the reader of this volume already has some familiarity with the term, but, for the purposes of this essay, and so that we can proceed with some clearer and common picture in mind, I offer this general description (written, somewhat prematurely, in the present tense):

Cyberspace is a globally networked, computer-sustained, computer-accessed, and computer-generated, multidimensional, artificial, or "virtual" reality. In this reality, to which every computer is a window, seen or heard objects are neither physical nor, necessarily, representations of physical objects but are, rather, in form, character and action, made up of data, of pure information. This information derives in part from the operations of the natural, physical world, but for the most part it derives from the immense traffic of information that constitute human enterprise in science, art, business, and culture.

The dimensions, axes, and coordinates of cyberspace are thus not necessarily the familiar ones of our natural, gravitational environment: though mirroring our expectations of natural spaces and places, they have dimensions impressed with informational value appropriate for optimal orientation and navigation in the data accessed.

In cyberspace, information-intensive institutions and businesses have a form, identity, and working reality—in a word and quite literally, an *architecture*—that is counterpart and different to the form, identity, and working reality they have in the physical world. The ordinary physical reality of these institutions, businesses, etc., are seen as surface phenomena, as husks, their true energy coursing in architectures unseen except in cyberspace.

So too with individuals. Egos and multiple egos, roles and functions, have a new existence in cyberspace. Here no individual is appreciated by virtue only, if at all, of their physical appearance, location, or circumstances. New, liquid, and multiple associations between people are possible, for both economic and noneconomic reasons, and new modes and levels of truly interpersonal communication come into being.

Cyberspace has a geography, a physics, a nature, and a rule of human law. In cyberspace the common man and the information worker—cowboy or infocrat—can search, manipulate, create or control information directly; he can be entertained or trained, seek solitude or company, win or lose power . . . indeed, can "live" or "die" as he will.

Now this fully developed kind of cyberspace does not yet exist outside of science fiction and the imagination of a few thousand people.[3] However, with the multiple efforts the computer industry is making toward developing and accessing three-dimensionalized data, effecting real-time animation, implementing ISDN and enhancing other electronic information networks, providing scientific visualizations of dynamic systems, developing multimedia software, devising virtual reality (VR) interface systems, and linking to digital interactive television . . . from all of these efforts one might cogently argue that cyberspace is "now under construction." Even popular computing's fascination with "windows" and 2+ dimensional graphic user interfaces (GUIs), together with the nation's burgeoning on-line newsgroups, electronic communities, and hacker subcultures can be seen as moves, however unwitting, towards the creation, someday, of the full-blown, public, consensual virtual reality that will indeed be *cyberspace*.

On the largest view, the advent of cyberspace is apt to be seen in two ways, each of which can be regretted or welcomed: either as a new stage in the *etherealization* of the world we live in, the real world of people and things and places, or, conversely, as a new stage in the *concretization* of the world we dream and think in, the world of abstractions, memory, and knowledge.

Both views are useful. But both are misleading insofar as they are both implicitly modeled on the historical processes of transformation, usurpation, and replacement rather than those of evolution, speciation, and displacement. With cyberspace the real world (let us grant some consensus here as to its physicality) does not *become* etherealized and thus, in the aggregate, less large or less real; nor does the "mental" world *become* concrete and thus, itself, less mental or spiritual. Rather, with cyberspace, a whole new space is opened up by the very complexity of life on earth: a new niche for a realm that lies between the two worlds. Cyberspace becomes another venue for consciousness itself. And this emergence, proliferation, and complexification of consciousness must surely be this universe's project.

Akin to the Teilhardian "noosphere," then, but not in any sense ideal, transcendent, or beyond reality, cyberspace unfolds in an expanding new landscape of ideational and electronic complexity.[4] But, just as printing did not *re*place but *dis*placed writing, and writing did not *re*place but *dis*placed storytelling, and just as movies did not *re*place theater, nor television movies . . . cyberspace will not replace either objective reality or dreaming and thinking in their historical modes. Cyberspace will not replace art museums, concerts, parks, or sidewalk jugglers; nor sex, books, buildings, or radio. Each of these earlier media and activities will move over a little, as it were, free—indeed obliged— to become more themselves, more involved in their own artistry and usefulness. Each will be dislocated in certain dimensions but freed in others, as Innis, McLuhan, and Carpenter so clearly saw.

Part One: On the Nature of Space, and Cyberspace

What Is Space?

We must begin with a large question: What is space? The existence and nature of space seems to be a truly basic, fundamental, and universal quality of reality; and if not of reality proper and entire, then, as Kant propounded, a necessary feature of our mind's operation in relation to it, and within it. Space and time, combined, appear to constitute a level of reality below which no more fundamental layers can be discerned, a field without natural parts, a universal attribute of Being that cannot be done away with, as much as Hume tried to do so.

As a rule we do not expect clear and final answers to metaphysical questions such as the above, questions among which we must include "what is the essence, or origin, of *space*?" Cast them as practical and empirical questions about physical reality, however, and we do expect some clear and helpful, if not final, answers from science. Can science, and especially physics, help us answer the question "What is space?" Is space a physical phenomenon, an *object* in some sense?

Insubstantial and invisible, space is yet somehow *there*, and *here*, penetrating, and all around us. Space, for most of us, hovers between ordinary, physical existence and something other. Thus it alternates in our minds between the analyzable and the absolutely given.

Or so it was until modern physics and mathematics revealed space's anatomy, as it were, showing its inextricability from the sinews of time and light, from the stresses of mass and gravity, and from the nature of knowing itself. The early part of the twentieth century saw post-Euclidean geometry and the Theory of General Relativity admit the concepts of curvature and higher dimensions, introducing "inertial frames," "manifolds," "local coordinate systems," and "space-time" to all informed discourse about space. These ideas had myriad practical consequences. Physical space, we learned, is not passive but dynamic, not simple but complex, not empty but full. Geometry was once again the most fundamental science.

With the techniques of differential geometry and algebraic topology we have come far in our power to reason mathematically about space's structure and "behavior." But our understanding of physical space's *actual* dimensionality, size, curvature, and grain at the macro- and microscales is less advanced, just as it is of physical space's relationship to the "mental spaces" of logic, representation, and the free imagination. Further, our understanding of the connections to the space and time of everyday experience—from mathematics and physics, through biological and social structures—has not improved very much in centuries. So if we can form neither a secure picture of physical space or space-time at the extremes of scale and velocity, nor a new picture of space in the ordinary sense, how can we intelligently, freshly, preferably scientifically, and with a view to creating cyberspace, answer our question "What is space?" Must we draw a blank?

Not entirely. We can define "space" in phenomenological, operational terms. That is to say, we can talk about how space appears/feels to us, and what both space and various concepts of space are "good for" objectively. We can ask: What operations does space permit or deny? What phenomena would be different if space were not "constructed" in this way or that? In what elusive physics are we so embedded that we cannot report on its laws? And if physical space *has* a discoverable and constrained topology, what of spaces of the imagination? Is not our ability to construe the latter precisely that which throws the former into relief?

With cyberspace, we ought not to feel dissatisfied to begin with a phenomenological/operational method, this rather than seeking after the objective *facts* of nature's infinitely subtle spatiotemporal constructions in the laboratory. Phenomenology, after all, entails nothing less than *taking appearances seriously*, and, containing no material objects, no energics, no physical dynamics, cyberspace is just such a realm of appearances to be taken seriously. Furthermore, cyberspace's spatiotemporal logic need only be consistent internally and locally. We do not presume to construct a permanent universe. Thus the criteria for success are pragmatic and human from the outset—workability, pleasure, and human empowerment taking precedence, always, over utter fidelity to nature at large.

Space as Freedom of Movement

Here is one phenomenological approach to a definition of space.

In almost all instances, and perhaps most irreducibly, space presents itself to us in the *freedom to move*, a freedom we "know" from the moment of birth.[5] Later in life—and in distinct stages, as Piaget showed—we come to reflect upon how the very possibility of movement depends on the preexistence of different and discrete locations for the same thing (including our bodies), locations between which continuous movement—that is, movement through all intervening locations—must occur over time.

So far so good. But notice: if the above is to function as a definition of space, a whole set of codependent terms must be grasped all at once: "location," "continuity," "identity," "freedom," "change" (and therefore, implicitly, time), terms which, alas, require that we already have some understanding of "space"!

Luckily, as evolved creatures, we have considerable knowledge of space "hard-wired" into each of us. This knowledge exists not just insofar as the laws of physics and chemistry "require" of all real things—brains no less than stones—that they be *in* and *of* space and time, but also as the set of everyday bodily sensations, reactions, and expectations, which appear to us as having immediately to do with the world's spatiotemporality.

For example, we simply *find* that can see and imagine innumerable discrete locations existing simultaneously: things in other places. With our eyes, and in our mind's eye, we can see innumerable paths and routes between locations; we see that most objects retain their *identity* as they move from one place to another. The phrase "the shortest distance between two points" immediately suggests "a straight line" to us, and we can easily extend this notion to the idea of a shortest time between two points, which may, or may not, be a straight line. We observe that objects do not really disappear here and reappear there in a disjointed or instantaneous fashion, though they may sometimes appear to, but must instead travel *between* here and there continuously, even if the route is concealed or very swiftly taken. Finally, we begin to assume that this arena of compounded, possible positions and reversible movements extends smoothly and indefinitely beyond our immediate perception in every "direction" (which is another whole story), retaining its local characteristics. This large set of phenomena, with its logical bounds and experienced character, we call *space*.

We must be clear about what is achieved when we "define" terms operationally and phenomenologically as we just have, especially when the definition seems causally or logically to antecede the defined. Defining A as "that which permits B, C, D . . . " (as in "space is that which permits identity, movement, size, etc.") is not a discovery of essences or of necessary grounds but a directive to examine carefully and critically the most immediate conditions, consequences, properties, and manifestations—*signs*, if you will—of a unitary A whose actual existence is, at best, sensed and presumed rather than seen and known directly.

The codification of such signs into a minimal and yet complete set often constitutes a set of *principles*, taken as necessary and sufficient conditions for, or "attributes" of, A. We will be looking at some of these shortly.[6]

We may need to employ advanced mathematical techniques in the design and operation of cyberspace. However, as esoteric as our thinking may become (and this quite aside from esoteric technical knowledge of hardware and software) we will have to ensure—in best computer industry tradition—that the ordinary user comes first. Even as we strive for higher dimensionalities or supernormal capabilities for the denizens of cyberspace, ordinary space and time must form the basis, the norm, *any departures from which* we must justify. Neither an advanced degree in math nor extraordinary powers of visualization ought to be necessary for a reasonably well-educated person to spend time productively in cyberspace.

Magic in Cyberspace: The Violation of Principles

As we know and will soon examine, in patently unreal and artificial realities such as cyberspace, the principles of ordinary space and time, can, in principle(!), be violated with impunity. After all, the ancient worlds of magic, myth, and legend to which cyberspace is heir, as well as the modern worlds of fantasy fiction, movies, and cartoons, are replete with violations of the logic of everyday space and time: disappearances, underworlds, phantoms, warp speed travel, mirrors and doors to alternate worlds, zero gravity, flattenings and reconstitutions, wormholes, scale inversions, and so on. And after all, why have cyberspace if we cannot (apparently) bend nature's rules there?[7]

But let us notice two things: first, that there is a limit to how frequent and severe such transgressions can become before credibility, orientation, and narrative power begin to be lost; and second, that myth and fiction do not contain violations of ordinary spatiotemporal logic but *descriptions* of such violations. Only today, and only on the display screen of a computer, can we find even modest spatiotemporal "miracles" actually performed in real time, routinely, and under our individual control.[8] One has only to watch an expert handling contemporary graphic user interfaces: exploding and collapsing icons; slip-sliding, disappearing, and reappearing panels and windows lapping each other six "deep"; buttons and sliders that are neither buttons nor sliders, that don't care where they are, and yet work; cursors that change what they are and do, depending on where they are; the stripping, skipping, scrolling, flying, popping, and gobbling . . . "effects," all, which are critically effective and have real world consequences, and yet when understood physically, at the level of the phenomenon, would call for a major rewriting of the laws of physics.[9] Similarly, the kinds of action "possible" within the fictional worlds of computer interactive fiction, video games, and consensual, real-time, nationally networked adventure worlds such as Habitat and Club Caribe, Carnegie Mellon's marvelous TinyMUD, and the latter's current spin-offs, defy physical constraints routinely. Magic!

In a way, cyberspace *is* the future of both graphical user interfaces (GUIs) and of networked text-games based on the place metaphor such as TinyMUD (more about this later). For cyberspace will not be just a description or staging of an uncanny reality—a matter of mental effect as in a novel, play, movie, or video game. More like GUIs, it will institute a virtual reality as a functional, objective component of physical reality. Cyberspace will provide a three-dimensional field of action and interaction: with recorded and live data, with machines, sensors, and with other people. Beyond consequences in cyberspace, these interactions will also have consequences that reach directly back into the physical world, from the efficient running of corporations, governments, and small businesses, to the enrichment of our individual lives with entertainment and communication . . . in short, to our real health, wealth, and happiness. After all, "cyber" is from the Greek word *kybernan*, meaning to steer or control.

What about interfaces beyond GUIs? With such new devices as data gloves, data suits, and head-mounted stereographic displays, a three-dimensional, electronically construed space may be entered sensorially. Is this vanished "interface," this envelopment by a computer-generated world already cyberspace? That is, *is the kind of experience afforded by virtual reality technology, ipso facto, cyberspace?* The answer has to be yes, and no.

The Independent Existence of Virtual Worlds

Cyberspace must be envisaged as a coherent and global virtual world independent of how it is accessed and navigated. There may be not one, but many ways to enter cyberspace, from simple, mouse-controlled animation of video monitor images, through VR (virtual reality technology being directed at re-creating the human sensorium as fully as possible), to direct neural plugs (as William Gibson imagined). Once in cyberspace, there may be many ways of getting around, from walking and crawling, to leaping through worm holes, from "bareback" riding or cyberBuick cruising, to floating and flying unencumbered. And there may be just as many alternative modes of action and manipulation. In other words, like a city, cyberspace is there to contain all these activities, happen as they may.[10] Therefore, although it depends on them technically, cyberspace itself is neither a hardware system, nor a simulation or sensorium production system, nor a software graphics program or "application." It is a place, and a mode of being.

These observations have immediate implications. For one, they help us see that the design and development of computer-human "interfaces," although a crucial and interesting complementary enterprise, is a separate project from the design and development of virtual or artificial worlds with which one would want to "interface" in the first place, regardless of how advanced the technology of "virtuality," of sensorium synthesis.

To illustrate the difference: few designers of today's GUIs, few journalists who observe today's GUI wars, and few GUI users are aware that two, separable systems are being created simultaneously: one is the space of window manager (WM) itself, and the other is the set of spaces in which both the WM and the data cooperate, spaces that can have their *own* value-laden lefts/rights, ups/downs, ins/outs, objects/voids. And of course, the two systems—the WM space and the data space— need not operate in lockstep. To make matters worse, they can be nested in each other indefinitely. For example, one will be as likely to find GUIs in cyberspace as cyberspace in GUIs.

Here is a simple example of the hidden valences of the WM space of a "desktop" GUI: why is the Macintosh trashcan icon—pale and ashen—positioned at the bottom right of the screen, while the rainbow-colored apple icon of the Apple system menu—happy and edenic—is positioned diametrically opposite, at the top left? Why have almost all GUI designers agreed that the top of the screen is icon/menu territory? These are vestiges of the organization of *pages*, which for thousands of years (even before there were "pages") have given different value to the top and bottom, center and margin, left and right, of things in general and then to fields of inscribed, textual and graphic information.[11] And whence *these* value assignments? From the body—with its eyes and anus, skin and heart—and from the earth, with its life, light, and view above and over, and its inert, dark, and blind things below and under. However subtle, these kinds of spatial/positional values will persist in the three dimensions of cyberspace.

But I am getting beyond the question. The point here is that the natural spatiotemporal principles suitably instated in the design of a vehicle/window manager need not be the same ones that govern the behavior of objects in the space of cyberspace itself, neither alongside the vehicle/window manager nor, further, inside/beyond each window. "Laws of physics" can apply differently to different classes of screen objects: this is already the case in all contemporary GUIs (though the Apple Macintosh's comes to mind most strongly here).

The ultimate *physical* basis for cyberspace resides, of course, in the actual construction and architecture of computers and communication links: in chips, circuits, crystals, cables, etc. Here, in the orchestrated flow of the myriad electrons and photons we expect no space-time miracles, no uncanniness. But cyberspace as such exists at a higher level evolutionarily and phenomenologically, that is, at the level of human perception and experience, thought and art. And it is at this higher level that the question of its "laws" comes into play. In cyberspace, I am saying, the distinctions that are already a part of GUI and game design will become keener, the number of permutations larger, and the necessity for a "natural order" and a consensus even greater.

In sum, then, it would be unwise to ignore the design of cyberspace itself while we are engaged in the myriad considerations of particular GUI and VR implementations. The design of cyberspace is, after all, the design of a another life-world, a parallel universe, offering the intoxicating prospect of actually fulfilling—with a technology very nearly achieved—a dream thousands of years old: the dream of transcending the physical world, fully alive, at will, to dwell in some Beyond—to be empowered or enlightened there, alone or with others, and to return.[12]

The Principles of Space and Cyberspace

Like the real world, cyberspace will continue to enlarge, to fill in, "complexify," evolve, and *in*volve, indefinitely. In time, the detailed and perhaps painful reexamination of the constraints, laws, and opportunities of the natural and physical life-world necessarily undertaken by cyberspace's first designers will become less frequent and less necessary. The second generation of builders will find that the new reality has its own, seemingly self-evident, rules.

Having gained some impression of what I mean by "cyberspace," we now turn to the task of considering its possible rules and principles. Using decidedly low-altitude mathematics, we will look at these in relation to the rules and principles of natural, physical space, and under five, essentially topological rubrics: *dimensionality, continuity, curvature, density*, and *limits*.

From this will emerge seven principles:

the *Principle of Exclusion* (PE)

the *Principle of Maximal Exclusion* (PME)

the *Principle of Indifference* (PI)

the *Principle of Scale* (PS)

the *Principle of Transit* (PT)

the *Principle of Personal Visibility* (PPV)

the *Principle of Commonality* (PC).

Each principle identifies a critical juncture in the system of possible correlations between the behaviors of physical space and cyberspace. These seven are neither empirically observed, as are the consequences of the various laws of physics, nor merely invented, as are those of the "laws" of fantasy worlds. Neither cast in stone nor wholly contrived, these principles, I will try to argue, are at the very least felicitous *conventions* for cyberspace, derived from the constraints and opportunities that physical reality seems to have chosen for itself, as well as the inherent limitations of computing and electronic communications. The models of cyberspace presented in the second part of this chapter are derived from, and are consistent with, all seven principles.

Dimensionality

We speak easily of physical space being three-dimensional. (As I mentioned earlier, I will treat time conventionally.)[13] But what is a *dimension* in our context? And what does it mean to say that space "has" three of them?

Though I will approach the problem as directly as possible, I must begin with a simpleton's view of the matter and hope that the reader will be patient when he encounters what is, to him, well understood. I will also answer the question backwards, beginning with consideration of abstract, numerical spaces (to which cyberspace is related) before coming back to real space.

Let us assume that we have before us some physical system, object, or phenomenon, dwelling and acting in ordinary space and time, such as an ecological system, a power plant, or any complex machine. We wish to *describe* and represent the system, and thus come to understand how it behaves and how we might control it. We might make some kind of measurements of the system's behavior through time; in fact, we would make a set of **N** different kinds of measurements, each of one aspect of its behavior. If these measurements are quantitative at all (if they have a largely numerical basis as in reading the control room's dials and gauges) or have any sensible linear ordering, then we can speak of the *state* of the system at a time t as the instantaneous value of all the measured variables, taken as a set, at time t. The behavior of the system is therefore reasonably described by the entire time-ordered set of states. Indeed, for some purposes, the system may be thereby defined.

Visualization If we are fortunate enough to need only two variables to describe the system's behavior then we can visualize the system's behavior in toto by plotting the two variables "against" each other on a chart. Each variable becomes one (pseudo)spatial *dimension*, which is commonly represented by one axis of a rectangular coordinate system laid out on a piece of paper, chalkboard, or computer screen. A *point* in this plane—in this space defined by the coordinate system—can therefore represent a certain state consisting, by definition, of the simultaneous, specific values of two variables. A continuous curve in this space can record the way these two variables covary smoothly in time—the system's "behavior"—or it can describe all the realizable states of the system, or both. A scattering of points has its own meaning too. It may represents some discontinuity of system behavior, or the action of a hidden, orthogonal variable such that the scattering of points manifests the peaks of a "submerged" surface, like island archipelagos.

A point in a three-dimensional coordinate system represents a single three-variable state, a curve a contiguous set of three-variable states, and so on. We may find—as we plot the states of our system in this abstract *data space*—that certain regions are more populated with points than others, representing the set of states the system is "in" most often and/or the action of a hidden variable. Surfaces may appear in our space, as may volumes of various shape, describing/recording the limits of the behavior of the system. These in turn may change over time . . . and so on.

All this is well understood. The role of computers in visualizing system behaviors in this abstract way is also well appreciated. But what if the system under review requires that we pay attention to five or more variables? Assuming that we still wish to visualize it, we face some design decisions: most of us simply cannot directly visualize four-or-more-dimensional spaces.[14]

We might simply *decide* which dimensions to work with, and drop the others. Working in round-robin fashion, we can then produce many representations of the system, no one of which will be complete in itself. Then we can look at all of these together, assembled in some way.

Or we may choose which dimensions to assign to "coordinate duty," and which to assign to the *character* of the point in the coordinate space, so that the character of the point and its position *together* describe a state of the system. This is the strategy I want to look at more closely. In a simple version, it is a strategy adopted by a number of investigators today (see Cox 1990, Ellson 1990, Brown, De Fanti, and McCormick 1987, for examples). But because of its profound importance to the development of cyberspace, here I attempt a general formulation.

Extrinsic and Intrinsic Dimensions   In Euclidean geometry, a "point" has no character: it has no size and no intrinsic, inherent properties. When one makes the statement "There *exists* a point such that . . . ," the point exists as pure position, and herein precisely lies its conceptual usefulness.[15] But in the physical world (and certainly in the world of actual representation) a point is always a "something"—a dot, a spot, a particle, or a patch of a field—a *point-object* to which one or more values can be attached that are descriptive of its character. In other words, unlike a true, Euclidean point, a point-object might have a color, a shape, a frequency of vibration, a weight, size, momentum, spin, or charge—some *intrinsic* quality or set of qualities that is logically (though it may or may not as matter of empirical fact *be*) independent of its position in space.

Now, any N-dimensional state or behavior of a system can be represented in what I would like henceforth to call a data space[16] of point-objects having **n** spatiotemporally locating, or *extrinsic* dimensions, and **m** *intrinsic* dimensions, so called because they are coded into the intrinsic character of the point-object.[17]

In sum:

$$N = n + m \qquad\qquad (N > 0\,,\, 0 < n < 5)$$

(# of) system dimensions = (# of) extrinsic dimensions + (# of) intrinsic dimensions.

Of course, if some of its assigned intrinsic dimensions are coded into perceptible, perhaps variable, size and shape, the point-object may no longer be able to be very pointlike geometrically speaking. Thus, a nontrivially sized object's so-called external dimensions, in the sense of some set of "external measurements" that depend on its shape belong, in our nomenclature, to its *intrinsic* dimensions because they belong to/characterize it as an object. By the same token, "internal dimensions," as of a room or car or pipe, are also intrinsic to the thing. In a data space, an object always contains within itself a privileged *address-point* (or, simply an *address,* given in extrinsic dimension values) that functions to identify accurately its position as a whole in the data space. Because intrinsic dimension data exist, technically, only *at* address points, an object may change its size and shape as its address changes—in other words, as it moves. Phenomenally, the moving object "taps" and reveals the data embedded in, intrinsic to, each address in the data space. One can see why *color coding* point-objects is so useful. Because (perceived) color consists in three, naturally intrinsic and non-space-occupying, orthogonal dimensions (RGB or HSV), color will not interfere when we wish to interpret the *configuration* of many closely positioned point-objects in reference solely to their positions in the data space, that is, their composite "shape."

Identity, Similarity, and Difference   If the behavior of a phenomenon or system can be cast as variables or "dimensions" at all, then these dimensions can also, in principle, be partitioned into two classes, namely, extrinsic and intrinsic to point-objects in a data space, as I have outlined. Given this, some convenient definitions follow.

Any two objects in the same data space can be said to be *identical* if they have the same values on the same, matching intrinsic dimensions; *similar*, if they have different values on the same, matching intrinsic dimensions; and *different* if they do not have the same intrinsic dimensions.

If we consider any two objects, each in a different space—that is, each having different set of extrinsic dimensions—then they can be said to be *superidentical* if they have the same values on the same, matching, intrinsic dimensions, *supersimilar* if they have different values on the same, matching intrinsic dimensions, and *wholly different* if they do not have the same intrinsic dimensions. (Notice that these definitions are meaningless for true Euclidean points, since points have no intrinsic dimensions. In other words, with a parallel argument made for true points' extrinsic dimensions, two points could be identical but not superidentical.)

The case of two, nonidentical objects having the same extrinsic dimensions and dimension values, whether at the same time, or including time as an extrinsic dimension at the outset, is forbidden, no matter what other comparisons may be made between their intrinsic dimensions and values. This constitutes the *Principle of Exclusion*[18] (PE) a commonsensical though very deep principle that in ordinary language says, "You cannot have two things in the same place at the same time." The case of two identical objects sharing the same place in space and time, of course, is solipsistic, since we could only be speaking of one self-same object.

A similar set of definitions can be made for single objects or point-objects that *move* through space in time. Any single object can be said to have *self-identity* if it preserves the same values on its intrinsic dimensions as it changes one or more of its extrinsic dimensions values (moves continuously in the space), and *self-similarity*[19] if one or more of its intrinsic dimension values are operationally linked to one or more of its extrinsic dimension values. Points in a "field" with gradient $\neq 0$ are self-similar by this definition.[20] If an object either switches or changes any of its intrinsic dimensions *in kind* as it moves (and not just, or in addition to, the values of such dimensions), then we can speak of its "transformation" as its *strange identity*.

There may be situations in which we may wish to distinguish whether a point-object has *super-self-identity*, or *super-self-similarity*. If it becomes "wholly different" from itself when in different spaces, then the question of "identity" becomes spurious: after all, in what sense is it the same object? The Principle of Exclusion does not apply to single objects.

Because there is some freedom in how to choose, partition, combine, and encode **m** intrinsic and **n** extrinsic dimensions for the representation of a any N-dimensional system as a data space, the shape and behavior of objects representing the system may appear different in each choice set. Provided that no information is lost, such representations are mathematically equivalent. But they are not necessarily functionally equivalent. In other words, when searching for a useful representation of a system of any complexity, it remains an empirical question as to which visualization—which partitioning of dimensions into intrinsic and extrinsic, which scheme of creating "objects" of character and the "spaces" of their inhabitation—will create a view most immediately intelligible and/or ultimately rewarding. In this sense, there is *more* information in a "good" visualization, and less in

Extrinsic Dimensions

	Same dims, same v's	Same dims, diff. v's	Diff. dims
Same dims., same v's	*self-same*	*identical*	*super-identical*
Same dims, diff. v's	*PE and PME excluded**	*similar*	*super-similar*
Diff. dims	*PME excluded**	*different*	*wholly different*

(Note: "dims "= dimensions, " v's "= values, "diff" = different
* discussed below)

Figure 7.1
Summary of relations of two data objects.

a "bad" one, and one must say that all representations of the data are *not* equivalent, even when all data points are in fact somewhere represented.

Given that there can be an enormous number of ways to partition the dimensions of a multidimensional system—and quite apart from any design decisions as to orientation, scale, and what sensible qualities should manifest them—are there any heuristics rules that can make the search for the best one shorter? I think there is one very fundamental one, namely the Principle of Maximal Exclusion.

Before we move on to this principle, we should now note that although all of the discussion above has been framed and conceived in terms of data spaces, it can apply as well to properly physical spaces. After all, there is no reason why the most suitable extrinsic dimensions to use in the description of any given system should not turn out to be the familiar X, Y, Z, and T axes of physical space and time, interpreted in their most neutral, Newtonian/Cartesian fashion, and merely *pictured* in cyberspace. Similarly, the most suitable intrinsic dimensions to use in describing a physical system may well turn out be the quite familiar momentum, charge, polarization, color, etc., or the myriad statistical compoundings and derivations of these that comprise the identity and behavior of relevant, separate parts of a physical system. This conception of things is carried out frequently, in pure and hybrid forms, in the fields of scientific visualization, geographic information systems, and so on. For example, it is common for one or more axes of two- or three-dimensional diagrams to map real distance or time, while the others have purely informational value, such as a geographic map in the X-Y plane where Z-values define a data surface of "land value" in units of dollars, or a map of the sky with stars where redshift, brightness, and a host of other spectral characteristics of stars are thought of as their intrinsic dimensions. Similarly, there may be occasions on which the neutral, uncoded, and unvalenced space "made" by Newtonian/Cartesian coordinates, interpreted as real space, serve as the best setting for objects whose shape, character, and solitary behavior—that is, whose intrinsic dimensions—*alone* are left to carry the entire responsibility for representing a system's behavior. In this case, space and geometry do not matter. There is no significance to up or down, left or right, closeness or distance. Satisfying only PE (and PME), objects form simple collections.

Some speculation: what if the very existence of (physical) "space," of "time," and of (self-identical) objects "in" space and time (the very terms and structure by which, as Kant observed, we comprehend reality) represents only a special case—a subset—of all the possible partitionings of the universe's variables into extrinsic and intrinsic dimensions? Then we might ask, Why did nature choose to divide her dimensions the way she did?[21]

We cannot of course really know why or even really how, but we can discern, I believe, a principle at work which, as designers of cyberspace, we may wish to emulate: the Principle of Maximal Exclusion.

The Principle of Maximal Exclusion (PME)   advises the following: *Given any **N**-dimensional state of a phenomenon, and all the values—actual and possible—on those **N** dimensions, choose as extrinsic dimensions—as "space and time"—that set of (two, or three, or four) dimensions that will minimize the number of violations of the Principle of Exclusion.*

This strategy minimizes information lost when multiple identical objects, without PME in operation, would be collapsed into single, self-identical ones, and multiple nonidentical objects would lose such self-identity as they (each) had. Let me give an example.

I have before me a two-color photograph, P. It consists of a large but finite number of point-objects $\{x, y\}$ whose extrinsic dimensions—X and Y—are the width and height of the photograph, and whose intrinsic dimensions are A and B, the two colors. The phenomenal color of a point is thus given by a pair of numbers $\{a, b\}$. At every definable position on the photograph (x in X, y in Y) there exists an object of certain color (a in A, b in B).

Now, it might occur to me to make a complementary photograph, P' of P, by exchanging the role of extrinsic and intrinsic dimensions, mapping one onto the other:

$$P = \{a, b\}_{x,y} \rightarrow \{x, y\}_{a,b} = P'.$$

That is, I make a "photograph" of dots whose extrinsic, spatial dimensions (X' and Y' of P') are A and B of P, and whose color dimensions (A' and B' of P') are X and Y of P. Thus the color of a dot in P determines its position in P', and the position of a dot in P determines its color in P'.

With a little thought we can see what would happen: as it is likely that there will be many dots in P that have the same color, these dots will wind up co-occupying the same position in P'. Conversely, because in P each and every dot's extrinsic dimension values are different, in P' no two dots will have the same color. This last transformation may not be a bad thing because information is not lost, but in the first, clearly, information is lost in violating the Principle of Exclusion. In mathematical terms, we have the problem of the many-to-one mapping. Thus, given data as a set of values on four unlabeled dimensions with which to construct a two-dimensional image, the Principle of Maximal Exclusion tells us to partition the dimensions such that dimension-pairs producing the largest number of unique tuples (or two-vectors) should serve as extrinsic. Computationally, this is a simple procedure for any real data set.

The Principle of Maximal Exclusion has a corollary that one might call the Principle of Maximal Object Identity (PMOI). It says this: *Choose as m intrinsic (object) dimensions that subset of n dimensions that minimizes the occurrence of unique m-tuples.* Together with PME, PMOI, I believe, provides a powerful heuristic for visualizing data to the best advantage.

Though something of a digression, let us look at these two principles, PE and PME, a little more closely.

From Democritus to present-day quantum physicists, the whole philosophical and scientific project that goes by the name "atomism" has consisted in pursuing a double goal: (1) the preservation of PE, and (2) the reduction of the number of permissible intrinsic dimensions of the physical universe, ultimately, to one; that is, the goal of producing a viable picture of reality consisting only of irreducible objects whose only quality is existence, and which cannot occupy the same space at the same time. Disallowed is the notion that atoms can have space, can be "internally open" and indefinitely so.[22]

In ordinary life there can be rooms within rooms, to be sure, literally and metaphorically, and this "anti-atomistic" fact constitutes one of the basic techniques of data manipulation in cyberspace, as we shall see. Though it obeys PE and PME, cyberspace is not consistently atomistic. Certainly no claim is made as to the objective, physical nature of the things of interest in cyberspace, nor is any privilege claimed as to the fundamentality of such space-within-space descriptions. There is no need. In cyberspace all entities are merely data entities, pure information, from the outset, picturing for us the gamut of life's information-generating systems at any number of levels and for quite specific purposes.

But a physicist might want to say that reality itself is just like that too! Fundamentally, he might say, reality *is* nothing but "data at play," a field of pure information. Experiments and theories are just instrument readings and interpretations of the data and nothing is as finally real as the atomists would want it . . . just as in our "cyberspace"! We reply, It may well be so.[23]

In a sense, PME acts as though interested in maximizing uniqueness and differentiation in the world, while making the atomist's project both possible and fruitful. But the consequences go further than commonly expected.

If I might wander farther off into cosmology here: PME tells us that the physical universe is as large, and only as large, as it needs to be to obey PE.[24] This implies that the size of the universe is a function of its information content: once small because it was simple, and now large, in part because we are here. Here also "largeness" must be understood not simply as the maximum value on some spatiotemporal dimension, but as the difference between the smallest and largest, physically possible units of spatiotemporal measurement that are informationally relevant. On this view, as immense as they are to us, space and time together provide the *smallest* realm/room needed for everything to be itself "exclusively"—the smallest arena necessary for the playing out of history in its most information-preserving configurations.[25] The citizens of another universe would call that set of dimensions that serves this function—that best supports PME—their "space" and "time."[26]

The reader, however, need not subscribe to this seemingly bizarre cosmology to see that, in cyberspace at least, where the preservation of information will be of practical importance and where there will be limits on how much information can effectively be displayed, the more information that is contained, the larger, in the above sense, the (apparent) space will need to be. Under PME, as cyberspace increases in complexity and content, there will be four methods to have it "grow" to accommodate. We might choose one or more of the following actions:

1. Increase in absolute size (area or volume) of cyberspace by adding more "territory," by landfill, as it were

2. Increase in the range of scales at which one can operate; the equivalent to increasing the resolution of cyberspace itself, the density of information per volume unit of (cyber)space

3. Increase the amount of information coded into the intrinsic dimensions of data objects; that is, more dimensions per object as well as more values per dimension

4. Invoke the information latent in considering the positional and temporal behavior and character of more objects together; that is, dynamic configurational behavior.

The theme of *information density* announced in these last few paragraphs will return. For now let us turn to the manner in which considerably more than three or four dimensions can best be visualized and manipulated given the parameters we have set ourselves. This will have direct implications for the method described in item (3) above.

Visualizing N Dimensions   In what follows we will assume that we are dealing with a cyberspace interface that allows us visually and directly to perceive data-objects in cyberspace as three-dimensional, changeable, and moveable through space and time in complete analogy to physical reality. In other words, consistent with what I have said earlier, the exact technology used for accessing/interfacing with cyberspace is not relevant. Thus we can imagine everything I will have to say here as happening vividly enough with available (if expensive) technology—on a high-resolution monitor connected to a very fast graphics workstation, capable of smooth, real-time animation of rather complex scenes. Stereo-optical techniques can be regarded as enhancements. Control by keyboard and mouse is sufficient.[27]

A tumbling arrow of variable length is (visually) an eight-dimensional object: 4 extrinsic (3-space plus time), and 4 intrinsic (3 angles for the direction of the arrow, and 1 for its length). A quivering, tumbling cube of changing color is a 16-dimensional object: 4 extrinsic again, and 12 intrinsic (3 for angular orientation, 6 (at least) for the amplitude and frequency of the quivering face-pairs, and 3 for the color (say, RGB values)). Clearly, it is possible to "see" a surprisingly large number of dimensions at play before the percept becomes unfamiliar. Considerable empirical research would be needed, however, to ascertain which kinds of dimensions, especially intrinsic ones, are easiest for us to see, remember, and attach value to independently. It is unlikely, for example, that a tumbling, quivering, multicolored cube in cyberspace would convey actually all the information it could potentially.[28]

Let us be conservative and assume that there are indeed quickly reached limits as to how much information can usefully be packed into intrinsic dimensions. It will turn out, perhaps, that only two encodings work reliably, say, color and orientation for a given geometrical object, and these only over a small range of values. Before accepting this, we might wonder: but what about object *size* and *shape*, those two most obvious visual attributes of physical things? Why and when can they not serve as intrinsic dimensions? This is worth considering in more detail.

Object size is not generally a good variable because when one object becomes large (relative to its address point, at the resolution of one pixel) then not only might it crowd out other objects in order to maintain PE, but we might be misled into reading subfeatures of its shape—say certain edges or corners—as having significance in terms of their own extrinsic coordinate values. I broached this subject earlier. Of course, this may be precisely what we want in certain situations, but then we would be dealing, in fact, not with one object with character, but with a set of small objects in some spatial relationship to each other, like a three-dimensional constellation, which is wonderful but not the same thing. As for shape, there may be limits to the distinguishability of object shape(s), especially when the object must be quite small to begin with. On top of this, many of the surfaces of a three-dimensional object are not visible (being turned away form the viewer) and these may be the very ones that we need to see.

What are our options? There are a few.

First, we can deal with many of the problems of size and shape I have mentioned by zooming in, by getting closer. The object, enlarged in our view, is isolated from the overall context. It might expand in inner detail, revealing complexity indefinitely.[29] Here we see intrinsic dimensions expand to become the extrinsic dimensions of the object now extended enough to have space within it, to *be* a space.

We may also rotate the object about its address point in any or all of three angular directions relative to the extrinsic coordinate system. If such rotation engenders change in the object, in color or size or shape, then up to three more dimensions capable of being encoded have been released.[30] We note that in ordinary, physical space, an object rotated on itself maintains its identity. In cyberspace this need not be the case, that is to say, an object rotated on itself may change as a function of that rotation. Similarly, in ordinary physical space, an object that is inspected from many angles retains its identity—that is, as it is revolved around *by* the viewer as a museum-visitor might walk around a sculpture. In cyberspace, again, this need not be the case, and the object can transform itself with our view of "it."[31] Things can get complicated here. Until research proves otherwise, I think it safe to assume that, in data spaces, both kinds of rotation—the object around its address, ourselves around the object—cannot effectively be used together to release information in six intrinsic dimensions, even if carried out sequentially and separately. I also suspect that, of the two, the first rotation—that is, of the object itself around its own address—is the more promising for the encoding of intrinsic dimensions. This is because, once carried out, the object stabilizes its identity, and any viewer motion around it becomes quite normal, revealing the object's whole shape.

This is all well and good. If, however, we wish to experience the object in its largest context—that is in (cyber)space, along with other objects—and if, therefore, it is important that the object not be so large as to obscure others, then we may need to adopt another method entirely for "releasing" intrinsic dimensions, namely *unfolding*.

Unfolding When an object unfolds, its intrinsic dimensions open up, flower, to form a new coordinate system, a new space, from (a selection of) its (previously) intrinsic dimensions. Data objects and data points in this new, unfolded, opened-up space thus have, as extrinsic dimensions, two or three of the ones intrinsic to the first, "mother" object. These objects may in turn have intrinsic dimensions, which can unfold . . . and so on, in principle, nested ad infinitum or until, at last, one has objects that have only one or two intrinsic dimensions and their self-identity left. At every occasion of unfolding, decisions are made as to the partitioning the remaining dimensions. At every occasion the Principle of Maximal Exclusion is applied.

Now, this hierarchical scheme ought to be familiar to anyone who has used the Macintosh, Open Look, Windows, or almost any contemporary GUI. When you click (or double click) on an icon, the "icon," effectively a two-dimensional data object, opens into a larger two-dimensional field or "window" within which new things can appear, including new icons. In cyberspace as we are beginning to picture it, three-dimensional "icons" unfold/open up into three-dimensional subspaces.

More than degree of dimensionality, however, the difference between today's GUIs and cyberspace is that in cyberspace the dimensions themselves (and values upon them) may not be neutral—that is, merely the actual dimensions of the monitor screen or the space simulated—but are themselves position-dependent state and behavior descriptors. It would be as though (in today's 2-D GUIs) all windows knew not only where they are on the screen but also where they "came from" spatially in the windows higher up in the hierarchy, all of which themselves know where they come from and all of which remain manipulable up, and consequentially down, the hierarchy. Right now, "windows" are almost completely undifferentiated bags of space.[33]

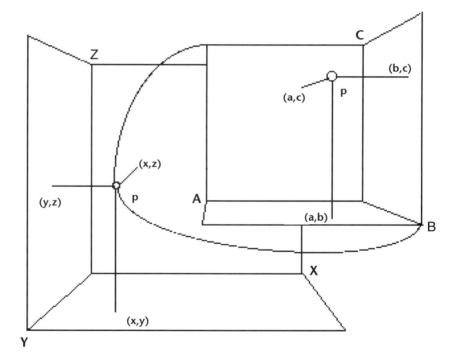

Figure 7.2
The intrinsic dimensions of the six-dimensional data object **p**, located at (x,y,z) in the (extrinsic) dimensional space of *XYZ*, are unfolded into the space of *ABC* and have the values given by the location (a,b,c).[32] If all objects p in XYZ are identical, then motion in XYZ has no effect in ABC.

Over and above the transition from two to three dimensions implied by the move from desktop to cyberspace, the difference will therefore also lie in how one keeps track of (extrinsic) positions in such nested coordinate systems, where observer and object positions are manipulated independently. This is a considerably richer and more difficult problem than that of arranging any number of flat and sequentially dead "icons" and overlapping "windows."[34] Will there be psychological limits as to how many levels "in" one can go, analogous to the limits Chomsky found for the nesting of clauses in sentences (and that I so frequently have tested in this chapter)? In the Macintosh GUI and most others there seem to be none. But this may well be because, as each new window opens, the previous one holds steady, falling away in significance. Will this be different when all the spaces together remain "live," positionally and operationally, and interactive? It remains to be seen.

We must conclude this subsection lest it swallow up all other concerns. But not without noting that in cyberspace the possibility exists very easily to make objects whose very existence, position in space, and character are a function (also) of the position of the observer/user in relation either to the coordinate system, or to some specified third object in the space. In real life, mirages and rainbows have this quality: that is, the quality of not being anywhere reachable in absolute, geographic space, but existing nonetheless visually, and always remotely, at a place determined by the invariant spatial relationship that obtains (in the case of the rainbow) between a given observer, the sun, and the water droplets (which, of course, themselves all have stable, reachable geographic positions). These kinds of objects travel with you, or appear and disappear as a function of your own motion and circumstance. Again, only empirical research will show whether this kind of entity will be of use in cyberspace.

Continuity

Physicists surmise that in nature there may exist a certain minimum length, the so-called Plank length, $L_p \approx 10^{-32}$ cm. Smaller than this, it is thought, nothing real or physical can exist; closer than this no measurement of position can be made. The existence of L_p effectively gives space itself a certain irreducible "grain" or resolution, making all length measurements properly only numerical multiples of the unit "10^{-32} cm."[35] But mathematically, conceptually, and certainly pragmatically speaking (most of the time) true space, physical or abstract, is understood to be smooth and continuous. That is, the X, Y, Z (and T) axes of a rectangular coordinate system are understood each to have the character of the real number line—to be infinitely divisible, monotonic, and supportive of the associative, commutative, and distributive laws that underlie ordinary arithmetic operations. This is the case regardless of practical limitations that may exist with respect to real acts of measurement and representation.

The number-line character of the coordinates of physical space generally forms the intuitive and functional basis for the representation of all other kinds of axes, dimensions, and coordinate systems: it forms the *substrate* upon which these are mapped. For example, (1) although certain transformations may be applied to any or all axes in a system (making them exponential or logarithmic in numerical value, say), (2) although an axis might be made of complex numbers or the ratio of two or more variables (as when one arrives at useful dimensions through statistical, "principle component" analysis of the data beforehand), and (3) although one or more axes might be subject to "quantization" so that regions of the number line are regarded as equivalent in position-specifying function, for the purposes of visualization these manipulations merely constitute recastings of the meaning, grouping, and scaling of the numbers in the rational, number-line coordinates of "true space."

Now, it is not clear that *all* information about the world can be represented spatially (let alone ought to be). Early in the previous section it was mentioned that in order to picture the behavior of a system in a "space" at all, whether in a fully developed cyberspace or a simple data space, one had to discover factors within the phenomenon that had the character of "variables." That is, one had to define features of the system's behavior that varied in a quantitative way, of which there could be "more" or "less," "many" or "few," of which it could be said that they were "weaker" or "stronger," "higher" or "lower," and so on. This was how we ensured that we could treat variables as number-line-based dimensions, and then arrange coordinates to create and govern a data space. Happily, this is a very powerful strategy: there are few factors for which we cannot conceive of there being varying extents, degrees, strengths, frequencies, and so on. Indeed, the differences *between* variables or dimensions are differences "in kind" precisely to the extent that they are not, and cannot be, differences "in degree."

Yet there exist many reasonable sortings of phenomena and systems into "parts" and "aspects," and many useful sets and lists of "things" that are not well characterized by one or two numerical variables, although they somehow belong together and adjacent to one another. Should we wish to represent and manipulate these sorted objects in a data space, we must declare them either (perhaps unfoldable) super-self-identical objects in some neutral, unvalenced space, or, if we wish to embed them in a space with some geometry and virtual physics, we must find in them, or apply to them, some relevant numerical ordering. For example, for all the other properties they may have, *goals* may be numerically ordered according to assessed "importance" or "difficulty," *furniture* by "size," "cost," "weight," "date of purchase," and so on. (Variables not adopted for the coordinate system are, of course, available for intrinsic object dimensions and characteristics, as we have discussed.)

And if worse comes to worst, we can avail ourselves of one or more of three linear/spatial ordering systems that are fail-safe, if not always optimal or interesting, as follows:

1. *Alphabetical*   Any group can be ordered by the letters of its constituent item names, which are arbitrary by the nature of language, and by the conventional recitation order of letters we call our "alphabet." (The same logic applies to alphanumeric pseudonumbering systems: for example, when stock parts each have an assigned, encoded "part number.") The alphabetic is an ordering system of last resort for almost all language-bound data, from customer lists and telephone books to encyclopedias, library catalogs, and dictionaries.[36]

2. *Geographical*   Since in reality everything has to be somewhere, any information about something real—any list of functional aspects, but more especially of parts or things—can be mapped into a data space so that its location mirrors where the thing actually is or could be, physically, in relation to other things, or to the earth itself. Atlases, city maps, machine layout diagrams, engineering plans, and all computer graphic "geographic information systems" present us information in this representational ordering; and again, like the alphabetic system, it is an enormously general and useful one.

3. *Chronological*   Events happen in time, often serially. A time-ordered list can be made of objects and events even if this aspect of their behavior is not truly salient. Manufacture dates, birth dates, times of arrival and departure, longevities, periodic sample values, annualized economic data, and so on, constitute in their diachronic and calendric ordering a very general mode for the construction of at least one coordinate of a data space. (A subspecies here might be termed "intervallic," that is, coordinates that map the time interval size between successive events. The "strange attractors" of modern dynamical systems theory, for example, exist in two- and three-dimensional data spaces whose axes are intervallic and twice phase-shifted from one-dimensional, time-ordered data.[37])

Arranging more or less "arbitrary" sets of aspects and lists of objects skillfully, of course, comprises much of the art of the database and database interface design; and insofar as cyberspace is just a gigantic, active, and spatially navigable database, these same skills will apply. Having thought the above thoughts, however, some future database designer with an eye to cyberspace may wish to add yet further criteria to the selection and design process. Indeed, he may approach the whole problem from a fresh perspective and bring into being an essential component of the new profession I ultimately have in mind: *cyberspace architecture*. But more about that later.

Now we return to the ideas of grouping and quantization mentioned earlier. Grouping and quantization can be used in combination. In the typical business "bar chart," for example, the vertical axis is often continuous, indicating, say, revenues in "dollars,"[38] but the horizontal axis is divided into discrete chunks, indicating, say, years. This is not to say that the data-generating (dollar-generating) phenomenon itself operates intermittently or only once a year, but that the only data measured and available has been summed (or sampled) at discrete intervals of time, namely, each year. Now, on the bar chart, a certain amount of space is simply annexed, its real horizontal dimension values divided, grouped, and banded together such that real leftness or rightness of position on the page (or screen) *within* such a grouping or bar is meaningless. All such positions are equivalent. But between bars the change is abrupt: a change in value occurs. By contrast, none of this is the case for the vertical positioning of a data point: every vertical change indicates change in value up until the quantum of the "cent" or perhaps "eighth-of-a-cent." The data space of a bar chart is thus discrete or quantized in the horizontal dimension, but (relatively) continuous in the vertical. (There are no rules here for how wide each band might be except commonsensical ones of legibility in relation to the amount of substrate, real space available, the amount of data, and so on.)

The same situation obtains for spreadsheets and matrixes in general. Here both axes are discontinuous and quantized. The location (address) and size of each cell constitute its extrinsic dimension values, the information contained in each cell space constitutes its intrinsic dimensions and dimension values. But the various positions and the amount of space *within* each cell are arbitrary and essentially meaningless. Thus matrixes and spreadsheets are only spatial insofar as their axes are number-line ordered, and only function spatially with respect to those aspects of the behavior of the system in question which lent their number-line ordering to the coordinates at the outset.

We have therefore two extreme cases to consider, *A* and *Z* (if I might use the alphabetic organizing system metaphorically!) and, between them, a quantized continuum of sorts wherein are situated the data spaces of cyberspace:

A: (representations of) physical spaces with their smooth and continuous dimensions and self-similar and self-identical objects, and

Z: pseudo data spaces that are so cellular and so arbitrarily ordered that, although they are mapped necessarily onto smooth, physical space substrates (such as screens and paper), they cannot and do not function spatially or geometrically, and whose "objects" can have little or no self-identity or self-similarity.

Intermediate kinds of spaces share to a greater or lesser degree both the fine grain and powerful monotonic ordering of natural space dimensions, and the simply pragmatic groupings of information classes, partially ordered, of structures such as spreadsheets.

In addition, workable data spaces can result from hybrids of these dimension types. For example, one might have a space whose coordinates map *ordinal* rather than cardinal numbers, creating thus an ordinal or *compacted* space. Why "compacted"? If, for instance, we have a set of data points whose extrinsic dimension values vary from some minimum to some maximum but such that not all, or not many, of the possible dimension values are represented, then the "unused" values can be omitted. The space collapses, so that what would have been a sparse scattering of points now becomes a more compact (and memory-saving) collection.[39] Again, one may thereby lose significant geometrical information—for example, information given by the angular spatial relationships of points to each other and the coordinate frame, information that would be revealed properly only in the fully extended, cardinally ordered space.

To all of these space types, however, and to all of these permutations of continuity and discontinuity, the cyberspace dweller will bring his inherited expectations of continuous, three-dimensional, physical space: real space with its natural ups and downs, left and rights, forwards and backwards, innernesses and outernesses, velocities and efforts, causes and effects. And it is out of a rapprochement of the two systems—physical space and data space—that cyberspace is born.

Limits

No discussion of continuity can conclude without a discussion of continuity's corollary: the idea of boundary or *limits*.

How large is our physical universe? No one can say; no one knows what lies beyond the most distant visible galaxies. It is not even clear that the question is legitimate. After all, what does it mean for every-thing-that-is to have a size?

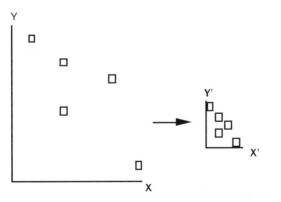

a) Extended (cardinal) space   b) Compacted (ordinal) space

Figure 7.3
(Note: these diagrams also indicate a space that is intervallic.)

One conventional view is to suppose that the universe is "finite but unbounded," like a four-dimensional sphere whose "surface" (our 3-space) is continuous and never ending, but is still of certain and finite size (volume). A more radical view is to hold that there is no meaning to the idea of boundary and therefore to the idea of the absolute size of space, except in the relationship between the "smallest" and "largest" measurements that can be made. Thus "size" is actually a sort of span, a ratio, a dimensionless number W, say, that can only be measured in principle from inside the system. It is measured in two directions, as it were, "up and out" and "down and in" from a particular position between macrocosm and microcosm occupied by the observer. The value of W in this physical universe, estimated from the value of the Planck length and the diameter of the universe, seems to be around 10^{70}.

To speculate further, it may well be that the value of W is a constant, no matter where along the macrocosm-microcosm spectrum the observer attempts to make his bidirectional measurement. Wherever he is, no matter what size he is, there is always a universe of things smaller and things larger than he. It may also be the case that the universe is expanding not only in its largest dimensions but in its smallest too, so that when we speak of the Hubble constant and of the "expanding universe," it is the value of W that is enlarging and not only its outer diameter. The picture created is not of a universe closed with respect to its "geographic" volume, embedded in and measurable with respect to some higher-dimensional and/or larger vessel space. It is, rather, something more fractal in flavor; topological closure, if any, exists in the meeting of the extremely large and the extremely small on some always-other-side from the observer. We stand, thus, always at a point on an expanding "circle of sizes" where in the "clockwise" direction things larger lie, and "anticlockwise," things smaller.

Whether or not this is an acceptable account of size in nature, we will nonetheless have to decide how cyberspace terminates. Will cyberspace have edges to blackness, walls of final data? Or will it be endless? If the latter, how? Like a planet, so that traveling for long enough in the same direction we find ourselves back where we started? Or might it be possible to present cyberspace phenomenally as a four-dimensional sphere, where striking out in any (three-dimensional) direction brings one eventually back to where one started? If we choose the latter, will we be able to conceptualize and navigate it? I think not. In the second part of this chapter I propose an intuitively satisfying way of making cyberspace boundless without depicting curvature or invoking intuitions of higher spatial dimensions; this by employing the idea from topology of a three-dimensional manifold whose horizontal edges are "abstractly glued" to form a "fat two-torus." Absolute size is calculable but not perceivable to the embedded user. Magical, but minimally so, an exact description of this cyberspace topology will have to wait.

If the two-torus is a good model of cyberspace's intrinsic boundedness and shape, it leaves open the problem of cyberspace's size and of its growth? Earlier, on page 148, I outlined a number of ways to respond to these problems, having to do with information density. In the next section, with the notion of information field density, **D**, we will find a plausible translation of nature's measure, **W**, into cyberspace. With it will come some surprising phenomena.

Density

How much space is there in space? Strictly speaking, which is to say, mathematically speaking, this question is either trivial or nonsensical. There are no more numbers in the real number line between 1 and 2 than there are between 2 and 3, no more distance in a 1-inch length here than a 1-inch length there, and no more area in a 4 x 6 region than a 3 x 8 region. True points are self-identical, have no interior and no size, and they can contain no space in themselves.[40]

It would seem, then, that to propose as interesting the issue of the "amount" of space there may or may not be "in" space must be to invoke a metaphor or analogy to some physical circumstance or process of containment, and thus not to be referring to space itself. We are likely to be thinking of space as some kind of compressible fluid, like air. A bottle may contain more or less air, for example, measurable by the air's "pressure," which is directly related to the gas's density (mass, or number of atoms per unit volume). Thus, in our analogy, more pressure is equivalent to more air per unit space, which means more space per unit space. And lest we believe that density/pressure is a matter of resistance to rigid containment, we can remind ourselves that, in the atmosphere, only loosely "contained," density can also vary for any number of reasons: heat, wind, humidity, altitude. . . . Thus we are supplied with a preliminary, if naive model for our notions of variable density of "space in space" without needing to think about containment and pressure in a literal sense.

To indulge this model any further is to conceive of two kinds of space: one, which is the space in whose varying amount we are so interested, and another, which is some absolute and homogenous underlying space whose linear metric forms the datum for the measurement. If the first is called the superstrate, **space$_o$** ("space-**o**ver"), and the second is called the substrate, **space$_u$** ("space-**u**nder"), then it is easy to see that we are concerned with the ratio or difference of the two.

Density of three-dimensional space-in-space = $D^{(3)}$ = **space$_o$/space$_u$**

Where $D^{(3)}$ is a dimensionless number (unless one wants to think of it in units of cm^3/cm^3).

Are there really two kinds of space? (And, if so, which do *we* inhabit?)

Rather than just say no—which is eminently sensible—let us remember the heady opportunities cyberspace offers us and pursue the matter. Perhaps we will be able to devise some interesting and useful experiences in our nonphysical universe with the idea of space-in-space; but the nature of the exercise is such that we may also find ourselves gaining insight into the possible nature of the physical universe.

There are two distinct and alternative ways of having "more space in space," or should I say, having more cyberspace in cyberspace. Both involve notions of density, but they do so differently.

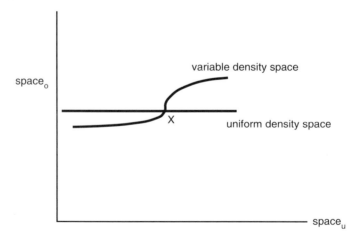

Figure 7.4
The observer is always at x, and imagines that space is universally uniform at the density value characterizing his own location.

Pixels and Voxels   In the digital world of the computer screen it is easy to create models that begin parallel to our "atmospheric model" of spatial density but then help us supersede it.

There are a finite number of finite-sized pixels on any video monitor screen, and it is not unreasonable to define the real (2-D) size/area of an object as the absolute number of pixels that comprise it. An object that is moving itself or is being "dragged" across the screen will retain its size and shape because the distribution of pixels over the screen is uniform. That is to say, grain aside, the superstrate array of pixels, **space**$_o$, and the substrate physical array of the screen phosphor molecules, **space**$_u$, map uniformly one onto the other. The ratio, $D^{(2)}$ expressed in pixels per square inch (or dpi^2) is a constant over **space**$_u$ (to which we give the more fundamental—because universal—status). Vary $D^{(2)}$, however, by physically (or mathematically) varying the size or distribution of the pixels, and the phenomenal shape-and-size of the object will vary depending on where it is. It may well shrink, swell, and change its shape in any number of topology-preserving ways. Furthermore, any traveling, one-pixel "particle" that, on a uniformly pixellated screen, would travel at a constant pixel/sec (and therefore inch/sec) *velocity*, would, to us, decelerate and accelerate as it traveled through areas of varying pixel density.[41]

More interesting is to consider two elaborations of this simple scenario: first, extrapolation into three dimensions, and second, the adoption of the perspective of an observer operating in **space**$_o$ rather than **space**$_u$:

1. *Into 3-D*: Three-dimensional pixels are often termed "voxels" (volume-pixel). Free of forces, all data particles or small data objects traveling through uniformly distributed voxels would behave close to the way they behave in everyday physical space (uniform **space**$_u$). Manipulating the density of voxels, $D^{(3)}$, will change the motion/behavior of the particle in a manner fully equivalent to the two-dimensional pixel case. In analogy to the Theory of General Relativity, for example, objects might reasonably be interpreted as following geodesics in "curved" space-time. (Indeed, one wonders what the notion of "curved" space(-time) could possibly mean—at least in popular physics—that could not be better visualized, if not explained, by the idea of space-in-space or spatial density gradients. Explaining

gravity using pictures of rubber sheets forming funnels for hapless marbles begs the question. After all, why should the marble spiral *down* the funnel if there were no gravitational force extraneous and additional to the geometry of the funnel that is supposed to, itself, explain/picture/cause the gravity?)

By way of comparison, the value of **W** for cyberspace is likely to reach no higher than 10^{24}. This estimate is based on a phenomenal measure of cyberspatial Planck length, $l_0 = 4 \times 10^{-6}$m, a voxel measure, v_0, therefore, of $l_0^3 = 6.4 \times 10^{-17}$m^3, and using a geographic ordering of the earth's land surface area 250m "thick." (Real space, over the same volume, has **W** = 10^{50}.) More interesting to the designers of cyberspace is the relative ease with which the observer can always be positioned such that macrocosm and microcosm are not themselves absolute "positions" but *directions* on the "circle of scale," a possibility I described earlier (p. 151). Thus, continuous or stepwise entering into smaller and smaller, unfolding objects-turned-spaces can circle us back to the macroscale view that contains our initial position, and vice versa. This Ouroboros-like, space-within-space circularity of scale can function independently of the two-torus topology recommended—or of any other topology—for the universal cyberspace.

2. Now let us enter and dwell in **space**$_0$.

But first, consider again what it is that really governs the dynamics of objects in a data space no less than what governs the pixels themselves. Both—objects and pixels—are *rigged,* as it were, like puppets. They dance their dance according to the results of an immense traffic of calculations being carried out in the background. Here, quite literally "behind the scenes," and invisibly, they are constrained, guided, and governed by the program residing in a microprocessor. How fast and how smoothly things happen on the screen (or in the space) is thus a function of the speed and control of these intricate bit streams to and from the screen (actually the frame buffer and video RAM) and not of the "natural" imputed properties of the objects we see, or of any direct, electromechanical, spatiotemporal interaction between them. Signals from object A to object B are not propagated across the screen at some pixel velocity; they are computed by whatever equations apply—taking no longer to compute large than small distances—and the results are "posted" on the screen as object behavior.

Of course, we can arrange it so that pixels—objects, texts, cubes, molecules, critters, whatever—move and behave in ways that seem natural, that is, that simulate the real objects they represent together with their physics. But this must be done deliberately, and, generally, it requires enormous computational power to do so. Furthermore, the speed of the process is apt to be critically sensitive to the *complexity* of the objects in question—to their detail, geometry, color, rendering, and so on—as well as the sheer number of objects involved at any one time. This is *quite unlike everyday reality* where an intricate object in a crowded room falls to the floor no more slowly than does a simple one in an empty room.[42]

This said, let us imagine a different system, one in which physical reality is more closely simulated. "Distant" CPU calculations are kept to an absolute minimum. The screen is bit mapped, the space is bit mapped; the video-RAM /frame buffer is "intelligent." All influences of one data object on another are *actually* propagated in the buffer, but are *phenomenally* propagated though the pixels or voxels of **space**$_0$ just as light and sound travel through space and air obeying the inverse square law. Objects move through space—and/or across the screen—dynamically constrained (1) by the rate at which their "information mass" can be "accelerated" and "decelerated," and (2) by the local value of $\mathbf{D}^{(3)}$.

Next we place *ourselves* subjectively in this superstrate space, **space**$_0$, much as Edwin Abbott's Flatlanders are embedded in theirs. We have become one of the objects. We can see and know only what comes across **space**$_0$ to us. We cannot see or know anything of the background calculations. We cannot rise off the screen, or out of the space, into a higher dimension so as to see the whole layout and thus have fore-knowledge.

We are embedded specifically, let us imagine, in a three-dimensional (cyber)space of locally uniform $\mathbf{D}^{(3)}$, and it is dark. All around hang fixed "stars" and seemingly fixed "planets."

Now we see an unidentified flying data object (UFdO) moving in the distance at a constant velocity. As we watch, quite suddenly it slows and shrinks, rapidly dwindling in size! What has happened? We cannot tell.

Three interpretations are plausible: (1) the UFdO has actually slowed and shrunk, (2) it has turned and is traveling away from us, or (3) it has entered a compacted region of space, that is, a region of increased density, $D^{(3)}$. Discounting interpretation (1) (that is, positing maximum object self-identity) we are unable to decide between (2) and (3). Indeed there is no method—not stereopsis, not radar—that can decide between them!

Now let us set out to catch up to the mysterious craft. Through the flat, dark field of **space**$_0$ we fly. Around us data stars and data planets move, flowing by more quickly to our sides than directly ahead or behind, and moving more or less slowly according to their distance from us, as is normal. Two minutes later we see the UFdO dead ahead of us! It seems to have stopped moving since it grows steadily in size, following the cotangent looming curve that is in agreement with our own velocity. As we draw near, it seems less a (cyber)spacecraft than an asteroid (". . . obviously intelligent, Number One!"). We approach, and as it looms we see more: like ordinary physical things, the closer we come, the more it reveals of its detailed structure.

We notice also, however, that *our* velocity has slowed; and, try as we might, we cannot accelerate towards the object! Only up, down, left, right, and away from it can we move any faster. In fact, moving away, we accelerate, and this without asking to.[43] It is as though we have fallen into a reverse gravitational field created by what now seems more like an asteroid, a small world. Either that, or the thing is expanding in size, and with it the space around it. ("This is no asteroid, Number One, this is a *planet*. And it's growing!") How can we tell? Looking around—as we now seem only to inch towards the ever more complex planet—we notice that all other objects in the "sky" are dwindling in size, seeming to recede. Indeed, we are entering a region of *more space*. What seemed small from afar turns out to be another world; and this not just in the normal, natural sense (in the sense that getting very close to any physical thing allows us in some way to "enter" its microcosmic structure and have it surround us), but in the more radical sense that space itself has expanded, affecting both our movement and the apparent position of the rest of the world. The planet is indeed—to all intents and purposes—surrounded by a reverse gravity field, one that, at the same time as it "repels" or decelerates that which would approach it, seems to "manufacture space" in ways analogous—if reversed—to what the General Theory of Relativity tells us is the case of natural gravitational fields.[44]

What could account for this scenario? Actually, the explanation in the context of computation and virtual worlds such as cyberspace is simple.

Assume a computer of finite computational speed. It must compute and display a data object at a fixed rate of around thirty frames per second. Assuming almost zero computing load in order to translate or enlarge an object on the screen—that is to redisplay without revealing/ recomputing new object information—and assuming a monotonic relationship between apparent data object size[45] and its revealed complexity, then it is clear that this finite computation speed will impose limits, alternatively or collectively, (1) on the rate of new-frame display, (2) on the level of detail and "realism" displayed, and (3) on the amount of the *increase in information* with each frame.

Since greater real motion on the part of the user generally reveals greater new information in his world, in case (3) we find the explanation of our UFdO scenario.

Of course, possibilities (1) and (2) continue to be theoretical and practical choices. They may also be functional options at the control of cyberspace travelers who could choose for themselves what to sacrifice for speed: smoothness of the animation, or detail/resolution and/or rendering niceties (such as shadows, color, etc.). Option (1), the rate of new-frame display, is a strategy common to animated video games and most "walk-though" CAD programs where "step size" can be specified; option (2), manipulating the detail or richness of the display, is the strategy of "adaptive refinement" first put forward by Henry Fuchs.[46] If, however, we wish to achieve maximum smoothness (of our own motion as well as the motion of objects "out there"), this in order to establish the phenomenologically fundamental **space**$_2$ status of cyberspace in imitation of nature, then we must forgo (1). No matter what quantization is present, inherent, in data objects themselves and in their environments, *the continuous substrate space of cyberspace is the fabric—the medium—through which* user *motion must occur.* This lends privilege to users as data objects among others: objects may perforce jump from slot to slot; but we, the travelers of cyberspace, always can glide on . . .

And if we wish to retain the maximum self-identity of data objects, we would be advised not to adopt the strategy of adaptive refinement (2) as a *norm.* For taken too far, this would violate one of the cornerstone principles of ordinary reality that would do well to remain in cyberspace and that I would now like to address (before returning to strategy (3)). Namely, the *Principle of Indifference.*

The Principle of Indifference states that *the felt realness of any world depends on the degree of its indifference to the presence of a particular "user" and on its resistance to his/her desire.* The principle is based on a simple phenomenological observation: what is real always pushes back. Reality always displays a measure of intractability and intransigence. One might even say that "reality" *is* that which displays intractability and intransigence relative to our will. This is why what is real always, ultimately, generates consensus. And science.[47]

Congruently, what is real always displays a measure of mysterious, even gratuitous, complexity, a complexity that does not adjust itself to our ignorance and that, in fact, exceeds what we know, always. More than merely something conceded to Art, this mysterious, "gratuitous" complexity, this extravagance, is reality's "calling card," its song of seduction. What is real always seems to have extra, more than we can use—more finesse, more detail, more possible uses, more reasons to be than for us alone—and we would be ill-advised, I think, to make cyberspace a place of complete knowing-through-appearances, a world where every object wears its explanation on its sleeve, and every inflection encodes some function.

The Principle of Indifference also implies strongly that, in a world we take to be real, *life goes on whether or not you are there.* This is a particularly powerful manifestation of the principle (perhaps deserving of its own name: "The Principle of Life Goes On"). To explain, the extent to which things freeze, go "on hold," simply because you are not there to keep them going, is the extent to which those things are not real. Most computer applications and games, for example, go on hold when you stop typing or issuing commands; some only when you log off. Upon your return, everything is where you left it. Things wait. In private spheres, real and virtual, this is probably how things should be, at least with nonliving/loving objects. But in public life and in common reality, the traffic of transactions, the flow of data and decisions, the movement of things, and the evolution of situations goes on relentlessly. A good part of the reason thousands participate in electronic BBS and newsgroups is because there is this kind of *life* there: life independent and indifferent, to large degree, on any single user's watchfulness and participation.[48]

Absence from cyberspace will have a cost.

The Principle of Indifference also underlies my earlier argument for the *independent* existence of virtual worlds (p. 132), for without this principle, cyberspace would rapidly devolve into countless roll-your-own realities, each as amenable to manipulation and as personal as a dream. As with a dream or an acid trip, the user himself would not believe in either the relevance of what happened there or its continuity with ordinary reality. Gone would be cyberspace's capacity to create a group of consistent worlds, let alone a single parallel universe for millions; instead, it would resemble William Gibson's *simstim* industry—VR game arcades and vacation parlors (à la *Total Recall*), home sensoriums, rock "virtuals" on CD instead of rock videos, and the rest. The level of fantasy is not at issue here. Nor is the desirability of a multi-billion-dollar VR/simstim business.[49] At issue only is the level of malleability that is appropriate to an ongoing, real-time, consensual public realm such as cyberspace.

On the other hand, too rigid an instatement of the Principle of Indifference would be stifling and alienating. Nothing would respond to our action; nothing would "know" of our presence. In cyberspace, the individual's ability to customize his environment and his experiences certainly ought to exceed that ability in the real world. A balance must be found, and finding it will constitute a large part of the art of designing cyberspaces both indifferent and responsive, both beyond the individual and yet *for* him.[50] We will look at some of these factors more carefully below, and then again in the context of the Principle of Commonality.

Let us return, then, to the strategy of adaptive refinement. Once in cyberspace, we might wish to govern the level of detail revealed by any particular object (thus releasing oneself from its sludgy, gravitational grip!), and this by direct command to the system. We might also find it useful to *choose* to move at a desired velocity, thereby draining the surroundings of detail and rendering richness in proportion to that velocity (. . . racing through wire frames, gliding through crowded museums . . .). However, only option (3), limiting the amount of new object information per frame, will create a consistent, if unusual, realm where *phenomenal immensity follows information density*, indeed, where the laws of information itself begin to create a new spatiotemporal physics. So convinced am I of the last argument, that I propose another principle for cyberspace, namely *The Principle of Scale*. This states that *the maximum (space$_o$) velocity of user motion in cyberspace is an inverse, monotonic function of the complexity of the world visible to him.*

Interestingly, the real world provides some examples of the manipulation of the relationship of information density to phenomenal immensity which I have named the Principle of Scale. Traditional Japanese gardens, for example, are miniaturized, their elements representing whole mountains, seas, rivers, and trees. Because the garden is real, closeness reveals detail everywhere; because it covers itself up—offering only partial views—it discloses new information constantly. In addition, however, the movement of the viewer is *slowed*—by bridges, stepping stones, roughness underfoot, obstructions, and other devices. Slow(ed) movement and an informationally dense environment ("bits" per steradian to the viewer, and "bits" per unit volume to the planner) combine to create the desired transformation of scale, the effect of spatial immensity.

The visitor to the garden nonetheless feels powerful: his every inertially registered motion makes a difference to what he sees—too much difference, really, and this belies the garden's true size, even as we enjoy its "immensity." By contrast, the enormous, empty halls favored by the Romans, and later, for similar reasons, by Albert Speer, have the effect—in their visual simplicity combined with their slow rate of visual change at normal walking speeds and near-zero information gain—of reducing the individual in size, making him feel impotent in even this most simple procedure: walking. We have immensity stripped of information. Similar effects, and other manipulations (such as light from below, tilted horizons) that strike us deeply, will be available to the designers of cyberspace environments from the histories of designed real environments.[51]

But it is too early to be adamant. In all likelihood, all three strategies—limiting (1) the rate of new-frame display, (2) the level of detail and "realism" displayed, and (3) the *increase in information* with each frame—will find their use, and we will devise others.[52]

The Principle of Scale forms a connection between the amount of *space* in space ($D^{(3)}$) and the amount of *information* in space. Following through on the idea requires us to look more closely at the concept of information *fields*.

The Information Field

In nature, space is not truly empty. "Empty space," that is, space without matter, is empty only in the classical mechanical view (before Michael Faraday). At the very least "empty space" consists of the active Higgs field in the quantum field-theoretical view, and of the electromagnetic and gravitational fields in the classical field-theoretical view. Thus the notion of a *field*—that is, of a space where every point contains, is, or has a *value* of energy, force, or information—is fundamental, especially when one believes that the existence of pure and empty space itself can only be inferred from the behavior of fields, which are themselves all that can be measured and perceived. In modern times, empty space—utter vacuum—is an abstraction. The set of laws of field behavior with respect to light and time is called *geometry*. For some, then, it seems reasonable to assert that "space" and "geometry" are equivalent terms.

Be this as it may, let us imagine a small, arbitrarily thin disc floating in Euclidean 3-space. (The space is materially empty, that is, empty of everything but the disc itself and, by implication, ourselves, the observers.) Now, the color of the surface(s) of the disc is changing over time such that we are able to say that the disc is emitting light-energy, and, further, that pattern of change of color constitutes (visual) information. Our question: *Where* is the information? To be sure, *on* the surface of the disc, but also *everywhere the disc can be seen from.*

In other words, the disc creates a *field* of the information originating on its surface extending indefinitely in all directions.[53] Or almost all directions. For there is a geometrical plane, coincident with the disc, from every point in which the disc is invisible—in other words, a plane in which the information field is null or empty. Solely from within this plane, there is no perceptible field, and thus no space, and thus no disc. Elsewhere, however, the information field is present and, unlike the energy illumination field, uniform or flat.

Now let us add a restriction. We will say that the observer has a lower limit on the solid angle he can detect or discriminate. At a certain distance from the disc that threshold is reached, and the field—now defined in terms of the character of the emitter *and* the observer—vanishes. What, we now ask, is the shape of the *boundary* of the field?

From the simple geometric theorem that states that the angle subtended by a chord of a circle on the circumference of the circle to which it belongs is a constant, we can see that the shape of our boundary is the surface of a sphere, a sphere of which the disc is a very thin layer. (Actually there are two such spheres, one for each side of the disc). Stated differently, the locus of all points subtending the same solid angle on a disc is a spherical surface containing the disc.[54] Differently again, the visual size of a disc—measured in solid angle subtended at the eye of the observer, **o**—is the same for all observers located on the surface of the same sphere containing the disc as a layer. The measure of this angle, α, is an inverse function of the square of the distance from the observer to the disc surface, and a direct function of the cosine of the angle between the observer and the normal through the center of the disc, β.

$$\alpha = \frac{\pi r^2 \cos \beta}{R^2} \quad (R >> r)$$

where r is the radius of the disc and R is the distance from the observer to the disc. Knowing one's own velocity, v, and *da/dt*, one can infer R, and without an absolute value for v, the ratio r/R is determinable.

The relationship of α to R is nonlinear, creating a field pattern that looks roughly like this for equal increments of the angle, α:

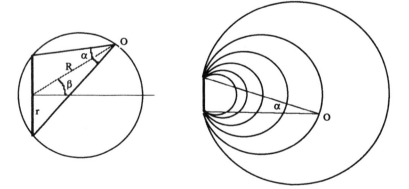

Figure 7.5
Information field of a disc in empty space.

The geometry of the field is a mapping of the geometry of the disc: different shapes form uniquely different fields; and from the field alone, in principle, one can reconstruct the fact of the disc, its size and its position. Indeed, this is exactly what we do when we see and locate anything, as J. J. Gibson pointed out (1950, 1986). The information field is thus holographic in the sense that its local parts, volumetrically, contain information about the whole. By the same token, a hologram is one kind of record of the information field, in particular, one that records/preserves/re-creates the field's original geometry over a sub-stantial area to one side—the "front"—of the hologram. (A regular photograph records/preserves/re-creates the original field only at a point in space, that is, at a privileged "station-point" where the picture projects perfectly onto the original scene.)

Let the amount of information broadcast into the field by the disc-source, s_i, be I_i, measured in something like bits/second. (I_i may also be a function of time, $I_i[t]$). A measure of the *information field density*, $\mathbb{I}_i$, of the disc at the observer's position $O = O(x,y,z) = O(R, \beta)_i$ is simply given by

$$\mathbb{I}_{i,o} = I_i / \alpha_i(O)$$

measured in, or, at this juncture, thought of *as*, bits/steradian. Thus the total information available at O is the sum of the information present in the field at O from all sources, $i = 1, 2, \ldots n$, and, concomitantly from all angles of view and all packings of a_i to the maximum 4π total. The information density of the field at a point, $\mathbb{I}_o$, is simply

$$\mathbb{I}_o = \sum_{i=1}^{n} I_i / 4\pi.$$

(We assume independent information sources, a rare occurrence, admittedly.) Should there be information density *limits* on the part of the observer, in terms of his/her/its processing speed and/or visual discrimination, then certain parts of the field, beyond those limits, can be "bad," while others, within, can be "good." Set operations on the fields of more than one source are thus possible. Patterns like this can describe the way people choose seats in a theater, tables at a restaurant, and so on.

Field density measured in terms of observers and their "optic arrays" (J. J. Gibson's term) can be given further and more conventional meaning. If we are trying to find a definition of information field density as information per unit volume of space, then we need only integrate I_o over Δx, Δy, Δz, being that volume of space that contains the information we are interested in, and divide by the **space$_o$** volume $\Delta x \cdot \Delta y \cdot \Delta z$, taking into account perhaps the density limitations of the observer and "shadows" created by obstructions. Indeed, the geometric pattern of obstruction-shadows, or occlusions, and the observer's geometric position vis à vis the sources of information in an environment is itself a useful class of information. This class of information is examinable in detail by my theory of *isovists*.[55]

We might now think back to the Principle of Scale and learn something more about its operation. For the total amount of information at a point in cyberspace is a sum, or, better, a union of two classes of information: the information available from other sources in cyberspace, I, as we have just been discussing, and the information latent to the location itself, that is, figured in the location's intrinsic dimensions, which we can denote I'.

If an observer/user happens to *be* at that point, then to both must be added a third class of information, namely, the information implicitly carried into cyberspace by the observer, I_o: his resident files, his programs, his display and image production software, any personal cyberspace or virtual world. Even his personal intelligence and memory might also be included in I_o, if we wish to go to the end of the argument.

When a user occupies a location in cyberspace, that-which-appears-to-him *in* cyberspace is either or both: the information content intrinsic to that point, I'; and the information pouring in from other cyberspace entities, I. It is up to the cyberspace system designer/programmer to make clear to the user which is which, and then also to distinguish both from the user's own, resident information, I_o. (In the second part of the chapter, a project that capitalizes on these distinctions will be described.)

The Principle of Scale, we might now reasonably recommend, applies only to I and not to I' or I_o information. If we allowed the Principle of Scale to apply to I' and I_o information, we would instate a mass equivalent in cyberspace. Space itself and users themselves would have a kind of inertia proportional to their private, internal complexities, to the "mass" of the information they contained. I say "kind of inertia"

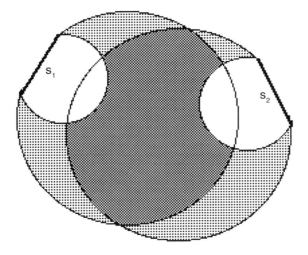

Figure 7.6
The shaded area denotes the part of the field in which the information of s_1 and s_2 is simultaneously available within the density limits set.

because affected would not be object-or-user *acceleration,* as is the case with physical inertia, but velocity. The effect would be more like *drag.*

Notice that users see each other as $\mathbb{I}$ information. Notice also that the $\mathbb{I}$-field is a **space**$_o$. Therefore, *the amount of (phenomenal) space in cyberspace is thus a function of the amount of information in cyberspace.* It varies objectively from place to place, is partially observer dependent, and governs motion in the predictable and yet uncanny ways we had begun to discuss.

This is not the place to elaborate further on the geometry, mathematics, and variants of (visual) information fields such as this.[56] My purpose has been theoretical and conceptual. For above all else, cyberspace is an information field most purely, and the boundary between cyberspace and real space can be designed so that the fields can extend maximally from one into the other, *geometry intact.* The intermingling of the geometry *and* content of two information fields, one electronically sustained, one materially and energically, is the goal, stated most abstractly, of all virtual reality technology: from stereo displays to the gloves, goggles, body suits, and force-feedback schemes of VR. Stated more practically: whereas standard, two-dimensional displays show the content of the cyberspace information field, $\mathbb{I}$, at some point of observation, replacing the geometry of the field with the geometry of the screen, the more advanced notion is to allow the information field of cyberspace as such to extend into—and to overlay—the information field existing in the user's real world as through a window, or as surrounding him entirely. And vice versa.

The Remaining Principles

So far we have discussed four of the seven principles that I proposed earlier for the design and nature of cyberspace: the Principles of Exclusion, Maximal Exclusion (together with Maximal Object Identity), Indifference, and Scale. These were developed, I trust, in a context that supports their plausibility. I would like now to turn to the remaining three principles—namely, the *Principle of Transit,* the *Principle of Personal Visibility,* and the *Principle of Commonality,* each of which has been more or less implicit in the discussion so far.

The Principle of Transit (PT)

This principle states that *travel between two points in cyberspace should occur phenomenally through all intervening points, no matter how fast (save with infinite speed), and should incur costs to the traveler proportional to some measure of the distance.*

This principle may seem at first to be an unnecessary restriction of freedom of movement in cyberspace, too real, and too conservative (in the literal sense of the word "conserve"). After all, one of the prime advantages of network computing, and computing in general, is the almost instant access possible to the files, documents, programs, and (soon) people one is interested in, regardless of remoteness or physical location. Furthermore, the user can easily "be" in two places at once with any multitasking or window-based GUI (especially X-windows), or remote video link.

In my view, the latter facility is not contra-indicated by the Principle of Transit, whose aim is to help build and maintain a sense of continuous spatiality (in accordance with the Principle of Indifference). Simultaneous presence in multiple locations is not as destructive of the mental geography that is cyberspace as would be cost-free, instantaneous travel. Like a child playing with a doll house, one ought to have no more trouble "being" in two places at once than a guard has watching several monitors, or anyone has, for that matter, scanning and looking into a number of ordinary spaces—rooms, cubicles, cups, refrigerators, roof gardens—more or less simultaneously from where they sit. The most likely problem to arise would be confusing the perceptions *others* might have of where/who/what/how many *you* are.[57] Potentially serious, this is a problem best solved by avoidance, that is, by limiting the proliferation of aliases or clones of oneself in cyberspace to two or three, and having instead "remotes"—something like cameras, microphones, sensors, or even agents—identified as such at the remote locations and limited in their capabilities. (I can hear the objections now: "No! no! I want the freedom to be, see, and be seen anywhere, everywhere, as anything, anytime, all the time. . . ." We will discuss some of this in a later section).

From here it seems a natural step to allow instant motion between remote locations. If you have the address, dial it up, and *be there!* Why not? (Why do the crew of the starship *Enterprise* ever need to *walk* to the transporter room? I have always wondered.) But consider:

1. Access is never *really* instant. It takes real time to search a disk, and it takes real time to send a message across the country. It always takes a few seconds or longer to locate and access new files and directories, and we don't seem to mind too much if the delay is reasonable and there is something else salient to see or do. Why should these delays not be made proportional to a determinate *distance* in cyberspace (and this does not have to be a simpleminded measure), and why should that time not be spent showing/experiencing the route between, however swiftly? What would one rather do? Look at the old screen, a blank screen, or watch two-second commercials?

2. But further, I hazard that *navigating* around file structures, *selecting* paths, *accessing* different and distant computers, and so on constitutes a good deal of the pleasure of computing. This navigating, waiting, and puzzle solving, which demands imagining and feeling the system *work*, succeeds in creating the very *environment* we computer fans find so addictive. These operations already build pictures in user and programmer's minds alike; and millions of dollars have been made by taking care of how consistently these mental geographies—as simple as they are—are designed, named, metaphorized, and physically managed. Hackers may be able to do without, but if cyberspace is to be the ultimate popular computing environment, why not build upon this orienting, world-building tendency in all of us? And why not preserve the fundamental concepts of distance and velocity? Without them, one

can have only disjointed places, like lit mailboxes tossed into the dark or conjured into existence from nowhere. Without distance and velocity as properties of our virtual world, one must rely entirely on quasi-spatial and abstract connectivity structures such as menus, hierarchies, and graphs, or on remembered alphanumeric codes, and manuals, hopping about from one to the other.

3. There is a good reason to *be* in transit for significant periods of time, and in relatively public areas. For it is *between* tasks, both spatially and temporally, that one is most open to accident and incident. In the real world, chance meetings in hallways, lobbies, airports, on sidewalks, and so on are essential to the formation of informal interpersonal networks. Browsing is essential to the acquisition of new information.[58] Without time in transit in cyberspace—open, spatiotemporally coherent, and free—one is imprisoned by one's discrete task domains, blinkered and locked to destinations. Perhaps most important, one has no presence to others. Indeed, without the utility (as well as pleasure) of relatively unstructured being-in-the-world, without the opportunity to soak in information, survey the field, and gradually define one's own degree of focus and interaction, cyberspace would hardly be necessary. Nor would be cities. And nor would have been the plains of Africa two million years ago.

4. If instant access to people and information were to become endemic to cyberspace, gone would be the process of progressive revelation inherent in closing the distance between self and object, and gone would be a major armature in the structuring of human narratives: the narrative of *travel*. Destinations would all be certain, like conclusions foregone. Time and history, narrativity and memorability, the unfolding of situations, the distance between objects of desire and ourselves—the distance, indeed, that creates desire and the whole ontology of eroticism (see Heim, chapter 5)—would be collapsed, thrown back, to existing in *this* physical world only, and only as lame, metaphorical constructions, here and there, in that one.

The concept of "cost" in the context of the Principle of Transit as enunciated is purposely left open to some interpretation. If it seems reasonable to identify the cost of cyberspace travel with *time* in some way, as we have, then this is because time is indeed the fundamental currency of computing, for it translates quite directly into data-processing capability and speed, and this in turn into the real-dollar cost of hardware, network on-line time, and access privilege. However, other "payments" for distance travel may be incurred as losses (lessened legitimate demand on the system). For example, loss of resolution (as with Fuchs's adaptive refinement technique), loss of range of view, of smoothness of motion, of the presence of certain data objects, of user capability, and so on. And we need not exclude direct, real-dollar charges similar to telephone long-distance charges. As long as these costs can be indexed to cyber-geographic trip distance and/or velocity, the spatiotemporal construct of cyberspace will be reinforced, and the Principle of Transit maintained.[59]

Now, none of this is to imply that free (instant) rides between two points in cyberspace should not ever be possible; only that these transgressions of the Principle of Transit should not become the norm at any and all scales, and that their special character be retained, even ritualized, by design, culture, and economic constraint.[60] For example, a zone or sector of cyberspace might have a number of designated *transfer stations*. These transport users to other sectors very quickly, blindly, and without time proportionality to cyber-geographic distance. As in city subways, orientation information is provided at these transfer stations in condensed form.

Extending the thought, we might wonder this: where does (should) one emerge when first entering cyberspace—that is, when logging in? Does each user have a place in cyberspace, somehow his or her own, that is "home," and from which they must start out and return? Or should one be able to enter cyberspace anywhere—boom!—upon merely specifying an address. Should there be distinctions, in terms of speed of accessibility, between frequently and infrequently visited places? Should leaving cyberspace—logging out—be a symmetrical reversal of the process of entering?

By way of negotiating among the competing demands and opportunities presented here, I offer these further suggestions:

It seems plausible to have a finite number of *ports of entry*, or simply, *ports*, which, like their real-world counterparts such as ship ports, airports, train stations, and bus terminals, function as landmarks themselves, while giving all travelers a concentrated geographic, cultural, and economic orientation to the sector of cyberspace entered. A user that logs into cyberspace finds himself at the port of entry he has designated, probably in a certain, standard position and orientation, and moves off from there in a self-guided, continuous manner. There should also be special kinds of ports, namely, *gateways*, that function to connect users in cyberspace to parallel, perhaps proprietary, cyberspaces and other electronically networked entities. The three systems—transfer stations, ports of entry, and gateways—each with its own set of protocols, could selectively overlap, coincide, grow, and connect as required.

If it is acceptable to be in two or more locations in cyberspace at once, as I suggested earlier, then it follows that one should be able to travel to, and enter, a second cyber place without relinquishing one's presence in the first.

It also seems plausible that frequent trips between the same locations could be truncated. Commuting can be tiresome, and the geography of cyberspace, once learned, may not be that enthralling to the jaded user, especially when he wishes not to be open to the eventfulness of "the between."

Finally, *leaving* cyberspace should not require a retracing of one's steps, literally or functionally. One should be able to exit more or less instantly, or at the very least, on "autopilot" back to the port of entry, and so on. The user can thereby witness the spatial logic of cyberspace even as he is swiftly extracted. (By the same token, autopilot may be a reasonable way to move anywhere one has the address for at maximum speed while still being able to "enjoy the scenery.")

Forgoing autopilot, allowing instant exit from cyberspace violates the Principle of Transit quite directly, but it does so for psycho-ecological reasons. Research has shown that on their daily rounds, all hunting/gathering animals in the wild, including bushmen, take a long, convoluted route out in to the field, foraging, tracking, finding food, etc., but the shortest and straightest route home with the haul, or at the end of the day.[61] This is a deeply natural and rational pattern of behavior, surviving, by and large, in urban "hunter-gatherers" to this day. Contradicting the pattern is always a disturbing affair, producing either feelings of entrapment or stranding, or Odyssean narratives of the nightmarish, protracted return. (The film *After Hours* is a nice, modern example). Thus the quick exit option in cyberspace—"option," because clearly one should be able to go back the way one came also.

We do not have the room here to continue discussing all the myriad enhancements, permutations, extensions, and subversions of schemes for getting about in cyberspace efficiently, profitably, and entertainingly. If we did, however, each should preserve in some way, and over significant portions of space, the Principle of Transit.

Navigation Data and Destination Data   Because it is a space, cyberspace seems to be very much "about" searching, finding and navigating data rather than the data itself—just as space, time, geography, and planning in the real world can seem for us to be "about" affording navigation to and from significant objects rather than about those objects themselves. Indeed, the cyberspace scenarios that occur to us most readily are of this kind—stories of marvelous travel, techniques of orientation and choice, and with not too much said about why. Taken too far, the emptiness of this "Top Gun" or "wizard-nomad" view of the spatiotemporal world—a view which holds that the world is there for us to *negotiate* both physically and informationally (in as much as reality is an endless chain of signifiers, a sequence of clues about clues)—serves to remind us that in fact there does need to be an end to traveling and a purpose to manipulation. There must be a point in time and a place where one can say "this is it," "this is what I came for," "this is what I needed to know." Cyberspace needs, therefore, to focus as much on arriving as it does on touring. It needs to provide the places, faces, voices, and displays of data beyond which there is no reason to go, that call for decision and action, that create "moments of moment." What follows is a short inquiry into this idea.

If one can imagine that the world is a single information field that comes alive, as it were, in the minds of sentient, motivated perceivers, then it becomes possible, I believe, to classify that information ($\mathbb{I}$) usefully into two classes: "navigation information" and "destination information." In the interests of euphony, I would like to rename these *navigation data* and *destination data*.[62]

Navigation data, as the name implies, is that class of information that orients us in time and space, in location and direction, that tells us of our progress in relative and/or absolute terms, that contains addresses, instructions to proceed, and/or warnings not to. Navigation data compartmentalizes and, in so doing, classifies, thresholds, and sequences potential experience. It is, in short, information that serves to organize us and the world in spatiotemporal terms.

Destination data, as the name implies, is that class of information which in some sense *satisfies*; it answers a question, delivers on a promise, rewards interpretation, engenders real-world (or cyberspatial) action. It may take the form of a text to be read, an image to be appreciated and judged, the face of a friend or colleague we need to speak to, and what they actually say; a price, a quote, a joke, a diagram, a piece of music or code. . . . Destination data, in short, is a body of information judged to be of intrinsic value, however arrived at.

Navigation data and destination data always appear together, and are often intermingled. With a change in our attitude and purpose, the one may even be transformed into the other. But typically, our attention is divided between them: now on navigation data, now on destination data; now on one aspect of the environment, now on the other. And, more often than not, those parts or aspects of the world that regularly serve these functions are designed differently. We must adopt a certain artistic or professional attitude, for example, to see a street sign for what *it* is, as a thing, in its own right. The covers, pages, layout, page numbering, marginalia, contents page, and indexes of a book or magazine, a system built upon hundreds of years of textual conventions, constitute its class of navigation data. Similarly, the stories, pictures, prices, addresses, advice, and so on constitute its class of destination data. So focused are we on getting where we want to go and getting what we want to get, that *buildings*—as works to be appreciated and read in themselves, as objects that reward questioning, as "destination data"—are effectively invisible to everyone but architects. This is because buildings and cities provide the physical world's most detailed navigation systems, and are widely perceived in terms of their navigation value alone. The distinction is similar to, but not the same as that of Marshall McLuhan between "medium" and "message."[63]

In the world of computing the distinction is even clearer; and here is one general way of looking at it.

The much heralded, punctuated evolution from command-driven to menu-driven to fully configured GUI-driven applications was, in fact, the stepwise bringing of implicit navigation data to the fore. Today it has reached the point where, as I remarked earlier, user and programmer involvement with navigation techniques—windows, buttons, icons, sound effects—has become equal to involvement with destination data techniques—visualization, interpretation, inference, decision support, and so on.

Can one have one class of information but not the other? Considered in a detailed way, no; but effectively, yes. In a navigation-data-dominant cyberspace environment, for example, one might cruise through electric grids and matrices in an infinite night, behold clouds of marvelous tinsel, fly over blossoming geometric solids, spin down spiral vortexes of color . . . without learning, knowing, doing anything. By contrast, one might stare at an old monitor, studying 23 lines of amber text with nothing navigational to do but scroll forward or quit. And the second may well be the more meaningful experience! A functional balance, clearly, is required.

Diagramming the evolution of interfaces allows us to extrapolate and illustrate what cyberspace offers as a next step:

Here, total information (available) to user = $\mathbb{I}_{tot}$ = $\mathbb{N}$ + $\mathbb{D}$ = navigation data + destination data. In the diagrams "navigation data" is normalized $\mathbb{N}$ = $\mathbb{N}/\mathbb{I}_{tot}$ x 100; similarly, $\mathbb{D}$ = $\mathbb{D}/\mathbb{I}_{tot}$ x 100. It is understood that the value of $\mathbb{I}_{tot}$ is a function of time also.

The cleft between navigation mechanisms and destination data are sharply drawn in command-driven and standard GUI-driven applica-

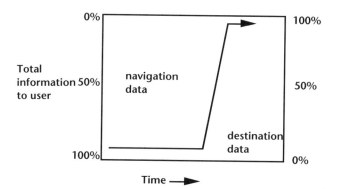

Figure 7.7a
Command-driven: a blank screen, a status line, and a few very significant commands ("navigation data") instantly give over to the document ("destination data").

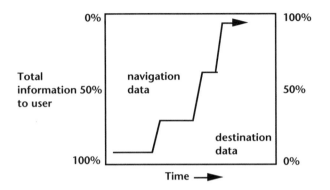

Figure 7.7b
Menu-driven: the user is given destination data in stages—its type and size, for example—by menus, each of which is orienting, sequencing, and thus aiding of navigation, until destination data fills the screen.

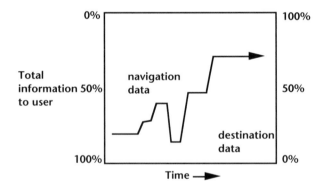

Figure 7.7c
GUI-driven: navigation and destination data coexist always, one sometimes dominating the other, but never completely.

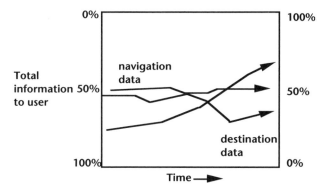

Figure 7.7d
Cyberspace (and super-GUIs): multiple destinations at different distances and different degrees of self revelation; navigation and destination data always available and balanced; tasks can involve "changing tracks of attention"; greater overall smoothness due to continuity of space in cyberspace (but steep changes—dI/dt and $dN/dt \gg 0$—still possible with destination unfolding and with step functions built into data objects).

tions. (We discussed some aspects of this earlier.) Cyberspace, on the other hand, offers a deep, spatially continuous environment rich enough for objects to be ambiguously navigational *and* "destinational"— switching, phenomenally, from one to the other as a function of user proximity, motivation, and attention, quite like reality. The red house on the corner is my home, but it is also the place to turn East if you want to get to the highway. *Seeing* in everyday life is also always *seeing-as,* as Wittgenstein pointed out; and a well-designed cyberspace will offer this kind of context dependency more or less naturally. Nevertheless, its seems likely that the earlier, fundamental distinction between extrinsic and intrinsic dimensions, including object-unfolding sequences, will map naturally onto functional distinctions between navigation and destination data.[64]

The Principle of Personal Visibility (PPV)
Based on the real world, but providing some significant enhancements, is the Principle of Personal Visibility. This states that *(1) individual users in/of cyberspace should be visible, in some non-trivial form, and at all times, to all other users in the vicinity, and (2) individual users may choose for their own reasons whether or not, and to what extent, to see/display any or all of the other users in the vicinity.*

Initially, the first provision of this principle seems to be a direct threat to privacy. But the kind of visibility I have in mind here is minimal.

A small blue sphere, say, for each person in cyberspace, indicating only his position, movement, and most simply and importantly, his *presence*, would suffice to satisfy the first provision of the principle. The user is then in complete control, via protocols too many to explore in detail here, of how to divulge any further information, either automatically or on request. There are no implied restrictions on the channels and media that might be used for interpersonal contact—voice, video, text, gesture, even VR touch—nor should there be any restrictions upon the reasons and content of the communication.[65] User-identity information need not be an essential part of the minimal presence (though it seems a good idea): anonymity is acceptable. The first provision of the Principle of Personal Visibility seeks only to prohibit individuals from "cloaking" themselves completely in cyberspace, from becoming entirely invisible.

But why? No doubt hackers will try to find ways to work in cyberspace without visible presence and without trace; certainly "sys-ops" will believe it their right, and voyeurs will feel it their need. Let us set aside the efficiencies and rewards of stealth; and let us let the watchers also lightly be watched. The Principle of Personal Visibility installs the belief that democracy, even in cyberspace, depends on accountability, and that accountability depends in turn on *count*ability, that is, on the obligation to "stand up and be counted," to *be* there, in some deep sense, for others. An open society requires the open presence—each to the other—of its people.

There are other, less ethically motivated reasons for PPV. In the real world many behaviors are guided by the grouping behavior—if you will—of others. We know "where the action is" almost instantly, and we can infer remarkably well from minimal, overt motor behavior what is happening in a social situation, "what's going down." Multiple and mutual individual adjustments of position can multiply into unpredictable molar patterns of group behavior, as researchers in "artificial life" often detail. In short, cyberspace must have street life. A good part of the information in cyberspace, as in the real world, is *in* other people, *is* other people. Further, when we step beyond the minimal presences—beyond small blue spheres, say—into the exfoliation of intrinsic dimensions presented as constructed, more or less fanciful *personae,* as will surely be the case and as Gibson so cannily depicted, the vitality of cyberspace will quickly begin to rival, if not transcend, that of the real world.

Personae, of course, need not have immediate, complete presence in cyberspace. That is to say, our "blue spheres," now strangely active, like tiny crystals or miniature flags, might unfurl, upon inquiry, into diaphanous images of beauty or power, straight from the pages of fantasy books. . . . Or they might not. In fact, it may be necessary to *limit* the scale of a given individual's presence in cyberspace. They might instead, upon querying contact, send oblique textual messages, steadily revealing information and opening channels only upon transactional agreement. The channels may open—voice, video, 3-D, color—people "gathering around" each other in twos and threes and fours . . . trading carefully in the data of human connection. (From afar, faint fractal lines might glimmer between them.)

Now let us look at provision (2) of the Principle of Personal Visibility. We may wish to feel alone, to work alone. We may wish to see some but not all of our fellow "cybernauts." Perhaps others obscure our view or behave distractingly. (I may want no self-styled, teenage mutant dragon to leap into my view when he chooses to.) In these cases there is no reason not to able to *select* who will be visible to us and who will not by various criteria: proximity, absolute identity, spatial grouping, task orientation, sex, origin, interest, and so on . . . in fact, by as many classes of information as are made available by others' public presences. This power to render others invisible (to us) is the other half of equation of privacy. For all its larger socioethical implications (for example, should we be able to screen out the suffering and protests of others?) I believe it is a provision worth transferring from daily life to cyberspace. The question revolves on the definition of public versus private domains of cyberspace as to where PPV does and does not apply.

The Principle of Commonality (PC)
Where the Principle of Indifference refers generally to the relationship between individuals and the elements of a virtual world, the Principle of Commonality extends our thinking to interpersonal communications and to the social dimension, with the virtual world—here cyberspace—acting as mediator.

Ordinarily, if you and I are in the same room, we assume that we are seeing pretty much the same things. We acknowledge, but set aside, our differences of perspective—although these may sometimes be crucial.

We acknowledge that the obstructions to our respective views are also different, that smaller objects may hide in my "view shadows" but not yours, and vice versa. But we do not deeply contest the reality or commonality of the features of the physical world so evidently *around us both*. Indeed we feel that commonality at this level is necessary to anchor, to root—and therefore to allow to grow—whatever differences in experience, feeling, and knowledge we bring to the situation. The physical environment functions as an objective datum; indeed, historically, it defines "objectivity." To refer to this by our earlier terms, environmental objects have self-identity and even super-self-identity.

In connected virtual worlds all of this can be done away with. Like people communicating on the telephone, one of us can be in a living room, the other on a beach . . . but now it will not be necessary that we really are in either. You might reach for a cigarette that in my world is a pen, I might sit on a leather chair that in your world is a wooden bench. She appears to you as a wire whirlwind, to me as a ribbon of color. While I am looking at a three-dimensional cage of jittering data jacks, you can be seeing the same data in a floating average, perhaps a billowing field of "wheat." These malleable data representations, worlds, and selves, seemingly so desirable, instantiating (at last!) our much-vaunted individual subjectivity and the late-twentieth-century notion that reality is nothing but a projection of that subjectivity, are, in fact, as much laid *against* each other as into each other. While the temptation to narcissism and deception are dismaying, the risks to rational communication are staggering. Even in the emotionally relatively neutral case of alternative data representations (let us say, of transactions and prices in the futures market) our co-respondent brokers in cyberspace, if they are to communicate effectively, must be able to co-witness and point out the salient features of all representations of the data.

The Principle of Commonality in cyberspace recommends *that virtual places be "objective" in a circumscribed way for a defined community of users*. More specifically, the Principle of Commonality requires that all comers to a given domain at a given time in cyberspace are to see/hear largely the same thing—the same place, the same objects, the same people—or at least some subsets of *one* "thing," and that the same direction considered as *up*.[66] More specifically still, the principle instates *self-similarity* as a norm for data objects (as "mere" objects) and for human presences. It is the idea of subsets here that allows us to achieve both the realness of the everyday world *and* the magical properties we would wish for cyberspace. In set-theoretic terms, the situation is easily represented in figure 7.8.

It becomes a simple matter to ensure that both A and B are proper subsets of the domain, D. We can further restrict communication between A and B to occur with and *through* the intersection of A and B only, that is, through something experienced that is both in D and that is experienced in *common*. Of course, no cyber-geographic reality is necessarily implied by this diagram, merely a logical/informational one. If Figure 7.8 suggests how our principle might apply abstractly (which is interesting in itself), it does not help us see how cyberspace as a place—a virtual world—can partake of both the depth of purely logico-semantic structures *and* the logic of being in a real space and time. Cyberspace is nothing if not the mapping of these two realms together. For this insight we require some consideration of the theory of isovists.

Isovists   In my definition of the Principle of Personal Visibility, I mentioned the vague term "vicinity" to indicate the limits of visibility. Although the actual range of one's vision might be individually adjustable, it seemed implausible that everyone in cyberspace at a particular time should, or could ever, be made visible to a single user. To make progress in understanding and implementing the Principle of Com-

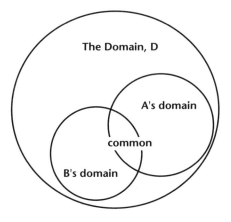

Figure 7.8
A, B, and their intersection are in D.

monality, however, we must make the notion of vicinity more precise; for commonality of experience in spatiotemporal, environmental terms entails "same-placedness" of some kind.[67] One means of reaching that precision is to use the concept of an *isovist,* and a part of the *theory of isovists* (Benedikt 1979, Benedikt and Davis 1979). Very briefly:

An isovist is defined as a closed region of space, V, together with a privileged point, x, in V such that all points in the space are visible from x. Whether an observer actually occupies the point *x* or not is irrelevant to the definition: isovists, like "views," are thought of as existing anyway, that is, objectively, anywhere and everywhere vision is possible. Thus one moves though the (visual) world, now "in" this isovist, next in that.

In the everyday world, with its walls and furniture and trees, isovists have varying shapes and sizes. (Only in a closed spherical room—in fact in any closed, convex room—do isovist shape and size remain the same. In this case, only the geometrical position of *x* relative to the isovist boundary changes.)

With this rather straightforward rendition of isovists, much can be done; and with a more sophisticated analysis of shape, even more. Here however, we can use isovists to clarify some definitions, specifically, the definitions of *concealment* and *isolation.*

Let A, B, and C represent three isovists in a domain D, with 0 denoting the null set, so that:

$$A = V_{x_A}, \ B = V_{x_B}, \ C = V_{x_C}, \ \text{and} \ D = \cup_x [V_x] = \text{all } x.$$

(Of course, if x_A was in B, then x_B would be in A, and vice versa, and there would *no* concealment since observers at these points would be visible to each other.) Isolation always entails concealment, but con-

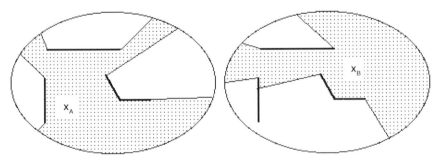

Figure 7.9
Two isovists V_{x_A} and V_{x_B} in the same domain.

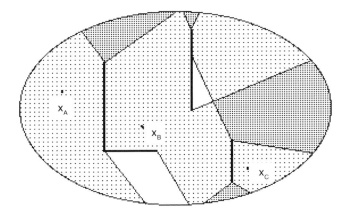

Figure 7.10
Three isovists in the same domain.

cealment does not always entail isolation. Neither isolation nor concealment are transitive relations. In experiential terms, isolation means that there is no space visible in common between you and another observer, while concealment simply describes nonvisibility.

Aside from making things clearer, these definitions from isovist theory also lend themselves to some quantification of vision-based phenomena: for example, minimal covering sets with respect to D, degrees of concealment based on area (volume) calculations, optimal trajectories to cancel isolation or to cover D, probabilistic measures on subdomains, and so on.

In the ordinary world, isovists are intricately interwoven, overlapping and excluding each other in ways that reinforce or frustrate social arrangements. Unlike our set-theoretic diagram, in real environments the intersection of n > 2 isovists can generate any number of disjoint common spaces, themselves distinguishable according to their com-mon*ness*.[68] When sets of isovists are put together and become *territory* for someone, asymmetries of concealment are common. Power is intrinsically associated with places from which one can see more than one can be seen, for example. Isolation, as the term was chosen to suggest, is something few of us seek out as a norm; sufficient privacy is usually achieved by controlling (self-)concealment. In fact, the seeker after privacy will try to maximize concealment but minimize isolation. And so on.

Many are the spatial behaviors and locational choices that are conditioned by the properties of isovists in a given, real environment. In cyberspace, the situation becomes more interesting yet.

Certainly if a domain in cyberspace is merely a re-creation of the real world, such as a virtual office or store, or some remote but real location as with the technology of telepresence, then isovists behave as they do in the real world. Vision is delimited by opaque surfaces in either realm. The fact that any object in a virtual world can be a *gateway* to somewhere else—unfolding into another space, or providing a passage to another universe[69]—does not affect the logic of isovists. No matter where they lead, these gateway objects must first be seen, as objects, from the space we are in.

However, should the domain and/or its contents be diaphanous, sparse, or largely transparent, then some conventions about boundaries, about isovist *horizons*, need to be instated. For with transparency our vision is indefinitely expansive and inclusive. Objects that are close merely overlay objects further away. With our X-ray vision we could see forever! A simple solution to the problem would be to instate spherical horizons: there would simply be a limit to the range of each user's vision. A diagram of the situation would look like Figure 7.8. However, accepting some notion of horizon, and even with arbitrary levels of transparency, there still are some important differences between the set-theoretic "bubble diagram" of the situation (Figure 7.8), the real world situation (Figures 7.9 and 7.10) and the situation in cyberspace. To wit:

1. The objects collected within data spaces are not arbitrarily positioned, as they are in a purely set-theoretic representation. Instead, they follow a spatial logic of position, proximity, size, and density and are governed by coordinate systems that orient, and indeed partially create, the data objects themselves. In addition to the topology of sets, we have the geometry of data space proper.

2. The shape of the horizon need not be a direct mapping of a radial range of vision of the observer. This gives us circles and spheres. Instead, it can reach variously to geometrically demarcated "private" domain boundaries that are opaque, as well as indefinitely across common, "public" space, through openings and gaps here and there, objects within the private domain being as transparent as the user wishes them to be.

3. Certain global aspects of cyberspace's navigational systems—beacons, cardinal points, grids—can be always be made visible through any

otherwise opaque surfaces; likewise, certain classes of objects can be opaque always—for example, the immediate surroundings of one's vehicle and tools.

Thus a user's isovist in cyberspace has a hybrid character: partially shaped concretely as it might be in reality, and partially shaped abstractly by the nature of the contents of the space he is in. This hybridization appears as a patterning of transparency, translucency, superimposition, and layering. This patterning is apt to change spontaneously over time and with the user's motion, the latter either as a function of the environment itself or as a "moving property" of the world *caused* by the user himself. I have in mind here the kinds of object rotations and self-revolutions I discussed earlier (pp. 143–144); and the way rainbows always stay out of reach, or how "columns" of space keep up with you as you drive past row-planted fields (p.146). Computationally, a record must be kept at all times of exactly what is in the user's isovist by virtue of where he or she is, and of what lies within his actual view and range—this being a subset of all cyberspace objects and features potentially available for the user to experience.

Perhaps most important, however, is how the ideas of isolation and concealment, and the much more intricate patterns of *common space* that are possible in cyberspace, can be used to give order to the social experience. Here are two recommendations that instate the Principle of Commonality in a general way—the idea, as always, being to map useful, if overlooked, features of everyday reality onto the strange reality of cyberspace.

Recommendation 1: *That users cannot see into domains of which they are not a part.* Modification 1: Owner/operators of *private* domains may control what and how much of their contents is visible from "outside," as well as other, related, boundary-transparency effects. Modification 2: Items in the *public* domain are always visible, in principle, at the discretion of the user, and within the range/power of his deck. (They may also not be allowed to *enter* certain domains—but this is another story.)

Recommendation 2: *That a monotonic relationship exist between the relative volume of space commonly visible to (any) two users and the bandwidth of possible communication between them.* More precisely stated: Let $V(A)_x$ be the volume of the cyberspace isovist available to A at x, and

$V(B)_y$ the volume to B at y. Let $R_{x,y} = V(A_x \cap B_y)/ V(A_x \cup B_y)$. Let $\kappa_{x,y}$ be the maximum communication bandwidth between users at x and y, then

$$\kappa_{x,y} = a(R_{x,y})^k + b, \qquad\qquad \kappa >= 0, a >= 0.$$

Here, **a + b** is a system maximum bandwidth, and **b** is the residual, default bandwidth available to users who are isolated, that is, who are in different parts of cyberspace. The value of κ is arrived at empirically: it "tunes" the rate at which getting together psychologically is getting together physically.

It is worth playing out some of the experiential dynamics of recommendation (2) in our imaginations. (The experiential implications of recommendation (1) seem clear.) We understand that "bandwidth" translates easily into something like the following, *cumulative* scale of communication channels and media:

. . . text (slow), text (fast), voice (lo fi), voice (hi fi), sound (hi fi), graphics (object), graphics (paint), video (graphics/stills), video (lo res.), video (hi res.), video (stereo), video (HTP stereo),[70] VR (head-mounted displays, earphones, gloves/suit), VR ("holodeck") . . .

The system maximum, **(a + b)**, may be reached anywhere in this scale,

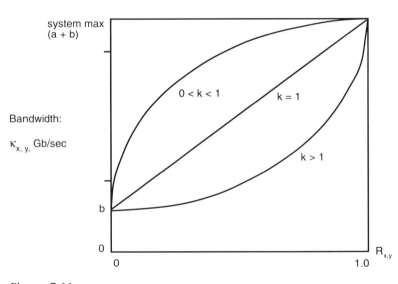

Figure 7.11
The relationship of communication bandwidth, $k_{x,y}$, to a measure of commonality, $R_{x,y}$.

depending on the whole system design, and reflects the least of either the capacity/speed of (1) the cyberspace network, or (2) the individual user's deck.

Now, the maximum value of R is 1. (I shall omit subscripts). It occurs when both user A and user B share the same space entirely. How might this happen? First, they might occupy the same point in space. This, however, is excluded by the Principle of Exclusion: they may only come "very close." Closeness is not required, however, if the space is convex and closed; that is, "being in the same room" may be enough to set R at unity. Let us imagine, however, that the situation is not quite so cozy; that with complex, phenomenal transparencies and different ranges, the value of $V(A \cup B)$ is rather large relative to $V(A \cap B)$, so that R is small. Four strategies are available to A and/or B: (1) deliberately curtail their range, (2) shut off certain views, (3) move toward each other; and (4) find locally convex regions. It is possible for two users to be in full view of each other and yet have R < 1. Conversely, it is possible for two users to have R very nearly equal to 1, and not be in view of each other, that is, to be concealed but not isolated from each other. The pair in the second case may have fuller communication (experientially). I believe this is as it should be.[71]

Let us say that two users are isolated from each other. Our cyberspace system has a residual communication bandwidth, **b**, and perhaps one or two media that can operate within that bandwidth. Our far-flung users *find* each other at the outset in ways similar to, but far sexier than, the ways we find each other now with telephone books and mailing addresses. Initial communications are confined to these channels. To open up fuller communication they must open windows to each other; in fact—without necessarily relinquishing where they are, as we discussed earlier—they must in effect *travel* to the other's place/space/domain in cyberspace, or to some third venue. The Principle of Commonality works in concert with the Principle of Transit, especially in the way it indexes, broadly, communication bandwidth to cyber-geographic distance.

To open full VR communications, with an individual or with a group, is to share a common reality. And vice versa. This is the very meaning of the Principle of Commonality.

We could go on. But hopefully, with this principle and all the others —individually and in concert—the reader is beginning to see how correlations between real life and cyberspace life can be maintained, and yet how much more magical and empowering the latter can be.

Part Two: Visualizing Cyberspaces

Remarks on Feasibility: The Symbolic vs. the Literal

By whatever hardware configuration, it is clear that cyberspace will require stupendous computing power and data communication speeds by present-day standards. Moreover, it will require that these capabilities be accessible to thousands if not millions of people.

When can we expect to see the requisite technology in place? Although such projects as installing the ISDN network (2 x 64 Kb/sec channels and 1 x 16 Kb/sec channel on telephone lines) and extending NSFnet (up to 2 Gb/sec, on fiber-optic cables) are being actively pursued, no one expects a super-high-speed networking system to be in place, in any widespread commercial way, before the early part of the twenty-first century.[72] Similarly, while significant computing power is increasingly affordable , the power required for color- and illumination-rendered, real-time, user-controlled animation of complex, evolving, three-dimensional scenes—around 1000 MIPS, and perhaps 4 million polygons/sec—is very far off as a common commodity.

Why, then, are we spending time, now, devising and divining the best principles for operating in cyberspace? Are we not premature? And why should we put any serious effort into visualizing mature cyberspace systems? Are we indulging in nothing more than fantasy, science fiction with an academic gloss?

In the introduction I suggested that "cyberspace" can be a motivating, unifying vision, one capable of directly coordinating, over the long term, currently disparate initiatives in computing and telecommunications. It can also motivate any number of research efforts in science, art, business, and, of course, architecture. Assuming some agreement on this point, I think there are two reasons for laying out the principles of cyberspace and attempting to visualize it now, so much sooner than it is likely to be fully realized. The second reason is less obvious than the first, and will take a little exploration. But first, the first.

Because the design, institution, and management of cyberspace will be a task of immense scale and complexity, it can simply be argued that "it is never too soon to begin." Like the early space programs of the United States and the USSR, the "cyberspace program" should begin experimentally, creating relatively crude, probably fragile, and certainly expensive cyberspaces, each with a limited number of users. As we work forward from these prototypes—improving them, connecting them—the lessons learned will be valuable ones. Spin-offs into many areas in computing—hardware, software, telecommunications, and interface design—will be plentiful. The experimental process will take decades, to be sure. But they are the same decades it will require for the technology to become affordable and for the whole enterprise to become profitable. With this strategy, at some point in the not-so-distant future there will be a happy convergence of means and ends, of capability and availability. In the meantime, the "cyberspace program" will be profitable for many: not only for thousands of engineers, programmers, designers, and managers, but for the companies and agencies that first use cyberspaces internally to increase productivity, the way Hewlett Packard now leads in the use of its own prototype networking and office automation products.

The second reason. As I have noted previously, most of the modern media and almost all of today's computer graphics and telecommunication systems—to the extent that they sustain consensual imagery, purposes, and discourses—can be seen, if not as cyberspace already, then as components of cyberspace in the making. Connected and coordinated properly, perhaps around digital interactive television, it can be argued that we have the essentials of true cyberspace in the palm of our hands already, and that we need simply to let these technologies evolve under market pressures.

Perhaps. But each of these technologies, from cellular telephones to TV shopping channels, represents an ideology and an economy with a life of its own; and the requisite coordination may never be mustered. In any event, again, we must wait.

As we await real cyberspace, however, whether it is assembled and evolved from existing networks and communication systems as just argued, or developed experimentally more or less from scratch as argued previously, we should not overlook the possibility that cyberspace as an *idea*, indeed as a *system* of ideas with rigor and purpose as well as a probable, future incarnation, can usefully inform the design of many computer applications *today*. Such applications need provide little of the direct, multisensory embrace we expect of virtual reality technologies and a mature cyberspace system, and yet they can function as cyberspace "generators" nonetheless. Just as one can, and will likely, experience cyberspace with screen-based, 3-D graphics and sound—that is, with less than full VR involvement and its attendant, enormous, processing requirements—so one can experience cyberspace "in the mind," created and sustained by programs hardly further along than today's GUIs, CAD programs, or even text-based on-line networks. Success lies in the consistency with which "cyberspace" as a functional metaphor, as a set of mental images and concomitant, real operations, can be propagated across platforms, applications, and networks. This depends in turn on the appropriate balancing of *symbolism* with *literalism* in what we design. If I might expand this last distinction:

Music can create spaces in the mind, as can mathematics. But the canonical example of symbolically created space, of course, is the space "found" in between the covers of certain books, in stories and in poems—the Siberian steppes in *War and Peace*, the desert in *Ozymandias*, the tangles of *Gödel, Escher, Bach*—in short, in the spell cast by words and numbers. If these spaces are not *cyber*spaces, it is not because they are constructed in the imagination, but because they are not constantly open to multi-use or to change, and do not themselves "know" that they are being read. But they can be "entered" by more than one person, and they can structure the discourse about, and therefore, in good part, "within" them.[73]

Symbolic systems in general—or, should I say, object systems understood symbolically—are triggers, lures, capable of eliciting rich "virtual experiences" from the mind's myriad depths with great efficiency. From the simplest of icons and abbreviations to the most enigmatic cosmological figures, symbols draw equally on our most recent training, on the decades of immediate and reported experience, and on millions of years of "neural programming." Thus symbols have enormous "leverage" compared to objects seen more purely as themselves, namely, in their literal, mechanical context. Moreover, symbolic objects themselves do not need to be very richly rendered or detailed. Drawn, crafted, gestured, mentioned, displayed on a screen certainly, symbols are energically, materially, and computationally "cheap." This efficiency, in fact, is the symbol's continued raison-d'être. The computational expense of symbol interpretation—of tracking and constructing the contexts in which they are relevant, of unfolding them into meaningful worlds—is borne by the experienced and educated individual mind in a living social context; in all, a perfect model of "distributed processing" with free hardware.

It is not symbolic but literal, "real," cyberspace that will have the insatiable appetite for MFLOPS, bus rates, pixel density, fiber-optical cable capacity, and the rest. Anything less than infinite computing power will deliver less sensory realism than ordinary reality, and less than we are apt to want. Therefore, *the question is not whether or not cyberspace should be symbol-sustained, but how much it should be so.* VR pioneer Jaron Lanier's dream of "post-symbolic communication" simply will not happen in any short term, if ever. (See Kelly 1989, Stewart 1991.) It may not even be possible. One must take what he says as a direction, a tendency, a preference, for sensory richness and literalism in virtual worlds. Indeed, rather than try to say just how literal cyberspace needs to be to deserve the name (as though we could accurately measure "literality" or "symbolicity" anyway!), I suggest the following double strategy: (1) with all technologies at hand, let us pursue establishing cyberspace as literally as possible: a multisensory, three-dimensional, involving, richly textured and nuanced virtual world converting oceans of abstract data and the intelligence of distant people into perceptually engaging, all-but-firsthand experience; but (2) let there be a sliding relationship between the symbolic and the literal, the first giving over to the second as technology and economics permit. Actually, better than "giving over": let the literal include and organize the symbolic, so that in the end *both* modes can intertwine to make one virtual world, a world that, like this world, is richer for the combination.

With this double scenario, the value of visualizing mature cyberspaces is clear. For with such images in mind, cyberspace can begin explicitly

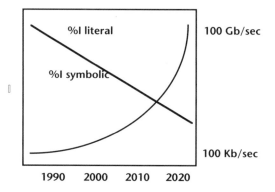

Figure 7.12
Note: here ▯ does not distinguish between information processing and communication rates.

to be constructed, if mainly symbolically, with today's resources. Time and technology are required only for converting cyberspace from a symbol-dominated, imaginary realm to a stimulus-dominated, literal one—a realm that, furthermore, will selectively retain, contain, and transform its earlier, more symbolic parts and incarnations. What begins as cyberspace in the mind, as carefully constructed as a good novel (but now interactive and encyclopedic), steadily transforms into the world that the novel *pictures*, cyberspace itself. (And in that world, experienceable by every user that jacks in, there are "novels". . .). Natural, which is to say psychological, limits as to what can be done literally and what can be done symbolically will become apparent, as will the unique opportunities afforded by each mode, but the construct of cyberspace both remains intact and evolves.

In the next sections I will describe some explorations of how the task of designing dynamic, three-dimensional, cyberspace structures—mainly databases—might actually be carried out with due awareness of cyberspace as an evolving rather than a revolutionary medium.

Visualization One: A Visual Database

This design exercise, done with graduate students in architecture Daniel Wise and Stan George, takes as its initial problem creating a cyberspace domain that allows browsing and access to a collection of over 300,000 slides of buildings, interiors, details, drawings, and landscapes, at my institution's Architecture Slide Library. The applicability of the design, I will hope to show, is general, extending not only to visual databases such as art collections and image banks, retail catalogs, photo archives, video libraries, and so on, but to countless other databases and facilities.

The slides are currently accessed in the typical way—a card catalog (now being converted to a text database)—and a room of metal cabinets with drawers of slides organized hierarchically: by *country* or *architect* (if modern), then *building type* (subcategorized by archetypes/projects, religious, civic, residential, commercial/industrial, interiors, landscapes, miscellaneous), *allied arts* (painting sculpture, products, posters, etc.), building *location*, building *name*, and then building *view* (drawing, exterior, interiors, details, environs).

A diagram of the abstract structure looks like Figure 7.13. At the top of the (inverted) tree is the set of all images in the collection; at the

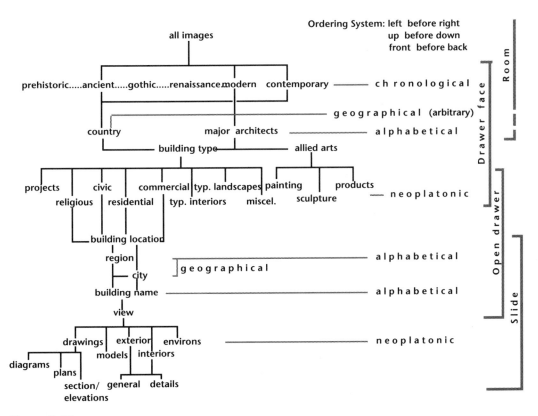

Figure 7.13
The abstract hierarchical structure of the School of Architecture Slide Library.

bottom is one image, one slide, presumably our destination. How is this hierarchical structure mapped into real space? Can one *navigate* by this structure?

First, as is typical, slides are kept in trays—metal drawers—these in cabinets, and these, in turn, in a temperature- and humidity-controlled room.[74] Access to the room is controlled by a librarian. However, the hierarchy of Figure 7.13 is nowhere to be seen, or hardly. Its upper levels only are visible on the face of the drawers; one must open drawers to see the "bottom" half of the tree. And even these are not shown in any spatial hierarchy. For in fact the slides are arrayed *linearly,* one after the other, in a long, folded, coiled row of 300,000 items, not unlike a strand of RNA or other macromolecule. (Straightened, the length of the row would be 1000 yards or 0.6 mile; side-by-side and thus directly visible the line would be 9 miles long.) The categorical hierarchy consists in markers along the length of row. To reach a destination slide, one must leaf through the neighboring slides, traveling "down" one limb of the hierarchy to the end, back up the next junction, perhaps across, and then "down" again. (The reader may begin to see the close affinity of the system with card indices and libraries in general.) Doing a search, say, for all the religious building interiors of a given period, is extremely difficult, requiring dipping into this drawer and then that. In fact, almost any search that does not coincide with the structure of the hierarchy is cumbersome. The hierarchy as a map, as navigation data, is of little use, and such usefulness as it has involves keeping it in mind and translating constantly into the physical reality at hand.

This is not to accuse our librarian, however, of arranging the slides poorly. In fact, a rather sophisticated system of orientations to navigate the tree is offered to the library user relative to the geography of the room and the cabinets, namely, the consistent if unremarkable spatial rule: left-before-right, up-before-down, front-before-back. These rules—very much prevalent in Western culture—"come" easily, and operate at the room level, the cabinet level, the drawer level, and even the slide (mount) and image level.

Interesting to note is the *variety* of ordering systems in the hierarchy of Figure 7.13 itself, each organizing a different level. Earlier we spoke of alphabetic, geographic, and chronological ordering systems (p. 148). All three are present here as well as one other—which I have named Neoplatonic, and all four read left to right as well as up to down.

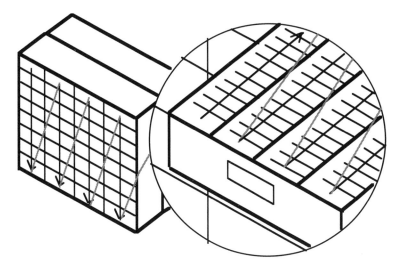

Figure 7.14a
The folded, linear physical organization of slides in drawers and trays.

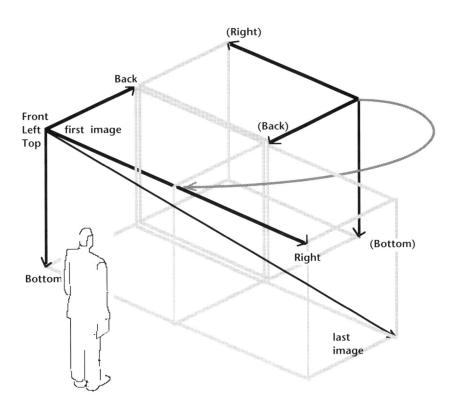

Figure 7.14b
The physical organization of the data in the room. (Note: in the real room two cabinet sets are back to back. The library user walks around to face the other way. In the diagram, the rear cabinet is swung around to complete the data space.)

What is this new scale? The Neoplatonic scheme places the spiritual above (or before) the material, the abstract above the particular, the potential above the actual, the eternal above the transient, the theoretical above the practical, and so on . . . forming a pyramid of sorts: at the top, the Forms, Truth and Beauty, God, the One; below this, angels and thought, true and beautiful things; below this, bodily existence, human and animal, impulses and instincts; below this, plant life, and below this, inert matter, earth, atoms. (Plotinus had read his Aristotle too). One can see this pattern—a value system, in fact, and one, I am sure, adopted unwittingly for being so "natural"—in the sequences in the slide library: projects-religious-civic-residential-commercial/landscape, painting-sculpture-products, drawings-models-building-environs, exterior-interior, and even diagrams-plans-sections and elevations. The fixed and timeless order of History dominates the highest level of organization, the detail of a particular eave or handrail occupies the lowest.

So thoroughly does it suffuse our culture, that I have no doubt that this Neoplatonic ordering system will manifest itself in subtle ways frequently, and throughout cyberspace.

Now, without the benefit of the notion of cyberspace and its technology, a software designer today intent upon making the library access system more efficient and intuitive is likely to want to do two things, one for efficiency and one for intuitiveness, or perhaps both. For efficiency, it seems natural to design a relational database with a textual "front end," one that can call up any image or sequence of images along any search criterion or set of criteria (from, say, an optical disc). In fact, our library is in the process of putting such a project in place. Ultimately, images will be copied rather than removed from the library, meaning fewer losses and allowing multiple "ownership" of an image. The system will be fast, will keep records, and so on, but it will be incapable of allowing soaking or browsing (see note 58) among the images, of creating a picture of a place or a period or an architect's oeuvre.

As efficient as the system may be, a more intuitive system, perhaps, would use a "hypergraphic" display of the hierarchy diagram as a map, a front end: users might click on a category and have subcategories open up, until they arrive at a series of slide names, and then images on screen to pick from. This display type, typical of more advanced GUIs, melds some of the panoptic logic of the diagram with the lattice-like, open logic, and instant, random access of a traditional database.

Now what if we extended our desire? What if we wanted to engage the library and its panoply of images more fully? After all, there are 300,000 fragments of worlds locked in our library. We want (1) to step into a virtual world where these buildings stand again, if not in true 3-D, then in good image projection, and (2) the way these are organized to be intrinsically spatial as well—a wondrous geography, a palace of places to wander through, corridors of History indeed. The basic distinction in this last sentence, of course, is between destination and navigation data respectively. Our design must somehow bring the two together seamlessly: with the ability, moreover, to navigate mixtures of chronological, alphabetical, geographical, and Neoplatonic orderings (and some forty subclassifications across these) and to construct or choose freely, according to one's interest, an *experience* of the images themselves that is repeatable, memorable, and yet unique.

I don't know that any of my students or I succeeded!

It was quickly evident that creating a *virtual building* of sorts, like a museum or "palace or places," that transcribed the hierarchical category diagram (Figure 7.13) into assemblages of rooms with doors, passages, etc., would not be efficient. Easy enough to do, coherent movement through such a model would offer none of the advantages of a database. Instead, we would be as locked into a pre-established scheme, and as locked away experientially from other rooms and the overall form of the collection, as we are presently in the real space of the slide library. Adding power and resilience to the simulation by allowing weird scaling, discontinuous motion, walking through walls, multiple presences, etc., all seemed destructive, a going-against what the notion of cyberspace and the technology of virtual reality "wanted" to do. The inverted tree had to go, and any "virtual museum"—based on it, or not—had to go too.

So a vast plain of images, the whole collection, flown over, divided cyber-geographically into landmarked regions and subregions mapping set-theoretically the hierarchical graph . . . this too would not allow reconfigurable or multidimensional navigation.

Daniel Wise's design employs the logic of higher dimensional representation discussed earlier and illustrated by Figure 7.2. This scheme was investigated quite extensively in proposal form, and indeed yielded a prototype for further work.

The user chooses a set of three dimensions to begin his search. Or there is a default set arrived at by application of the Principle of Maximal Exclusion. For example, "Date of completion," "Name of architect," "View," with the implicit ordering types: chronological, alphabetical, and distance,[75] respectively. These are displayed in a unique way, which is to say, pairwise in a three-dimensional volume, rather than truly three-dimensionally. Let me enlarge on the uniqueness here.

It is common in scientific visualization to present three- and four-dimensional data points in a coordinate system space where three of the dimensions are extrinsic, that is, are reflected in/as position in the space, and one is intrinsic (usually color).

Using this method, the space itself is often opaque with its own data, filled with a solid fog. Indeed, it is more like a solid object than a space.

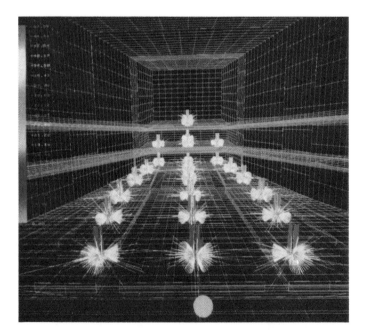

Figure 7.15
Nearly a dozen parameters displayed in the simulation of the acoustic qualities of an auditorium. Image by Adam Stettner and Donald P. Greenberg, Cornell University, 1989.

It must be peeled and sliced to see within it, examining its anatomy, and the facilities to do just this are common to all advanced data-visualization graphics packages. But how can one *inhabit* such a thing? Imagine a room so densely hung with strings of colored beads that one could see no farther than one or two beads away in any direction no matter where one walked. Nonfunctional, and suffocating. This is what it means to collapse too soon the navigation data carried by the presence and valence-structure of the "room" with the destination data, the colored beads themselves—the fog—in the room.

In our scheme, this problem is avoided. The walls of the room—we called it a "cell"—correspond to the three planes constituting a conventional, rectangular, and Cartesian coordinate system, namely, the horizontal plane, the "floor" (X-Y), and two vertical planes, the "walls" (X-Z) and (Y-Z). (It will be apparent later why Z is chosen for the vertical dimension.) The space in the cell itself is almost empty: "almost," because it contains *us*, our *vehicle*, and a *probe*, more about which soon.

Now, the walls of the cell are not blank but display quasi or partial destination data, namely, the *amount* of information, **H**, in units of megabytes, in the entire slide library that is selected by that particular value of X and Y, or X and Z, or Y and Z as the case may be, and under one of two user-definable conditions: (1) *regardless* of the currently selected value of the remaining third dimension, or (2) *given* the currently selected value of the remaining third dimension. This amount, **H**, is represented in one of three graphical forms depending on the dimension-type combination: continuous/continuous, continuous/discrete, or discrete/discrete, either as a colored and/or low-relief field, a set of ribbons of variable width[76] or a plane of discrete rectangles of variable size. Technically speaking, any one of the walls may in fact be ignored: the three dimensions are adequately determined by any pair. But the destination data a wall displays, uniquely, might well be missed.

We have a rectangular cell of an absolute size we can sense only by the application of the Principle of Scale: the speed of our subjective movement within it. And we *can* move within it, flying, floating, looking at any part from almost any perspective, from our vehicle: a virtual "pod" of some design that, with us always, provides us with interface and motion controls, navigations aids, communications devices, and many of the common features of GUIs. For the vehicle, the boundaries of the cell are not the walls however. They extend a determinate amount beyond, and with a specific geometry, shown in Figure 7.16. This permits a view of two and a half full walls (and about the equivalent vertically, involving the far wall, the floor, and ceiling) with an angle of view maximum of around 75 degrees. (Experiments have shown that this envelope is about as far outside of a room as one can choose a viewpoint and still feel inside the room, and the maximum angle of view representable without significant distortion. The near wall is transparent always.) Notice that within the envelope any amount of horizontal rotation of view is possible, except as curtailed at the corners "p" and "q," and at the boundary itself, where fuller rotation would cast us outside the cell. (Later we consider what exactly *is* outside.) However, keeping the line of regard parallel to a cell wall might simplify the computations involved, as one generates "one-point perspectives" only in this way, and it makes targeting easier in general since one's position in the X-Z plane is inscribed on the far wall "dead ahead" as in View from A of Figure 7.16.

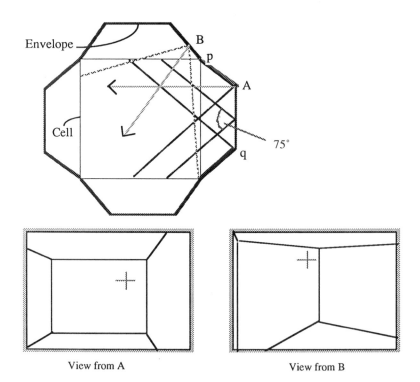

View from A   View from B

Figure 7.16
Implicit vehicle position envelope around prototypical cell for cell interior views.

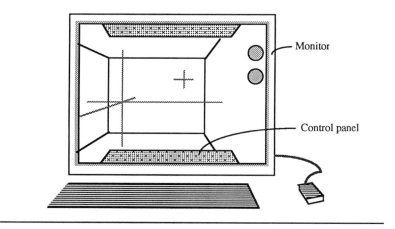

Figure 7.17
Screen-based deck system (very schematic).

One might imagine that VR technology such as head-mounted displays would obviate many of these framing considerations. Certainly the vehicle envelope could be coincident with the cell boundaries, but constraints on angle of view and freedom of rotation—especially in the vertical plane with pitch, yaw, and roll—would still need to be instated for the sake of avoiding vertigo. As well as providing a much-needed functionality, the metaphor of the vehicle gives us a legible system of motion constraints. In fact, and in whatever way it is designed in detail, the literal frame created by the vehicle and its control panels provides an essential, intermediate frame of reference between us—in this real world—and cyberspace. It belongs ambiguously to both.

The "probe" is imagined to be a satellite of the vehicle, controlled by its pilot, the user. It functions as a three-dimensional cursor within the space of the cell. The probe itself is located at the intersection of three cross hairs. It is moved by "clicking" on it and "dragging." (The mouse may be any multidimensional controller, or simply a two- or three-button mouse). Thus the vehicle's position relative to the cell and the probe's position relative to the vehicle (and the cell) is user controllable.

Now the actual information available at the x,y,z coordinates selected by the probe—the image we wish to see or the set of images we wish to browse—is not yet visible. This data is intrinsic to the probe and must be unfolded. As shown in Figure 7.2, this is easily done. Up to three more dimensions of search become available simultaneously, active in a subspace that unfolds from the probe that has its own subprobe. Upon the far wall of this subspace—or anywhere the user chooses, really— the image or set of images appears. The user now sees not only his destination, but also the structure and "geography" of the space wherein the information is stored. Both spaces and probes remain hierarchically active: moving the first probe changes entirely the contents of the subspace, while moving the subprobe changes the display only in the subspace. This continuing "live-ness" of the spaces is precisely what was called for earlier in my discussion of GUIs and windows (p. 145). The subspace as a whole can be moved forward and backward, left and right, up and down, as is convenient. It can also almost fill the screen, this in one of three ways: (1) by bringing the vehicle up closer to it (since the subspace itself is floating in the cell), (2) by bringing it up close to the vehicle, and (3) by direct screen enlargement.

A full description of the interface and guidance system would take up more space than is reasonable here. The system is rich in potential, as I hope the reader can begin to see. Almost all of today's two-dimensional GUI facilities are available "inside" the vehicle from its control panels—for example, windows and menus can pop up from the lower, or slip down from the upper, consoles. These facilities are thus not so much superseded as included in a more evolved system.

Finally, a vital aspect of operating in cyberspace is the presence of, and interaction with, other users. By the Principle of Personal Visibility, any user can be made at least minimally visible. His/her vehicle location and probe behavior is public. Every user has communication facilities, of course, including text, voice, video, and even VR, acting under the Principle of Commonality, as was earlier described. Illustrating the presence of others is not attempted here, however.

Plates 1, 2, and 3 show Daniel Wise's 1988 rendering of the ideas. Indications of the user's vehicle and control panels (the shaded frame on Figure 7.17) are omitted. A functional computer prototype of this model is currently being undertaken.

Extending the Model

The idea of a data cell is quite general. Owned and maintained in some way, it corresponds to the idea of *property*, of real estate, and we can expect some part of the economic system of cyberspace to revolve on dealings in such property. There are little or no restrictions on what can be experienced in a cell, and because it is cyberspace, there is no real restriction on its phenomenal, visual *size*. This, by the Principle of Scale, is a function of the amount of information visible in the cell, and there is no restriction but a technical and economic one as to how much and how rich this information can be. (Its auditory size is another and very interesting matter). In addition, there is no limitation on the number of *subcells* that can the opened and/or entered, except this: all subcells are entered from above. The reason for this limitation will become clearer shortly.

We imagine the cell multiplied: hundreds of them, thousands of them. We lay them out on a plane, like squared paper, their volume below the level of the plane, somewhat like a honeycomb. This is a rendition of the Matrix (Gibson 1984). Although, indeed *because*, the cells can be so different in phenomenal size once entered, their size on the surface can be more or less equal. I say "can be" because they need not be. Just as in the real world, the size of a "plot" of cyberspace is itself information: about the power and size of the institution that owns and operates it.

Some institutions (businesses, corporations, individuals, services . . .) may own several sets of cells. When these are adjacent, a transparent superstructure, a structure above the plane of the Matrix, may be erected. These are custom-designed within a pyramidal envelope to maximize visibility of them and past them simultaneously, and to correlate height with coverage.

To enter a cell is to descend into it, once permissions have been obtained. Not everything within is visible until one enters: hovering above, preparing or deciding whether to descend and enter, one might see only outlines of what lies within the cell, or one may see clear down, or one may see a special display . . . all this depends on the owner of the cell and his architect.

The cell displays default information along default dimensions. This is the state one finds it in. The user may, however, reconfigure these as required just before he descends, or while within.

Some cells might contain the kind of wall markings described earlier (in connection with Wise's visualization of the slide library) and might operate in similar ways to allow users to search them. The system is quite general. Others may use the opposite wall for data display directly—a grid of real-time video images, say, of the people who work for the institution—and use the floor and one side wall for choosing three-dimensional data points with their probes. Then again, objects and scenes of interest may simply materialize in three dimensions in the space of the cell, transparent or opaque. All the while, the vehicles and probes of others hover and move around like fireflies; and, looking up, one sees their traffic above the plane—the plain—of the Matrix.

How does one travel from cell to cell, from one landmarked or colored region of cyberspace to another without violating the Principle of Transit?

In general, one cannot simply break through to a neighboring cell. This would be an unnecessary violation if it were owned by someone else.[77] (Nor can subcells invade that space perceptually; hence, in part, their restriction to the cell floor.) Large-scale motion is carried out in two ways: by instant teleportation between transfer stations (cf. pp. 172), and by smooth flight. This flight is unique, however, in the way it amplifies everyday expectations. The "sky" over the Matrix is layered by *velocity zones*. The higher one travels, the faster one can move. The higher one travels the less there is to see: the tops of distant sector and region beacons, the iconic crowns of the superstructures of the largest institutions, the Matrix far below.

Actually this is not quite true. One may see as much information at high altitudes as at any other level. The difference is that at high altitudes the objects that *are* seen are at considerable cyber-geographic distance. That these objects are also likely to be fewer in number, and simpler, is coincidental. Here we see the power of the Principle of Scale: for the enormous sky becomes small by virtue of our speed, and yet does so "naturally," no matter how full it is visually. Thus, travel between distant parts of cyberspace can be all but instant, and the Principle of Transit is not violated.

As one descends, and slows, the Matrix fills in with detail, until one is cruising gently over the cells, looking down into their glowing, teeming, glittering interiors: here a performance, there a busy market, here a library, there a cell mysteriously empty, large, blue. With resolution on "full," we let our vehicle accelerate and decelerate as it will as it rides through the density waves of **space**$_o$.

And what is the *topology* of this cyberspace, so closely modeled on a city? Does it simply peter out into desert darkness?

Certainly, the vertical direction is accounted for. It is open-ended, and it maps the global dimension of "generality-specificity" from above to below. Indeed it does so into and through the primary cell floors. (Phenomenally, the floor of one cell can stretch a great distance, creating a surface almost as large as the whole Matrix above). In the horizontal plane, however, the plane of the Matrix folds back on itself in every direction, so that traveling in the two o'clock direction (and here clock directions make more sense than cardinal/geographic directions) finally brings one around to the same position, approaching from the eight o'clock direction. The Matrix itself can have absolute coordinates and addresses, somewhat like longitude and latitude but which never converge at poles. The overall topology of the "plane" of the Matrix is thus, technically, that of an *abstractly glued two-torus*, as mentioned earlier (p. 153). No cyberspace traveler however, sees the torus. She sees only a terrestrial geometry of plain, horizon, and sky.

Finally, in this all-too-brief sketch, it may seem that the "air" above the plane is rather empty—just translucent geometric forms all too reminiscent of office buildings. This is not the intention. One must imagine this space, first, alive with traffic—traces of hundreds of cyberspace travelers on free trajectories, clouds of sparks, perhaps glowing more brightly as they enter the pyramidal envelopes, as though entering a spotlight, or perhaps, in some, blinking out—and, second, alive with entities licensed to inhabit this public realm, floating like ribbons, hot-air balloons, jellyfish, clouds, but in wonderful unlikely shapes, constrained only (1) to represent information systems in the public interest, and (2) to be mostly transparent. There would be a thousand words and images, a din of voices and music (and, yes, advertisments are possible) . . . all to be tuned in or not, at the traveler's discretion. And, of course, the geometric forms themselves can be vastly elaborated and individuated even within the constraints mentioned.

Plates 4, 5, and 6 were developed by Stan George in 1989, using Arris CAD software. They begin to describe graphically only the framework of what we have been discussing. Much of the richness of content is missing; this was a preliminary exploration. Nevertheless, the reader should begin to get a feel for the whole.

Some Variations and Alternatives

What follows are very brief introductions—extended captions, really—to the visualization efforts of a few more of my students at The University of Texas at Austin. Many were created using computer graphics, but none were created solely with computer graphics. Like the previous projects, they are themselves far more suggestive than they are descriptive, more defining than definitive of cyberspace's possibilities. Needless to say, none of them literally work in a real-time, computational sense. They are what they are: partial visualizations, "artists impressions," if you will.

James Rojas, 1988: This visualization addresses the problem of the architecture slide library also, providing a model perhaps a little less generalizable than the one already discussed. It is based on the observation that *styles* in architecture are historically cyclic—from classical through mannerist through baroque, rococo, primitive/romantic, "gothic," and back to classical, which revives its earlier expressions and values, and so on. The process is one of codification and purification, tinkering, elaboration, decadence, exhaustion, return to "beginnings," development, and new codification . . . a process without beginning that may take centuries or decades to repeat. In cyberspace, this cycle is directly rendered as a helix, spiraling upward in time and enlarging (hence the spiral form) as more works of architecture are/were built and recorded. The whole "theory of cycles" was adopted as a conceptual, navigational scheme with full knowledge of its limitations, but with the hope that it would, nonetheless, produce a useful and memorable structure in cyberspace.

Rojas's visualization shows the overall form of the database floating, one can only say "somewhere," in cyberspace. Once within, other users are seen as liquid figures moving along the ramps of the spiral. Helped by agents curiously like Vanna White, one passes through the color-coded panes of a panel set across the ramp to enter a subcategory of the search, perhaps a geographical one. One arrives finally at a display of the desired building image and its textual support. This is the system's destination data.

Perhaps most interesting about this visualization is the creative acceptance of *noise* and *distortion* in the image. Indeed, the appeal of the exercise is partially dependent upon it. It suggests a strategy of adaptive refinement, of "graceful degradation" in the way the system works, that is more akin to radio operation—especially shortwave radio—than it is to standard computer graphics. Artists are well aware of the role of the forgiving, partially out-of-control medium in producing wonderful accidental qualities that become essential to the work. Rojas's artistry here shares that quality. It remains a challenge to computer hardware designers as well as programmers to create systems that behave generally in this "soft" way, giving users the initiative and the means to "drive" their systems as hard as they choose, with consequences that are visually apparent, nondestructive, and progressive . . . and yet not entirely predictable. (See Plates 7 to 12.)

Clyde Logue, 1990: This visualization of how a sales convention might happen in cyberspace takes the idea of graceful degradation a step further. Although, like all the previous examples, it creates a world that could easily be rendered in full, three-dimensional, virtual reality mode, in this screen-based model we see a number of concurrent and different screen resolutions; information density ($D^{(2)}$) is thus allowed to vary visibly. Figures of desire emerge from the noise. Cyberspace itself is imagined as a kind of three-dimensional static, images forming and dissolving from and into a deep video snow.

The convention structure first appears as a flock of panels in tight formation. Among these one drifts at will, immersed in a hubbub of voices and faces and products, each one bearing a coded label, each one a manipulable, hypergraphic element, and each class of elements resolving and unresolving as the user changes his degree of attention towards them. (See Plates 13, 14, and 15.)

Gong Szeto, 1990: Inspired by submarine graphics, this scheme shows visualized data as a sort of underwater topography. The attempt is to give form to the idea of "immersion" in data that is implicit in most VR and cyberspace discourse. The concept of a cell is present but somewhat relaxed. The vehicle itself—the cyberspace deck—provides a rich overlay of screens and analytical devices, together with navigational information as to one's location. Interesting is the tilt of the vehicle, indicating a sort of freedom of user motion relative to a more stable landscape, and the treatment of "on-board" video and data windows as semitransparent and rotatable and yet strongly reinforcing the necessary intermediate frame of reference between user and cyberspace. The system is reminiscent of the head-mounted graphics of VR, flight simulators, and real piloting systems now in use. (See Plate 16.)

Daniel Kornberg, 1990: A "cyberspace video store" hardly describes what is proposed here. The model goes one step beyond the rather easy-to-come-by notion that in cyberspace one should be able to find and preview movies (or musical performances) before downloading them, in compressed form of course, for home play. It shows the interior of a hexagonal structure along whose surfaces hundreds of movies, or parts thereof, are continuously playing on myriad screens. These are arranged in the depth dimension according to a user-chosen variable, defaulting to time/age. As at a market, the user collects images like flowers, flowers-become-screens, floating along with him. These screens in turn can be linked to other parts of structure, and/or other films, chosen for having the same actor, theme, director, location, or writer . . . the possibilities go on. Thus the user becomes a creative force through the act of research and collage, effecting a unique passage not only through the cyberspace structure as such, but through the passages and places depicted in the films themselves. And he need never download. This browsing, this "shopping," may well *be* the destination experience (and we are reminded of the fine line between shopping and consuming in general when it comes to information). (See Plate 17.)

By Way of Conclusion

How tempting it would be to claim that the two parts of this chapter correspond to the distinction Theory/Practice. In fact, both parts are Theory—some might say speculation. It will be some time before anyone can write a section fully deserving of the name Practice.

And yet, I hope to have shown at least how extending the logic of graphic user interfaces into truly three-dimensional realms might be accomplished, and how we might begin to organize consensual and public worlds out of the vast networking power of today's and tomorrow's computers and electronic media. To help make progress in these directions one does not need to be an advocate of cyberspace as such, with its attendant cultural imagery, nor of virtual reality technology with its current promises and rhetoric. My hope is that there is sufficient

material in what I have presented— from the observations and suggestions I have elevated to the status of principles, to the impressionistic
descriptions and visualizations I have shown as possible outcomes—
to contribute to the continuing evolution and speciation of our infor-
mation-age culture though the computer.

I have tried not to be too partial toward, say, abstract, commercial,
scientific or generally academic applications of cyberspace, but there
are those, I think, who will find my treatment of persons—personae—
and of what virtual worlds offer by way of self-portrayal and interpersonal communication, somewhat lacking. I have not given the full
treatment one could to the science and art of telepresence and the
usefulness of this technique for experiencing staged or ordinary, if
remote, worlds directly (such as "going" to Mars through a telepresence-
equipped robot connected to a VR console at home). I have largely
ignored the technical difficulties in actually programming and implementing what I have recommended. And I have been rather sanguine
about the possibly less-than-salutary political and economic effects of
a fully deployed cyberspace.[78]

What can I say?

It is traditional to end a scientific paper with a list of acknowledged
shortcomings followed by a statement of how essential further work
will be to addressing the issues raised and to resolving remaining
difficulties.

Consider the tradition upheld, but, with our topic, and my efforts,
more so.

Acknowledgments

I would like to thank my colleagues David Emory Campbell, Larry Doll,
Don Fussell, Marcos Novak, and Robert Swaffar for their readings and
for the valuable conversations we had while I was writing this chapter,
all my "cyberspace students" at The University of Texas at Austin for
challenging me to be clear and for their creative and critical input over
the last two-and-a-half years, and all the participants of The First
Conference on Cyberspace for their confidence, inspiration, and energy. Special thanks are due to Bob Prior, a courageous acquisitions
editor indeed, and to my wife, Amélie Frost Benedikt, for making the
paper readable at all.

Notes

1. Indeed, the economy of material *property,* which is inherently spatial and which dominates the classical economic theory, in cyberspace is subsumed by the economy of *information,* and with it the idea of *time* as the only true scarce resource. The economy of *capital* is seen as a stage in this subsumption.

2. For an account of the change being wrought in our society by the "informating" process, see Zuboff 1989. There is hardly a problem mentioned in this book to which cyberspace is not the solution. For a discussion of the quality of *realness,* see Benedikt 1987.

3. Gibson 1982, 1984, 1987, 1988; Brunner 1975; Vinge 1987. See also *The Mississippi Review* 47/48, (vol. 16, numbers 2 & 3, 1988); *Mondo 2000 #2* , Summer 1990, whole issue *Metropolis,* Sept. 1990, pp. 40ff; *Whole Earth Review,* Fall 1989, pp. 108ff., and *Smithsonian,* vol. 21, #10, January 1991, pp. 36ff.

4. What George Gilder (1989) calls *microcosm.*

5. And we must often be tightly swaddled to ease the shock.

6. On the matter of principles: many would leave unanswered the question of which of the two—the felt, somehow-known-as-unitary *phenomenon,* or the set of operable *principles*—has the deeper, more originary status. In practice, some are guided by a sense or vision of the phenomenon itself, even as they investigate and encode it by rules and principles, while others are guided by the play of rules and principles among themselves and their unwinding interaction in long, symbolic chains, without visualization of what, as a whole, these rules and principles together constitute or signify. At best, such wholenesses are intuited.

Certainly it is the skill to work in this second way that has allowed mathematicians to develop many exceedingly complex theorems in topology, algebraic topology, and functional analysis that may involve many more than three dimensions and/or deal with manifolds of complex numbers.

7. I think the case can convincingly be made that religion in the form of a church/ temple was historically the first information business, and that the coordinated space of gods, the sites of their interplay, and the world of sacred texts, images, accounts, and accountings sustained by religions since time immemorial, constitute the forerunners of cyberspace. Much that was done, and of what simply happened, in the real world of ordinary things and ordinary people was seen as the direct result of goings-on in this sacred virtual world, access to which by mortals was limited to shamans, priests, heroes, and the dead. (Today we might want to add hackers to this list, or, at least, hackers might want to add hackers to the list.) In any event, there will be room for considerable scholarship someday into what will be seen less as the *invention* than as the *evolution* of virtual worlds into the electronic medium. I hinted at all this in my introduction to this volume.

8. In movies we have "user-uncontrollable" illusions of such; another sort of writing, and another sort of reading on our part.

9. I have always thought that the modern computer screen resembles less a "desktop" than the patch of sidewalk used by a street-corner conjurer.

10. Though clearly one should design the former to suit the latter. This observation was also at the heart of widespread objection to the co-option by Autodesk Inc. of William Gibson's word "cyberspace" as a trademark for their VR interface system with AutoCad. See Sterling 1990.

11. Part of this has to be cultural, of course. Hebrew is written right to left, Chinese top to bottom, and so on. But within these cultures, top is "elevated" over bottom and center over margin, always. I expect that all systems of literacy using a *field* of inscription or markings to record information—be it a stone block, the wall of a cave, or the side of a building—vivify that field with polarities and values thought of as *belonging* to the field more or less intrinsically.

12. It is probably best not to think about the task in these grandiose terms too often however ... except to notice how every development in computer technology, from chips to monitors, from connectivity to games, every step that moves towards realizing this essential, if unstated, vision is greeted with such unreasonable enthusiasm. Examine the hype, examine the names chosen for companies and products: the entire computer industry, it can be argued, is drawn along at maximal velocity by the fantasy of virtual realities such as cyberspace, even as it is currently sustained by the enormous profits to be made in the rather dull business of processing business and scientific data and computerizing the second and third worlds.

And there is this complementary movement: a nostalgia for working with our hands, for being craftsmen, for honest, real labor. Cyberspace, in the way it reifies information and operations on information, offers symbol workers—from executives and academics—feeling guilty in some way that their work is abstract and invisible, intangible, and yet highly paid and displacing of their fathers' more "honest" labors, a return to or recovery of the procedures associated with tangible things and with visible creation. On this theme, see Zuboff 1989.

13. To think of the dimension of *time* as equivalent to a dimension of space and therefore of reality as profoundly and somehow symmetrically four-dimensional is less easy, even though a library of books has been written about the nature of this "equivalence" and the superior scientific status of the four-dimensional space-time world description. I expect that the reader, like myself, has read no small part of that library!

14. I am assuming that we can always deal with 4-space by constructing a model in 3-space that changes over time.

15. In the language of algebraic topology, a point can also be *open* or *closed*.

16. Rather than "phase space." Phase spaces are always conceived of as n-dimensional. The whole state of a system is thus described by one mathematical point in the phase space. What physicists are apt to call "phase space," mathematicians will often call "measure space," econometricians and engineers, "state space," and so on. To my mind "data space" is the most generic and inclusive term for these and other representations such as matrixes and "spreadsheets," and will lead most easily on to the notion of cyberspace as a kind of superrealm of living, breathing data spaces with which you can interact.

17. I like "intrinsic" and "extrinsic" rather than the terms "internal" and "external" because the last pair of terms—especially "external"—too easily leads one to think of the "outside" shape and size of the object as such rather than its spatiotemporal location and kinetic behavior, which is what I mean by "extrinsic."

18. So named because of the similar postulate in quantum mechanics called the Pauli Exclusion Principle. This states that no two electrons (or fermions) belonging to the same atom can have the same quantum numbers. This simple restriction on an "object" generally regarded as a point spatially, combined with the basic level structure of quantum theory, is responsible for the structure of the periodical table, the nature of chemical bonding, and most of the properties of matter including the macroscopic exclusion principle referred to here. (Citation from Cooper 1970, p. 395.)

Pauli's principle is weaker, however, not only because it applies to electrons in one atom only, but also because it excludes two "objects" only from having the same value on all dimensions, only some of which are spatiotemporal. For example, two electrons can have the same momentum and level as long as they have different (opposite) "spin." Our exclusion principle says that intrinsic nonidenticality forbids extrinsic (spatiotemporal) identicality, while intrinsic identicality and extrinsic identicality devolves to the simpler case of self-identity or singularity.

19. I am aware that this is not the definition of the term "self-similarity" today.

20. Indeed, a "field" in mathematical physics is broadly defined as a function, g, in which every point, extrinsically specified, has a mapping to an intrinsic value or values of some quantity: $F = F_{x,y,z,t} = g(x, y, z, t)$. F may be scalar (single value) or vector (two-value), or tensor (matrixed multivalue).

21. The reader may be getting jumpy at my presumption at this point, but let us remember the presumption implicit in being Creators of cyberspace in the first place.

22. This is why hyperdimensional grand unified theories (GUTs) in physics, such as supergravity and superstrings seem illegitimate to the true atomist. Coiling or twisting or otherwise packing all the world's remaining fundamental physical dimensions (there are apparently 7 more) into tiny regions of regular space-time, it seems to the atomist, is a little too easy, a sort of sweeping of the problem into the rug. "Have problem? Hitch another dimension to space-time. But better make it intrinsic, that is, invisible and/or non-space-consuming; otherwise we'll have to explain where it went, why it isn't out here. And if this intrinsic dimension *must* be spatial (after all, in this trend towards the geometrization of nature we don't want to end up with anything like Cartesian *stuff* with inherent character) then coil it up in an inaccessibly small labyrinth at the heart of the tiniest 'particle.'"

I am being flip here, of course. But in essence this seems to be the situation. Viewing nature's doings ultimately as geometry, as a "condition of space-time," may be a worthy goal (it is the true atomist's goal at that), but having then to conceive of seven such "spatial" dimensions not acting spatially—not "out here"—but locked up in uncountable microcosms is surely forcing the meaning of the word "spatial." This in turn leads to interpreting the fact that many behaviors of particles lend themselves to visualization in geometrical formalizations—indeed, in data spaces—as *proof* that such spaces exist in physical fact. (Of course, thinking of intrinsic dimensions as spatial, and representing them this way as an aid to visualization, is perfectly legitimate.) The universe may indeed be "eleven dimensional" at root, but either only three are space and one is time, or we need a darned good explanation of how and why the other seven got kidnapped.

I am not the first to wonder about this, by far. Popular expositions of this territory are legion. This comes from Davies 1984 (p. 162):

"In their search for a reason why seven dimensions should spontaneously compact themselves, theorists have been working on the assumption that physical systems always tend to seek out their lowest energy state. . . . This suggests that a shrunken squashed seven sphere is in some sense the lowest energy configuration of space-time."

Perhaps. I suggest that it may be an informational matter, that intrinsic dimensions (if they exist at all physically) are utterly nonspatial, and that all the whole phenomenon is the work of the Principle of Maximal Exclusion. This interpretation makes sense of the surprising mathematical fact, for example, that *only* 4-dimensional space-time (not 3, or 5, or 6 . . .) allows for an uncountable infinity of differentiable structures in mathematics and stable orbits—among other phenomena—in physics. Cf. Stewart 1987 and Barrow and Tipler 1987 (pp. 258–276).

23. The evolution and exploration of cyberspace will shine light not only on the questions of physics but on the meanings of "evolution" and "exploration." Fundamental connections between cyberspace and reality may only become understood as, perhaps, they draw parallel in the far distant future. Until then, they are best thought of as separate.

24. It also tells us why the world seems to consist of persisting objects and field types in spatiotemporal "motion": space-time is the largest container, informationally speaking, for the "complexification" of the world to "take place" at maximum "extension." (Extension: the classical word for *space*, used by Descartes, Spinoza, Leibniz, and their contemporaries.)

25. For a comparable set of ideas, cf. Barbour 1989. Barbour employs a graph-theoretic connectedness measure of identity such that the "distance" between two vertices is a measure of the difference in their connectedness. Maximal variety is the principle whereby the number of distances in a network is maximized. The suggestion is that just such a mechanism can "cause" space-time.

26. It is appropriate to make further note now of some of the similarities and dissimilarities between the formulation offered here and that of the notion of *phase space* in physics and scientific visualization.

In classical physics, the state of a system consisting of **n** unconstrained particles—each particle with a position (3 dimensions), and a momentum (another 3 dimensions)—can be represented as a point in a phase space of 6**n** dimensions. The evolution of a system is described by the Hamiltonian equations and can be visualized as the trajectory of the point through phase space.

Notice that, in our terms, all the dimensions are here treated as extrinsic; the idea of the "point" is thus saved, as it were, from being nonmathematical, that is, from having any quality or character other than pure existence. The idea of a point in phase space goes atomism one better: the state entire universe can be represented as a single point! The fact that no one can actually visualize a space of more than 4 dimensions—much less a phase space several thousand dimensions that even a simple, atomically considered real system would generate—is regarded philosophically: that is, as sad, but true. Nevertheless, several ideas and several theorems can be elegantly represented with the idea of phase space. One such example is Joseph Liouville's deduction that the volume of phase space occupied by Hamiltonian-governed systems must be preserved over time.

In phase space, the Principle of Maximal Exclusion does not apply, and this by default. Why so? Because in phase spaces—for all their similarity to data spaces—the complete state of a system is imagined contracted to, and coded as, a true point's simple position in a reference frame. There is no possibility that *one* entire system can, at the same time, be in *two* states. Of course, if there is uncertainty as to the state of the system, one may proliferate points and attach to each a probability of being the case. Alternatively, one may map and regard simultaneously all the possible states of the systems, thus viewing the behavior of the system as itself a system (in phase space). In both of these cases PME might indeed apply if interpreted as a "desire" of the system to differentiate its possible states, to wander ever further and in more complex ways through its phase space. Indeed this is the character of so-called ergodic systems and chaotic systems, but I am unclear as to how to make firmer claims that I can only intuit for PME in this context.

For us, PME only becomes useful—operative—when partitionings are being made between extrinsic and intrinsic dimensions.

If phase space casts all dimensions as extrinsic and if, because of this, it is difficult to visualize complex or highly populated systems in full or accurately, then phase space's logical opposite (which I will call briefly *h-space* for no good reason) is slightly less of a problem. For in h-space all of a system's dimensions would be regarded as intrinsic, and only one or two, extrinsic. (Having *no* extrinsic dimensions seems to defy existence, *pace* Kant). The situation is one we can visualize more easily. It would consist of arbitrarily small and complex objects in a 1- or 2-space such that the position of the objects, their trajectories, and geometrical interrelationships made no difference and meant nothing. PME would similarly have no meaning. In fact, the spatiality of such a space would be largely gratuitous.

(Phase space is a common notion in physics, both classical and quantum [where it is called Hilbert space after the mathematician David Hilbert]. For a recent and illuminating rendition for the nonspecialist, cf. Penrose 1989 (pp. 176ff.).

27. Stereo sound is, of course, the easiest extra channel of information to add to the purely visual cyberspace we are designing. It may well be a crucially important one, as may well be voice communication. However, this chapter will deal with the opportunities and problems of sound no more than tangentially. I will not specifically refer to hardware and software technology again until the second part of the chapter.

28. It is likely, for example, that the sound of an object is the most suitable medium for carrying large amounts of intrinsic data rather than sight. In the real world, the sounds of things—noise, music, voice—seem to issue from within them and express their character without affecting their look or spatial behavior. Thus sound seems to be a natural model for carrying intrinsic data of all sorts, as I remarked earlier.

Take the "cybercube" of the previous paragraph. We could easily hear its faces quivering, and even its pattern of tumbling.

Another approach is to use our natural ability to decode certain complex social stimuli such as facial expressions. The shape and condition of facial features, for example, can be made to reflect the simultaneous values on a large number of dimensions, having, of course, nothing to do with faces. These would be its intrinsic dimensions, the place of the face as a whole—or its address point—being its extrinsic dimensions. See Chernoff 1973.

29. Cf. the work of Ken Perlin, Department of Computer Science, Columbia University, on "fractal windows." "Scratch and Pad Demo," videotape, 1990.

30. This in essence was the way the visualization of a hypercube was achieved in the 1970s by a number of researchers. See papers by David W. Brisson, A. Michael Noll, and Thomas F. Banchoff and Charles M. Strauss, in Brisson 1978.

31. Of course, at a lower level this happens necessarily as the computer itself computes which view of a stable non-self-transforming object to put up on the screen every nth of a second.

32. Actually there is still some "graphic redundancy," or overdetermination, in this visualization. Only two surfaces are needed to locate a point in 3-space, that is, one pair from the set {(x,y), (x,z), (y,z)}. It is possible therefore not to need to unfold a *second* space but to choose complementary pairs of "walls" to locate data points in the *same* space (phenomenally speaking). For example, in this figure, **p** in XYZ could be located by the left "wall" and the "floor," and **p** in ABC by the "back" and the right walls. Thus the behavior of a line segment can represent the action of 6 variables, while in theory, if we bring the "ceiling" and "proscenium wall" into play, the behavior of a triangle in a single phenomenal "room" can represent the action of a 9-dimensional system. Now add color to the triangle's surface . . .

The reader will also realize that the unfolding of intrinsic dimensions into subspaces can go on indefinitely. The possibility of *circularity* also presents itself, that is, where the third or fourth unfolding, for instance, unfolds to the first "mother" or matrix coordinate system!

33. I say *almost* completely because the user can arrange his icons in the window to reflect some of their properties in their position—for example, recency and size—and it is conceivable that the computer can do this for the user intelligently; and certainly there are already implicit rules one can follow having to do with the Western tradition of how text belongs on a page such that the top and left of a page are privileged over the bottom and right, the center over the margins, and so on. See also the chapter by Alan Wexelblat in this volume.

Of course the undifferentiatedness of the space of "windows" is part of their usefulness: they are very forgiving. Like real space you can put things/icons more or less anywhere in a "window" and the window almost anywhere you please on the screen (and the computer almost anywhere you please in the room, and yourself almost anywhere you please in front of the screen . . .).

34. One small example of the further options that appear with this notion, and using not only extrinsic but intrinsic dimensions as controls: an object may not

bloom unless it has been rotated into a certain position . . . or moved, or has turned to a certain color . . . thus making certain states prerequisite to others.

35. Cf. Chris Isham, "Quantum Gravity," in Davies 1989.

36. Praise for the alphabet as an invention too often focuses on the fact that is a limited set of symbols of tremendous combinatory power, and not frequently enough on what a universal precedency system the absolutely tight *order* of the alphabet provides together with the gift of naming . . .

37. For a very accessible treatment of this, see Abraham and Shaw 1982–1988.

38. Strictly speaking this axis is not continuous but has a "Plank length," or quantum, of one cent.

39. The technique is not dissimilar to standard data compression techniques, especially for graphics.

40. Ways to finesse the ancient, concomitant dilemma of how *any* number of size-less things can amount to something of finite size remain problematic for most of us. Typically we simply declare that there is as much space in any defined space as the volume measure, V, of that defined space indicates, measured in the metric of the coordinates of that space.

41. It will also shrink and expand in $space_u$. In fact, any two-or-more pixel particle could exhibit all the laws of refraction, where **D** is akin to the refractive index, and of gravitation, where **D** is a dual of mass density.

42. Actually this point is debatable, especially if one follows Edward Fredkin (see Wright 1988).

43. The reader may wonder how it is that we *know* that we have slowed, embedded as we are in $space_o$, unless we also have access in some sense to an inertial $space_u$ relative to which we have truly slowed. To expound on this fully would be to attempt to explain the phenomenology of General Relativity, a task in analogizing that I leave for the future. I hope it suffices here to suggest that (1) indeed, we have no way to sense acceleration noninertially except by the variability of the size of space around us, and (2) we make the continuous assertion that $space_o$ is uniform, at the current value, everywhere, that is, time is the constant metric, and $d\mathbf{D}3/dt$ is the variable of phenomenological import.

44. But in reverse of course. Light bends around a star not because of curvature, but because of the rarefaction of space-time, a lowering of $\mathbf{D}^{(4)}$, in the region of the star. Perhaps gravity—natural "attractive" gravity—is nothing more or less than the "simplicity in things". . . calling to each other. What makes black holes so very attractive, then, is their extreme, and total simplicity. (Notice the tie back to PME here.) But this is to take perhaps too seriously the notion of the real world as an *information field, and information as the ultimate substance of the universe.* At any rate, the idea is deeply relevant to cyberspace since cyberspace is explicitly, and by design, nothing other than a field of information (and very crude compared to physical nature). The fruitfulness of the comparison to me must by now be quite evident to the reader of these last footnotes.

45. Essentially screen size in pixels or "physical closeness in cyberspace," measured in steradians.

46. See Bergman et al. 1986. There is also the well-known strategy of selective or adaptive detail rendering as a function of gaze direction, and techniques such as Incremental Radiosity, where only changed or new objects in a scene are calculated.

47. The equation "reality generates consensus" has its converse too: "consensus generates reality." To some extent, the same can be said of the phrase "reality generates science . . ." To go much further with these statements is to embark on a debate that has been one of the major themes of philosophy, art, and science, East and West, for thousands of years. . . .

48. The lesson has yet to be fully learned by most arcade video-game designers. On the Principle of Indifference (and on this corollary of it, the Principle of Life Goes On) the game should go on even without the player. The situation in the machine should develop with play, of course. But any one player's situation may deteriorate simply out of his neglect to play. Things may be taken "behind his back," overnight. Messages may await him. As in real life, he must return: return to fight against others, to influence the World that will be others' to deal with, and just to fight against entropy and aging. (There are a few text-based PC and Mac adventure games that have the capability of changing the situation if you think too long—the situation usually changing to your disadvantage. Also, most arcade games keep a record of top scorers' initials. This list can change in your absence.)

One might reasonably have moral qualms here. The Principle of Indifference, in the form of "Life Goes On," will make addicts of all of us! Another name for the principle? The Principle of Mortality.

49. This, I fear, is the implicit promise of today's virtual reality technology, even though VR's proponents have loftier goals. One cannot but be aware of the psychedelic cast Timothy Leary and others have put upon the area of VR.

Now, being a "child of the sixties" as they say, I personally do not wholly disapprove of the psychedelic project. But cyberspace, if it is to be a public and democratic realm, will have to seek consensus as to what is privately mutable and what is not, similar perhaps to the kinds of social restrictions placed on real-space architects and public-space sculptors and muralists. As an architect, I dread *this* possibility taken too far also (zoning and aesthetics committees in cyberspace?—please, no!), but it must be realized that without true physics, nature, evolution, or tradition, without clients, economics, or others with other*ness*, without even a necessary landscape such as a horizon and such as verticality and some definition of territory and address as backdrop, the alternative will be a chaos of images useless to everybody. In fact, cyberspace might not even get "off the ground."

50. It would seem at first that the ultimate intransigence/nonmalleability on the part of a computer system is its sometime "refusal" to do the user's bidding: after all, this is the system being independent, saying no, saying "I am what I am."

However, with ordinary reality there is always something that can be done: the "natural system" never really *freezes*—as computers often do—and, so far, the natural system never completely *crashes*—which computers sometimes do. Thus, holding to the Principle of Indifference does not entail advocating these ultimate intransigencies for cyberspace. Cyberspace entities and locales ought to remain responsive to manipulation and navigation always, even if truncated by equipment failure, user malevolence, or user ignorance. In fact, the ability to freeze or crash the whole or part of a cyberspace system deliberately, or even by mistake, is an indication of the system's relative fragility and therefore unreality. You cannot break what is indifferent to you.

One implication of all this is the nonadvisability of having cyberspace be located or controlled centrally, or its communication links constrained only to one "system of wires," say ISDN. Rather, cyberspace processing should be distributed, and its communication channels many and alternative: phone lines, satellite, HDTV cables, even radio and power lines. The technical aspects of all of this are daunting: we have not really begun to solve the problems of distributed databases and distributed processing on the scale required by cyberspace. And, of course, there are myriad political, economic, and power-related questions involved in this notion of decentralization and redundancy . . .

51. For a beginning compendium, see Rasmussen 1959, Harbison 1977, Nitschke 1966.

52. For example: as far as I know, another strategy is that of software-based, user-controlled differential or *multiple resolution* on a given screen, not implemented but as illustrated in Part Two of this chapter—that is, the arrangement whereby a particular data object or set of objects is assigned a display resolution different to others, and different to the background.

53. The loss of light energy by the inverse square law is inconsequential; our observers have indefinitely sensitive light detectors, let us say.

54. This is an approximation that breaks down slightly when r, the radius of the disc, is comparable to r, the distance to the observer.

55. Benedikt 1979, Benedikt and Davis 1979. ISOVIST, a computer program written for the Unix OS and SunView, available from the author and The University of Texas at Austin.

56. See my "The Information Field: A Theoretical and Empirical Approach to the Distribution of Information in the Physical Environment " M. E. D. Master's Thesis, Yale University School of Architecture, 1975. See also Koenderink and van Doorn 1975, 1976.

57. Throughout this chapter, it has been taken for granted that just as data exists visually, aurally, in cyberspace, so do its *users* for each other. There is much to be discussed here about the nature of personal presences in cyberspace. Cf., for example, Stone (this volume).

58. Mark Heyer provides the following taxonomy of increasing engagment with information: *grazing, browsing,* and *hunting.* Cited by Brand (1987, p. 43). In my teaching I like to expand the taxonomy thus: *soaking, browsing, watching, hunting, fetching/getting, making.* Only cyberspace is able to provide a suitably rich and indifferent environment for the first four stages, and perhaps the sixth. If one only wishes to *fetch* or *get* information, whose type and location one knows, one does

not need cyberspace, nor, for that matter, any GUI. With a connection to Internet and an FTP (File Transfer Program) on your own machine, for example, you need only to know the Internet address of the host machine, the name of the file, and perhaps a password. On request, the information in that file will simply pour onto the screen (or into your computer's memory). The file is pure destination data; navigation data is minimal, or rather, is maximally compressed and encoded into a string of digits and characters, absolutely critically ordered. With requests for user and file directory information—word lists—a certain amount of *hunting* can be done this way also. See Stoll 1989 for a hacker's account.

59. It should not be overlooked that the dollar-cost of travel to users may be one of the economic engines that drives cyberspace as a money-making enterprise (no matter who "runs" it or parts of it). Other sources of income to the owners and maintainers of the system are easy to see: outright purchase of real estate in cyberspace, the leasing of such, advertising time and space, connect-time charges to the system and to individual presences, innumerable hardware purchases and upgrades, cabling systems, satellites and so on, access software, endless enhancements to this, etc.; and all this in addition to the value of the information bought and sold as such within the system.

More broadly, one should note again that *time,* and not spatial/physical property, is the new and salient capital and currency of the "information age." Allocating it, saving it, investing it, buying it, selling it; in minds, in computers, on tape; on line, on air, on screen; at home, at work, in transit . . . We have not yet begun to formulate the new economics of time and information, as I alluded in the introduction to this chapter. It is being invented "on the fly," with sporadic progress reports in *Newsweek* and *USNews.*

60. In analogy: a public hypertext in which every word led to another document or another part of the text would be disorienting and close to useless, unless *tracks* of some sort are laid down and are visible to later users. See Nelson 1990.

61. See Isaac 1980. Glynn Isaac, an anthropologist, reports finding the same pattern in the daily movements (tracked from the air) of a variety of African mammals, including the Bushmen of the Gobi desert. From home base the day is spent searching, tracking, gathering items here and there. When the goal is reached (an animal or enough food) one comes home as directly as possible. Actually, arboreal primates such as chimps and gorillas do not maintain the pattern as strongly since they range through the forests continuously and do not make permanent homes.

62. Conventional usage quite correctly follows the hierarchy: "data"/ "information"/ "knowledge/ "wisdom," and technically I do indeed mean navigation and destination *information.*

63. Instantly, of course, we must remember his slogan too: the medium *is* the message (actually he liked to say *massage*). Be this as it may, the distinction between navigation and destination data holds up to all but the most determined post-structuralist deconstruction, and along these lines: no information does not lead to other information, therefore no information can be a true destination; all information is navigation data and destination both, and neither, intrinsically, and by the nature of signs. For a more detailed look at this argument, cf. Norris 1982, Benedikt 1991.

64. The possibility also exists, however, for any or certain objects in cyberspace to be under a modicum of [temporary] user control in this regard, as when an object is "asked" to bias its self-presentation to the individual user toward either its navigational or destinational aspects. This would accord well with Fuchs's adaptive refinement, except that here the semantic category and not just the net amount of information processed/displayed is controlled by user intention/attention. More elaborate selective display mechanisms are a possibility too, and in the section following the next we will look at this question again.

65. The problems of communication privacy and security will be no better and no worse in cyberspace than is the case with today's computers and telephones. What can one say here: there will be no "wire tapping"?

66. This is a consideration that at one time I had elevated to a principle in its own right: the Principle of Universal Verticality. Millions of years of life on this planet has ensured this most fundamental of agreements, and as the experience of astronauts testifies, it has become deeply necessary for perception and orientation for there to be a consistent vertical orientation in the environment. More specifically: it seems obvious that insofar as negotiating cyberspace will involve perceiving symbols, pictures, people, faces, texts, and so on, that it would do no good for them to appear out of nowhere in random "gravitational" orientations. Also, cyberspace as a whole might have gravitationlike valences, different meanings, capabilities, and destinations tied to "altitude" and to vertical motion in the large.

67. Abstract commonality as in ". . . you and I have a lot in common: we're both from Minnesota, have red hair, like guava . . ." is not intended here.

68. The property of no pair of sets having more than one disjoint intersection can define the meaning of set *convexity*.

69. This passage from *The Lion, the Witch, and the Wardrobe*, and by way of interlude, captures it all. From C. S. Lewis 1950, (pp. 5, 6, 7):

Everyone agreed (that) this . . . was how adventures began. It was the sort of house you never seem to come to the end of, and it was full of unexpected places.

(After looking into many rooms) they looked into one that was quite empty except for one big wardrobe; the sort that had a looking glass in the door. To (Lucy's) surprise, it opened quite easily . . . She immediately stepped into the wardrobe and got in among the coats and rubbed her face against them, leaving the door open, of course, because she knew that it is very foolish to shut oneself in any wardrobe. Soon she went further in and found that there was a second row of coats hanging up behind the first one. It was almost quite dark in there and she kept her arms stretched out in front of her so as not to bump her face into the back of the wardrobe. She took a further step in, then two or three steps . . . pushing the soft folds of the coats crunching under her feet . . .

(It) was soft and powdery and extremely cold. "This is very queer," she said, and went a step or two further. Next moment she found that what was rubbing against her face and hands was no longer soft fur but something rough and even prickly. "Why, it is just like branches if trees!" exclaimed Lucy. And then she saw that there was a light ahead of her; not a few inches away where the back of the wardrobe ought to have been, but a long way off. Something cold and soft was falling on her. A moment later she found she was standing in the middle of a wood at nighttime with snow under her feet and snowflakes falling through the air.

. . . "I can always get back if anything goes wrong," thought Lucy. She began to walk forward, crunch-crunch, over the snow and through the wood towards the other light.

One might as easily have chosen a passage from Lewis Carroll's *Alice in Wonderland* or *Alice through the Looking Glass,* or the myth of Orpheus and the Underworld.

70. This is a technology currently under development by the author and cannot be divulged until patenting is secured.

71. The reader may enjoy interpreting the behavior of lovers in terms of the strategies (1) to (4) just outlined.

72. Cf. Markoff 1990 for a nontechnical overview of the national effort.

73. A symbolic cyberspace based on the symbolic space of literature, as it were, was mentioned earlier; namely, TinyMud, the interactive, text-based, networked, city-metaphored venue for hundreds of participants that functioned in 1989 and 1990 out of Carnegie Mellon University. ("MUD" is an acronym for Multi-User Dungeon). More graphic was/is Lucasfilm's Habitat of 1985, also based on the model of the city, which ran on the Commodore 64 computer. (See the chapter by Morningstar and Farmer, this volume). Both systems created consistent imaginary environments for hundreds of concurrent participants through rather simple means, computationally speaking. Although both were essentially entertainments, there is no reason why good amounts of useful information could not have been located, accessed, and manipulated within the structure of their imagined, and therefore doubly unreal, cyberspaces.

TinyMUD has spawned many "MUDs" since, most of them running out of university workstations, for example, at the time of writing: Islandia (Carnegie Mellon), Club Mud (U. of Washington), Chaos (U. of Oklahoma) TroyMUCK (Rennselaer Polytechnic), Pegasus (U. of Iowa) Mbongo (U. of Calif. at San Diego); and many more. See the newsgroup *alt.mud* on the USENET for updates.

74. The room is also equipped with slide viewing racks, that is, large rear-illuminated, sloping panels with horizontal tracks to arrange slides upon. Certain mechanical actions are demanded: physically opening the drawers, reading labels, holding slides up to the light for preliminary viewing, taking them over to and setting them upon the viewers without dropping them, leaving behind special markers, mounting the slides in projection trays, etc. . . . these cumbersome and time-consuming activities would easily be superceded by any computerized optical disc storage and retrieval equipment. But we are here more interested in the data itself and its spatial organization.

75. There could reasonably be a clean division here between *exterior* and *interior,* and a further one between *day* and *night* views. These could be disposed linearly within the dominant dimension, or with enough images, could become orthogonal dimensions in themselves.

76. One has the opportunity here to encode two independent values instead of one—that is, one for left-of-centerline ribbon width, one for right-of-centerline width, like the grooves on a stereophonic vinyl record; and yet another dimension or two is available if the *color* of the ribbon is varied and encoded. These options were not pursued.

The reader may also have noticed that, since there are two plane surfaces per orientation in a real room (floor and ceiling, left and right walls, fore and aft walls), *two* points in a cell, perhaps with a line joining them, can represent *six* dimensions at once. Each point's position is referred to, and governed by, a different set of three walls. In fact, four walls—two *pairs*— are sufficient for six dimensions because of the redundancy inherent in the partition X-Y, X-Z, Y-Z.

77. One alternative is for neighboring cells to be dark, empty—perhaps just rendered in outline—if they are "broken into" from the side. If the new cell were also transparent, one would thus find oneself in an immense, largely empty, cubic framework, stretching to infinity. If the cells are partially "content-full," and translucent, however, one would find oneself embedded in a universe of deeply overlayed images. Assuming sufficient rendering and computing power, the second option could be most pleasant and stimulating. But lost would be the navigational data available and inherent in the *over*view and the *over*flight scheme that I am suggesting should be, if not necessary, then the norm.

78. After all, William Gibson was portraying a future he did not and does not condone—indeed, one that he dreads—even as he was relishing its narrative potential. We can expect his future books to make this clearer.

Bibliography

Abraham, Ralph, and Shaw, Chris, *Dynamics: The Geometry of Behavior,* vols. 1–4 (Santa Cruz, California: Aerial Press, 1982, 1983, 1985, 1988).

Barbour, Julian, "Maximal Variety as a New Fundamental Principle of Dynamics," *Foundations of Physics*, vol. 19, 1989, pp. 1051–1073.

Barrow, John D., and Tipler, Frank J., *The Anthropic Cosmological Principle* (New York: Oxford University Press, 1988).

Benedikt, Michael, *Deconstructing the Kimbell* (New York: Lumen Books, 1991).

Benedikt, Michael, ed., *Collected Abstracts of the First Conference on Cyberspace* (School of Architecture, The University of Texas at Austin, 1990).

Benedikt, Michael, *For an Architecture of Reality* (New York: Lumen Books, 1987).

Benedikt, Michael, "Architecture and the Experience of Reality." In Long, E. (ed.), *Knowledge and Society*, ed. E. Long (Greenwich, Connecticut: JAI Press, 1986), pp. 233–250.

Benedikt, Michael, and Burnham, Clarke A. "Perceiving Architectural Space: From Optic Arrays to Isovists." Chap. 6 in *Persistence and Change*, ed. W. H. Warren and R. E. Shaw (Hillsdale, N.J.: Lawrence Erlbaum, 1984).

Benedikt, Michael, "On Mapping the World in a Mirror," *Environment and Planning B*, vol. 7, 1980, pp. 367–378.

Benedikt, Michael, "To Take Hold of Space: Isovists and Isovist Fields," *Environment and Planning B*, vol. 6, 1979, pp. 47–65.

Benedikt, Michael and Davis, Larry, "Computational Models of Space: Isovists and Isovist Fields." *Computer Graphics and Image Processing*, no. 11, 1979, pp. 49–72.

Bergman, Larry, Fuchs, Henry, Grant, Eric, and Spach, Susan, "Image Rendering by Adaptive Refinement," *Computer Graphics*, Proceedings of SIGGRAPH '86, vol. 20, #4, 1986, pp. 29–37.

Brand, Stuart, *The Media Lab: Inventing the Future at MIT* (New York: Viking Penguin, 1987).

Brisson, David W., Ed., *Hypergraphics*, AAAS Symposia Series (Boulder, Colorado: Westview Press, 1978).

Brown, M. H., De Fanti, T. A., and McCormick, B. H., "Visualization in Scientific Computing," *Computer Graphics*, vol. 21, no. 6, 1987.

Brunner, John, *Shockwave Rider* (New York: Ballantine/Del Ray, 1975).

Chernoff, H., "The Use of Faces to Represent Points in k-Dimensional Space Graphically," *Journal of the American Statistical Association*, vol. 68, pp. 361–368, 1973.

Cooper, Leon N., *An Introduction to the Meaning and Structure of Physics;* Short Edition (New York: Harper and Row, 1970).

Cox, Donna J. "The Art of Scientific Visualization" *Academic Computing*, March 1990, pp. 20–40 passim, and see References, p. 58.

Davies, Paul, ed., *The New Physics* (New York: Cambridge University Press, 1989).

Davies, Paul, *Superforce* (New York, Simon and Schuster, 1984).

Ellson, Richard, "Visualization at Work," *Academic Computing*, March 1990, pp. 26ff.

Gibson J. J. , *The Ecological Approach to Visual Perception.* (New Jersey: Hillsdale, 1986); *The Perception of the Visual World* (Greenwich, Connecticut: Greenwood Press, 1974, 1950).

Gibson, William, *Burning Chrome*, (New York: Ace Books, 1982); *Neuromancer* (New York: Ace Books, 1984); *Count Zero* (New York: Ace Books, 1987); *Mona Lisa Overdrive* (New York: Ace Books, 1988).

Gilder, George, *Microcosm* (New York: Simon and Schuster, 1989).

Harbison, Robert, *Eccentric Spaces* (New York, Knopf, 1977).

Isaac, Glynn, "Casting the Net Wide: A Review of Archeological Evidence for Early Hominid Land Use and Ecological Relations." In *Current Argument on Early Man*, ed. L. K. Konigsonn (Oxford: Pergamon Press, 1980), p. 236.

Kelly, Kevin, "Interview with Jaron Lanier," *Whole Earth Review*, Fall 1989, pp. 108ff.

Koenderink, J. J., and van Doorn, A. J., "The Singularities of the Visual Mapping," *Biological Cybernetics*, 24, 1976, pp. 51–59.

Koenderink, J. J. , and van Doorn, A. J., "Invariant properties of the motion parallax field due to the movement of rigid bodies relative to an observer," *Optica Acta*, 22, 9, 1975, pp. 773–791

Lewis, C. S., *The Lion, the Witch, and the Wardrobe* (New York: Macmillan 1950; Collier Books 1970).

McLuhan, Marshall, *Understanding Media: The Extensions of Man* (New York: McGraw-Hill, 1964).

Markoff, John, "Creating a Giant Computer Highway," *The New York Times*, Sept. 2, 1990, pp. F1, F6.

Nelson, Theodor H., *Literary Machines 89.1* (Sausalito, California: Mindful Press, 1990).

Nitschke, Gunther, "MA," *Architectural Design*, March 1966, pp. 116ff.

Norris, Christopher, *Deconstruction: Theory and Practice* (New York: Methuen, 1982).

Penrose, Roger, *The Emperor's New Clothes: Concerning Computers, Minds, and the Laws of Physics* (New York: Oxford University Press, 1989).

Rasmussen, Steen Eiler, *Experiencing Architecture* (Cambridge, Massachusetts: MIT Press, 1959).

Sterling, Bruce, "Cyberspace (TM)," *Interzone*, #41, November 1990, pp. 54–62 passim.

Stewart, Doug, "Artificial Reality: Don't stay home without it," *Smithsonian*, vol. 21, #10, January 1991, pp. 36ff; and "Interview: Jaron Lanier" *Omni*, vol. 13., #4, January 1991, pp. 45ff.

Stewart, Ian, *The Problems of Mathematics* (London: Oxford University Press, 1987).

Stoll, Clifford, *The Cuckoo's Egg* (New York: Doubleday, 1989).

Vinge, Verner, "True Names." In *True Names . . . and Other Dangers* (New York: Baen Books, 1987).

Wright, Robert, *Three Scientists and their Gods* (New York: Times Books, 1988).

Zuboff, Shoshana, *In the Age of the Smart Machine* (New York: Simon and Schuster, 1989).

8 Liquid Architectures in Cyberspace

Marcos Novak

Introduction

What is cyberspace?

Here is one composite definition:

Cyberspace is a completely spatialized visualization of all information in global information processing systems, along pathways provided by present and future communications networks, enabling full copresence and interaction of multiple users, allowing input and output from and to the full human sensorium, permitting simulations of real and virtual realities, remote data collection and control through telepresence, and total integration and inter-communication with a full range of intelligent products and environments in real space.[1]

Cyberspace involves a reversal of the current mode of interaction with computerized information. At present such information is external to us. The idea of cyberspace subverts that relation; we are now within information. In order to do so we ourselves must be reduced to bits, represented in the system, and in the process become information anew.

Cyberspace offers the opportunity of maximizing the benefits of separating *data*, *information*, and *form*, a separation made possible by digital technology. By reducing selves, objects, and processes to the same underlying ground-zero representation as binary streams, cyberspace permits us to uncover previously invisible relations simply by modifying the normal mapping from data to representation.

To the composite definition above I add the following: Cyberspace is a habitat for the imagination. Our interaction with computers so far has primarily been one of clear, linear thinking. Poetic thinking is of an entirely different order. To locate the difference in terms related to computers: poetic thinking is to linear thinking as random access memory is to sequential access memory. Everything that can be stored one way can be stored the other; but in the case of sequential storage the time required for retrieval makes all but the most predictable strategies for extracting information prohibitively expensive.

Cyberspace is a habitat of the imagination, a habitat for the imagination. Cyberspace is the place where conscious dreaming meets subconscious dreaming, a landscape of rational magic, of mystical reason, the locus and triumph of poetry over poverty, of "it-can-be-so" over "it-should-be-so."

The greater task will not be to impose science on poetry, but to restore poetry to science.

This chapter is an investigation of the issues that arise when we consider cyberspace as an inevitable development in the interaction of humans with computers. To the extent that this development inverts the present relationship of human to information, placing the human within the information space, it is an architectural problem; but, beyond this, cyberspace has an architecture of its own and, furthermore, can contain architecture. To repeat: cyberspace *is* architecture; cyberspace *has* an architecture; and cyberspace *contains* architecture.

Cyberspace relies on a mix of technologies, some available, some still imaginary. This chapter will not dwell on technology. Still, one brief comment is appropriate here. A great number of devices are being developed and tested that promise to allow us to enter cyberspace with our bodies. As intriguing as this may sound, it flies in the face of the most ancient dream of all: magic, or the desire to will the world into action. Cyberspace will no doubt have physical aspects; the visceral has genuine power over us. And though one of the major themes of this essay has to do with the increasing recognition of the physicality of the mind, I find it unlikely that once inside we will tolerate such heavy devices for long. Gloves and helmets and suits and vehicles are all mechanocybernetic inventions that still rely on the major motor systems of the body, and therefore on coarse motor coordination, and more importantly, low nerve ending density. The course of invention has been to follow the course of desire, with its access to the parts of our bodies that have the most nerve endings. When we enter cyberspace we will expect to feel the mass of our bodies, the reluctance of our skeleton; but we will choose to *control* with our eyes, fingertips, lips, and tongues, even genitals.

The trajectory of Western thought has been one moving from the concrete to the abstract, from the body to the mind; recent thought, however, has been pressing upon us the frailty of that Cartesian distinction. The mind is a property of the body, and lives and dies with it. Everywhere we turn we see signs of this recognition, and cyberspace, in its literal placement of the body in spaces invented entirely by the mind, is located directly upon this blurring boundary, this fault.

At the same time as we are becoming convinced of the embodiment of the mind, we are witnessing the acknowledgment of the inseparability of the two in another way: the mind affects what we perceive as real. Objective reality itself seems to be a construct of our mind, and thus becomes subjective.

The "reality" that remains seems to be the reality of fiction. This is the reality of what can be expressed, of how meaning emerges. The trajectory of thought seems to be from concrete to abstract to concrete again, but the new concreteness is not that of Truth, but of *embodied* fiction.

The difference between embodied fiction and Truth is that we are the authors of fiction. Fiction is there to serve our purposes, serious or playful, and to the extent that our purposes change as we change, its embodiment also changes. Thus, while we reassert the body, we grant it the freedom to change at whim, to become liquid.

It is in this spirit that the term *liquid architecture* is offered. Liquid architecture of cyberspace; liquid architecture in cyberspace.

Part One: Cyberspace

Poetics and Cyberspace

Well then, before reading poems aloud to so many people, the first thing one must do is invoke the *duende*. This is the only way all of you will succeed at the hard task of understanding metaphors as soon as they arise, without depending on intelligence or the critical apparatus, and be able to capture, as fast as it is read, the rhythmic design of the poem. For the quality of a poem can never be judged on just one reading, especially not poems like these which are full of what I call 'poetic facts' that respond to a purely poetic logic and follow the constructs of emotion and of poetic architecture. Poems like these are not likely to be understood without the cordial help of the *duende*.

—Federico Garcia Lorca, *Poet in New York*

The *duende* is a spirit, a demon, invoked to make comprehensible a "poetic fact," an "hecho poético." An "hecho poético," in turn, is a poetic image that is not based on analogy and bears no direct, logical explanation (Lorca 1989). This freeing of language from one-to-one correspondence, and the parallel invocation of a "demon" that permits access to meanings that are beyond ordinary language permits Lorca to produce some of the most powerful and surprising poetry ever written. It is this power that we need to harness in order to be able to contend with what William Gibson called the "unimaginable complexity" of cyberspace.

How does this poetry operate?

Concepts, like subatomic particles, can be thought to have world lines in space-time. We can draw Feynman diagrams for everything that we can name, tracing the trajectories from our first encounter with an idea to its latest incarnation. In the realm of prose, the world lines of similar concepts are not permitted to overlap, as that would imply that during that time we would be unable to distinguish one concept from another. In poetry, however, as in the realm of quantum mechanics, world lines may overlap, split, divide, blink out of existence, and spontaneously reemerge.[2] Meanings overlap, but in doing so call forth associations inaccessible to prose. Metaphor moves mountains. Visualization reconciles contradiction by a surreal and permissive blending of the disparate and far removed. Everything can modify everything: *"Green, I want you green / green wind, green boughs,"* writes Lorca, *"Over the green night / the arrows / leave tracks of warm / lilies / / The keel of the moon / breaks purple clouds / and the quivers / fill with dew"* (Lorca 1989).

If cyberspace holds an immense fascination, it is not simply the fascination of the new. Cyberspace stands to thought as flight stands to crawling. The root of this fascination is the promise of control over the world by the power of the will. In other words, it is the ancient dream of magic that finally nears awakening into some kind of reality. But since it is technology that promises to deliver this dream, the question of "how" must be confronted. Simply stated, the question is, What is the technology of magic? For the answer we must turn not only to computer science but to the most ancient of arts, perhaps the only art: poetry. It is in poetry that we find a developed understanding of the workings of magic, and not only that, but a wise and powerful knowledge of its purposes and potentials. Cyberspace is poetry inhabited, and to navigate through it is to become a leaf on the wind of a dream.

Tools of poets: image and rhythm, meter and accent, alliteration and rhyme, tautology, simile, analogy, metaphor, strophe and antistrophe, antithesis, balance and caesura, enjambment and closure, assonance and consonance, elision and inflection, hyperbole, lift, onomatopoeia, prosody, trope, tension, ellipsis . . . poetic devices that allow an inflection of language to produce an inflection of meaning. By push and pull applied to both syntax and symbol, we navigate through a space of meaning that is sensitive to the most minute variations in articulation. Poetry is liquid language.

As difficult as it may sound, it is with operations such as these that we need to contend in cyberspace. Nothing less can suffice.

I am in cyberspace. I once again resort to a freer writing, a writing more fluid and random. I need to purge a mountain of brown thoughts whose decay blocks my way. I seek the color of being in a place where information flies and glitters, connections hiss and rattle, my thought is my arrow. I combine words and occupy places that are the consequence of those words. Every medium has its own words, every mixture of words has a potential for meaning . Poets have always known this. Now I can mix the words of different media and watch the meaning become navigable, enter it, watch magic and music merge.

Cyberspace Navigation, Synthesis, and Rendition

Some initial definitions are necessary before we proceed: *cyberspace navigation* refers to the traversal of information spaces; *cyberspace synthesis* refers to the reconciliation of different kinds of information into a coherent image; *cyberspace rendition* refers to the production of high-quality graphic presentation of that image. These are separate tasks.

The Hypermedia Navigator

Navigation through cyberspace is achieved by interacting with a *hypermedia navigator,* a virtual control device that follows the user and always remains within arm's reach. It is possible for the user to circumvent the cyberspace synthesizer and enter and traverse the space of the navigator, riding the links, as it were.

Every paragraph an idea, every idea an image, every image an index, indices strung together along dimensions of my choosing, and I travel through them, sometimes with them, sometimes across them. I produce new sense, nonsense, and nuisance by combination and variation, and I follow the scent of a quality through sand dunes of information. Hints of an attribute attach themselves to my sensors and guide me past the irrelevant, into the company of the important; or I choose to browse the unfamiliar and tumble through volumes and volumes of knowledge still in the making. Sometimes I linger on a pattern for the sake of its strangeness, and as it becomes familiar, I grow into another self. I wonder how much richer the patterns I can recognize can become, and surprise myself by scanning vaster and vaster regions in times shorter and shorter. Like a bird of prey my acuity allows me to glide high above the planes of information, seeking jewels among the grains, seeking knowledge.

Just as hypertext allows any word in a normal text to explode into volumes of other words, so a hypergraph allows any point in a graph to expand to include other graphs, nested and linked to any required depth. We may, of course, extend this idea to other media to arrive upon corresponding hypermedia. We can now make some further distinctions: static and dynamic, passive and active, pure and hybrid. A static hypermedium is one in which the links are fixed and can only be changed manually; a dynamic one is one where the links are in some way variable. While the distinction of static/dynamic applies to the links of a hypermedium, the distinction of passive/active applies to the nodes between the links. A passive hypermedium is one in which the information nodes themselves remain stable though the links may vary dynamically; in an active hypermedium the information nodes themselves can change. Pure hypermedia remain within the confines of a single medium, hybrid ones roam freely. The hypermedia navigator, or *navigator* for short, is a virtual device for traversing vast hybrid hypermedia spaces that have both active links and dynamic nodes.

Sometimes I wander out of my world into the larger spaces. I travel along pathways mostly empty. The passages I traverse are not still, however. Along their boundaries processes sparkle, information flows like water on a moist wall, schools of data swim around me curiously, and lattices of fact and fiction tangle and untangle. The ones I touch open out into texts and images and places.

Every node in a hypermedium has a dimensionality. Hypertext, for example, occurs in a one-dimensional space, but we can easily envision hypermedia with higher dimensions. While the dimensionality of a node is fixed, the dimensionality between nodes need not be: a word in a text can open to a hologram, a point within the hologram can open to an animation, a frame in the animation can return to a text.

Every node in a hypermedium is therefore an information space, a space of potential information, and the "text" of the node is the actual information within that space. In an active hypermedium the information within the information space, as well as the links among spaces, may change according to internal or external conditions. These conditions may be user-controlled or user-independent. The processes within which these conditions are embedded may be defined explicitly by the user or may be autonomous.

My point in space is given by my navigator. Its forms can vary but the idea remains the same: I control a point in an n-dimensional space, say a cube. I assign meaning to each of the axes, and to any rotational parameters, material parameters, shape parameters, color and transparency parameters, and so on, that describe the "reality" of my icon. By moving my icon in this abstract space I alter the cyberspace I occupy. My navigator follows me at all times, and my position within it is fixed while I move within the cyberspace I have defined. Should I decide to search through a slightly different "reality" all I need to do is reach out for my navigator and alter a parameter. Otherwise, for more drastic navigation, I can alter a dimension, or even the number of dimensions. Finally, I may choose an entirely different coordinate system. In every case my deck is responsible for synthesizing the requested information in a new cyberspace.

If a point in one cyberspace is an entry into another cyberspace, then a new navigator is spontaneously created and pushed onto my stack of navigators. I can now maneuver in the new system without losing my place in the old one. Since the stack of navigators constitutes a pathway, I can be reached by anyone who encounters one of my icons, opens a channel, and sends me a message.

There are no hallways in cyberspace, only chambers, small or vast.

Chambers are represented as nodes within my navigator. The topology of their connections is established by the settings and nesting of the coordinate systems of cyberspaces within my navigator. Chambers allow different users to share the same background, as well as encounter and interact with the same objects.

Motion   There are two kinds of motion: motion within my navigator, and motion within the cyberspace pointed to by my navigator. If my place in my navigator is fixed, the reality I experience is also stable, though I can move in it and interact with the entities that inhabit it. If, however, I concurrently change places within my navigator, then the reality I experience is no longer stable, but fluctuating. As my location within the navigator changes, new, perhaps distant, realities are brought forth to replace the old ones. Entities, landmarks, and landscapes appear and disappear, time and space become discontinuous, and the increment of my motion changes from one scale to another to fit the current reality.

Others   Initially, a user has a personally configured cyberspace, and maneuvers through it by manipulating the navigator. If this cyberspace is configured to accept signals from other users, or simply *Others*, that is, if communications ports have been defined and enabled, then an appropriate indication is made within the cyberspace that another user is hailing. At that point the users must agree on a manner of interaction, ranging from simple text to full interpresence. For full cyberspace interaction the coordinate systems of all interacting users' navigators must be configured along identical orderings. Partial coincidence will result in hiding of information.

This world of mine has ports: through them I gather and give. This world has windows: through them I can see and be seen. Through ports other worlds are accesible; through windows other worlds are simply visible. I can open and close my ports and windows, naturally, and I can also have curtains and filters that only permit some information to enter or escape. Sometimes the information itself controls which aperturess are open, how much, and how.

Within my navigator traces of light move from node to node, indicating the presence of others in the chambers of cyberspace; outside my navigator, in the chamber I am currently occupying, I encounter some of them directly.

Interaction with others depends on the degree to which they share information coordinate axes and orderings. If users wish to coexist fully in the same virtual space, they must set their navigators to the same settings. Communications established between users sharing information axes partially will result in "ghosting," that is, inexplicable and unpredictable "appearances" and "disappearances" as aspects of the cyberspace of one user engage and disengage aspects of the cyberspace of the other. In this respect cyberspaces are "consensual," since any complete exchange requires a sharing of settings of the participants' navigators.

Using my deck, I enter the cyberspace. At first the world is dark, but not because of an absence of light, but because I have not requested an environment yet. I request my default environment, my personal database. From it I choose my homebase, or workbase, or playbase. I am in my personal cyberspace, and I am not yet in contact with others. This is my palace, and it is fortified. Only guests can visit my "fortress of solitude," and in here I can be Superman to the Clark Kent of my real-space self. Sometimes I organize my information around my armchair and navigate through it at a glance, extracting what I need by effortless exercises of will; other times, for the sake of exercise or play, I scatter it around my globe and fly across immense distances to recover minute recollections with the most strenuous "physical" effort. Sometimes I use a single surrogate, other times I divide into a legion.

I sense the presence of others. I see the traces of passage, the flares of trajectories of other searches. Those who share my interests visit the spaces around me often enough for me to recognize the signature of their search sequences, the outlines of their icons. I open channels and request communication. They blossom into identities that flow in liquid metamorphosis. Layers of armor are dropped to reveal more intimate selves; otherwise, more and more colorful and terrifying personifications are built up in defense; but true danger is gray.

The world opens and others flood in. Now there is congestion and noise, interference, but also excitement, risk, and challenge. I travel with the constellation of my possessions, and barter and trade information. I can scan the horizon and avoid what is busy, enjoy what is free.

Cyberspace Synthesis and Rendition

Visualization is the task of a *cyberspace deck*, or more precisely, a *cyberspace synthesizer*. The function of this device is to receive a minimal description of the cyberspace, coded and compressed, and from it to render a visualization of that space for the user to navigate within. The quality of the rendition will vary according to the particular technology used and the parameters set by the user. Thus, one user may possess a very powerful, photorealistic cyberspace synthesizer and use it to produce a detailed, hallucinatory blending of images in cyberspace being traveled through, while another user may possess a less advanced synthesizer and use it to produce a rendition that is less detailed but more conventionally "real."

The *cyberspace protocol* is a communication standard. Cyberspace decks are virtual reality synthesizers whose quality varies with available technologies. The cyberspace protocol establishes a description language for virtual realities, a user-configurable interface standard, a list of primitives and valid relations among them, and operations upon these. The overriding principle in every case is that of minimal restriction. As with audio synthesizers, increased capacity is the responsibility of the deck, with higher number of concurrent channels, active senses, degree of realism, layering, and effects all set by the user independently of the reality of the cyberspace description.

Rendering cyberspace is a separate process from synthesizing cyberspace. Cyberspace decks are primarily responsible for virtual reality synthesis, with the actual rendering performed by current graphic supercomputer workstations.

Ultimately, the synthesis of cyberspace occurs in the mind of the user as a mental space, a spatialization of the sum of affordances into a series of worlds of opportunity and restriction, promise and constraint.

Underlying Considerations

Minimal Restriction and Maximal Binding

The key metaphor for cyberspace is "being there," where both the "being" and the "there" are user-controlled variables, and the primary principle is that of *minimal restriction*, that is, that it is not only desirable, but necessary to impose as few restrictions as possible on the definition of cyberspace, this in order to allow both ease of implementation and richness of experience. In addition, *maximal binding* implies in cyberspace anything can be combined with anything and made to "adhere," and that it is the responsibility of the user to discern what the implications of the combination are for any given circumstance. Of course, defaults are given to get things started, but the full wealth of opportunity will only be harvested by those willing and able to customize their universe. Cyberspace is thus a user-driven, self-organizing system.

Multiple Representations

Cyberspace is an invented world; as a world it requires "physics," "subjects" and "objects," "processes," a full ecology. But since it is an invented world, an embodied fiction, one built on a fundamental representation of our own devising, it permits us to redirect data streams into different representations: selves become multiple, physics become variable, cognition becomes extensible. The boundaries between subject and object are conventional and utilitarian; at any given time the data representing a user may be combined with the data representing an object to produce . . . what?

Digital technology has brought a dissociation between data, information, form, and appearance. Form is now governed by representation, data is a binary stream, and information is pattern perceived in the data after the data has been seen through the expectations of a representation scheme or code. A stream of bits, initially formless, is given form by a representation scheme, and information emerges through the interaction of the data with the representation; different representations allow different correlations to become apparent within the same body of data. Appearance is a late aftereffect, simply a consequence of many sunken layers of patterns acting upon patterns, some patterns acting as data, some as codes. This leads to an interesting question: what is the information conveyed by the representation that goes beyond that which is in the data itself? If a body of data seen one way conveys different information than the same body of data seen another way, what is the additional information provided by one form that is not provided by the other? Clearly, the answer is *pattern*, that is, *perceived* structure. And if different representations provide different perceived information, how do we choose representations? Not only do different representations provide different information, but in the comparison *between* representations new information may become apparent. We can thus distinguish two kinds of *emergent* information: *intrarepresentational* and *interrepresentational*.

I substitute the characters on this and the next page of this text with grey scale values; two images emerge, pleasingly rhythmical. The gray tone of the letter e stands out, forming snakes along the pages; I apply spline curves to the snakes, and, in another space, my text itself is changing—what will it say? Now I combine the two pages, and convert the result into a landscape, using the grey values to represent height. What snakes were left after the combination of the two pages become Serpent Mounds. The Others who were reading my text with me a while ago are now flying over this landscape with me, but only I can command it to change. Today.

Attribute-Objects

There are no objects in cyberspace, only collections of attributes given names by travelers, and thus assembled for temporary use, only to be automatically dismantled again when their usefulness is over, unless they are used again within a short time-span. Thus useful or valued objects remain, while others simply decay. These collections of attributes are assembled around nameless nodes in information spaces. Travelers and processes can add, subtract, or modify collections of attributes. By specifying a set of attributes, travelers define temporary attribute-objects, and by ordering each attribute in any desirable way they create their own dimensions for navigation. Three such dimensions result in a three-dimensional space, and motion along this space results in transfigurations of the environment that enable a holographic browsing. X, y, z, roll, pitch, yaw, color, material, size, all the parameters that define my point in space are indices into dimensions of attribute space. My motion makes my environment melt from one image into another, and my navigation becomes a knowledge dance.

In order for it to be possible to direct data into alternative representations it is helpful to have an open and extensible high-level manner of structuring that information. Object-oriented programming has shown the usefulness of interacting with objects that respond to "messages" with "methods." We can envision a world of attribute-objects, objects assembled as collections of attributes, where each attribute is itself an attribute-object, to some limit. Messages and methods are now attached not to objects, but to attributes of objects, in such a way that if a series of objects share an attribute, they can be expected to behave in a certain way. This implies a connection of attributes to affordances. For example, any attribute-object that has a weight-attribute, and whose weight-attribute is greater than some minimum value, can at any time be used as a paperweight.

Furthermore, if these objects are not placed in any permanent system of categorization, that is, if as few assumptions are made about what these objects are, beyond their collection of attributes, then it becomes possible to envision cyberspaces created along the parameters of users' needs. Attribute-objects can be gathered and sorted by attribute or combination of attributes, and these sorted collections can then be mapped onto coordinate axes. An information space can thus be formed, and motion through that space can imply "browsing." Consider for example a point in Cartesian space, whose coordinates are indices along the three axes that define the space, and whose rotational parameters and material and other attributes are indices along additional dimensions. Let us say that along the x-axis we have some ordering of "objects that are blue," sorted by degrees of blueness. Moving the point along the "blueness" axis, we scan these objects as if browsing through the shelves of a library.

Other attributes can be mapped along the other axes. Moving the point along those dimensions, or any kind of diagonal motion will alter the reading of the cyberspace correspondingly. The point thus becomes a control point in one of infinitely many potential cyberspaces, a navigation tool through actual or inferred information.

Attribute-objects are locations in attribute-space, and can themselves open to reveal other spaces or information.

Attribute-objects are saved as code strings containing genotype information. Selected kinds of attribute-objects can evolve autonomously by random mutation and cumulative change processes (Dawkins 1986), the main survival criteria consisting, first, of operation within available resources, and second, of interaction with other attribute-objects.

It will not be possible to assign all pertinent attributes to all objects, or to foretell what attributes a user may request for a particular journey through a cyberspace. It is therefore necessary to provide the ability to infer new attributes from those that are defined. This inference implies intelligence and expertise. Each user must therefore develop inference rules and a knowledge base with which to scan the environment and extract from it those objects that are pertinent to the task at hand.

Cellularity and Distributedness

As far as possible, the computational requirements of cyberspace need to be distributed among its users. By necessity, cyberspace itself must be reduced to a digital communications standard that allows users to exhange information in the most compact form possible, with cyberspace synthesizers and renderers taking on most of the work. The topology of perceived interconnected cyberspaces need not have any direct connection to that of the array of support computers, since the cyberspaces are perceived, not actual spaces. Like a cellular telephone system, a neutral grid of computers will permit access and communication within an *apparent* hierarchy of cyberspaces.

So far, most discussions of cyberspace and virtual reality have focused on the current computational paradigm of Von Neumann computing. The emerging paradigm of parallel distributed processing, though still not developed enough to be an immediate contender, offers an alternative view into cyberspace. In a neural net simulation, information about objects, relations, laws, and other components of a world, is not encoded in an explicit way. Rather, it is encoded implicitly, as weightings on connections between simple computational cells. Reality is an emergent property of the cell, a particular statistical setting of interconnection strengths that changes, ever so slightly, every time it is used, and manages to learn that which we are not yet able to articulate by observing the patterns in our data or our usage.

Mathematically, parallel distributed processing can be described in terms of surfaces with local and global maxima and minima, and convergence is given as a traversal along the surface to a state of minimum "computational energy." The actual contour of the surface depends on what the network has learned. In other words, in neural networks information is already directly spatialized, according to the networks' "learning." Thus, an information landscape emerges through usage. The implications of this for cyberspace is that the "unimaginable complexity" implicit in the idea of moving through interlinked information spaces, global networks, and so on, can be "learned" by a neural cyberspace component within a hybrid system. Subsequent to this learning and the resultant spatialized visualization of it as a landscape, we can use our own navigational skills to find the local extrema we are interested in.

Intelligence

Visualizations of information spaces can be understood in many ways: static or dynamic, as in photography and cinema; passive or (inter)-active, as in television versus interactive computer graphics; direct or hyper-, as in normal text versus hypertext; and now, with cyberspace, normal or intelligent. These sets of categories are orthogonal to one another; they are, in essence, attributes of the space itself. Some or all may exist at the onset of a session; new ones may be added; old ones may be deleted or modified.

We need a provisional definition of "intelligence" in order to proceed. Intelligence, as used here, refers to the capability of a subject, object, space, time sequence, or process, to detect, respond to, and modify its own or any other pattern. Patterns can be nested within patterns and may be invisible if the data within which they are found is represented in a manner that does not correspond to the pattern-recognition capabilities of another entity. Higher "intelligence" can detect and operate upon patterns more deeply nested, while simpler "intelligence" is restricted to surface patterns. A knowledge base, consisting of collections of "pattern-recognizer" attribute-objects, specifies what constitutes pattern for a cyberspace entity, and under what conditions, and in what ways, such a pattern may be modified at any given time. Pattern recognizers need be very general, based perhaps on simple but general methods for detecting unexpected correlations in data streams. Once again, the capacity of neural nets to act as correlation mechanisms for information that is only implicitly connected makes them suitable for many such tasks. While training networks is time-consuming, the performance of hardware implementations is virtually real time (Anderson and Rosenthal 1988).

Of course, not all "intelligence" will be operating at such a fundamental level. Different attribute-objects will use different methods of encoding responsiveness using procedural, or declarative, processes, as well as other present and future artificial intelligence techniques.

Subject, object, space, time, and process form the basic elements of cyberspace. These, and any others that are added subsequently, need to be considered not only continuously, but also discontinuously. Not only may a subject detect a pattern of light and darkness and call it "day/night," but the pattern-object "day/night" may itself detect the recurring presence of certain subjects or objects at times called "Monday," and may therefore adjust itself, or them, as required by the nature of its intelligence.

Overlap, Transparency, and Identity

In physical space two objects cannot occupy the same space at the same time. In cyberspace such a restriction is not strictly necessary. The identity of objects does not have to be manifested physically; it can be hidden in a small difference in an attribute that is not displayed. The motivation for this is two-fold: first, to allow a poetic merging of objects into evocative composites, and second, to keep the implementation of cyberspaces as simple as possible.

More precisely, the standard that implements cyberspace communications need not be assigned the overhead of deciding when two objects meet precisely, for example. It may be sufficient to place them so that their bounding boxes touch. It is then the common task of the decks that are viewing the two objects to resolve the situation according to their capabilities and the representations chosen.

What seems most likely to occur is that entities will behave as "ghosts," passing through each other freely and interacting on the basis of a list of operations vaguely reminiscent of their real-world analogues. As we become acclimated to this new ecosystem, our nostalgic desire for the vestiges of real-world physics will for the most part disappear. We will adjust to cyberspace far more easily than cyberspace will adjust to us.

This blending is possible because in the computer each entity is indicated by a token-value pair. Surface appearances are results of particular values assigned to tokens, and even if two entities have identical attributes, they can be distinguished by tracing the final, absolute identity of their tokens, ultimately the physical address allocated to them at the time of their creation. It is, nevertheless, possible to envision a situation where, through appropriate subsumption rules, if two entities have identical attributes they can be made to collapse into one, but this seems to limit us to a reality that is surely familiar but not necessarily appropriate to cyberspace.

Animism and Empathy

In cyberspace animism is not only possible, it is implicit in the requirement that all objects have a degree of self-determination, or are controlled by an Other. Thus a measure of empathy is required in order to comprehend the behavior of the entities one encounters. To the extent that any object may act as a front for a real person, its motives will have to be considered.

Furthermore, it is not simply the animism of objects that provides information, but the animism of the space itself. More specifically, any entity in cyberspace, including space itself, and "cybertime," can provide several levels of information: information about something else, information about itself, information about the observer, information about the surrounding environment, and global information. Beyond this physical atomism, any entity can provide information about structures of information that it is allowed to have knowledge of. Instead of atomism, organism. Within a region of cyberspace, time itself may pulse, now passing faster, now slower.

Lorca's *duende* is therefore not only a poetic description of an attitude toward the construction of cyberspace, it is also a tangible reality in the sense that every entity in cyberspace, and cyberspace itself, is somehow animate. The Platonic difficulties of this can perhaps be made most clear in the following example. Societies of mutual interest emerge spontaneously in cyberspace, as entities meet and interact. Friendships and associations of surrogates that only meet in virtual worlds are inevitable and have already been observed in on-line networks as well as in early implementations of consensual worlds. Suppose now that two entities meet in cyberspace and choose to produce offspring. Assuming that behind these two entities are humans who are in a very real sense the "souls" of these entities, we are now faced with the question of where to find a "soul" for the offspring. Will this be a friend of the humans in the real world? Will it be an applicant on a waiting list to cyberspace? Will it be an artificial intelligence? This problem may of course be avoided by forbidding such frivolities from taking place in the first place, but the issue has nevertheless been raised, and it bears a striking resemblance to Plato's idea that the souls of children exist prior to their births, and that love and friendship in the physical world were made possible by the memory and recognition of similar relations among souls in a metaphysical space prior to 'birth.'

A grand paradox is in operation here: even as we are finally abandoning the Cartesian notion of a division of mind and body, we are embarking on an adventure of creating a world that is the precise embodiment of that division. For, it is quite clear that our reality outside cyberspace is the metaphysical plane of cyberspace, that to the body in cyberspace we are the mind, the preexisting soul. By a strange reversal of our cultural expectations, however, it is the body in cyberspace that is immortal, while the animating soul, housed in a body outside cyberspace, faces mortality.

As we move farther from metaphysics to metafictions, more paradoxes become visible. Cyberspace is a dream of escape from a mortal plane even as it is an acknowledgment of that plane. In a comparison between reality and cyberspace, we can make the following observation: what nature is to us and to our creations, some technological ground is to cyberspace; what we call "nature" is simply the technological and theoretical field that we operate in.

Consider the problem of a cellular automaton sufficiently complex to exhibit not only the characteristics of life, but even a degree of self-consciousness and intelligence, and a scientific and ontological curiosity. What could such a creature be able to discover about its "reality"? The automaton could initially observe its own behavior and the behavior of other inhabitants of the cellular universe within which it finds itself; it could observe personal and social patterns; but, like the creatures in Flatland, it would have no knowledge of higher dimensions. It would be unable to discern the cellular grid itself, the field, within which it existed; but even if it managed to infer that its universes consisted of an array of cells whose values were assigned according to certain laws, it would not be able to comprehend the technology by which it was "displayed," nor the intellectual and physical structures behind that technology, and certainly not the "motives," if any existed, behind the existence of any of the above, not for any theological or mystical reason, but simply because those aspects of its existence were orthogonal to its experience and because the *grain* of that reality was finer than its own.

Similarly, *from the perspective of an entity within cyberspace,* the laws
of physics of cyberspace are layered; the uppermost, and most visible,
are the laws implicit in the software of cyberspace and the interactions
thus permitted; at a second level are those laws that constitute the
conceptual structure of cyberspace; next, more deeply hidden, are those
laws that pertain to the hardware that cyberspace operates on; finally,
are the laws that pertain to the physics within which the hardware is
operating. Far, far removed, are the motives and preconceptions of
those who set up the cyberspace itself.

Here cyberspace leads us to question our own attempts to comprehend what is "real." Not only are we limited by the grain we are built
upon, but even more so by the possibility of the existence of processes
that are orthogonal to our experience. For, it is evident, such processes
would not be bound to have local causes and local effects *in our realm,*
though they might in their own frame of reference. The effects of
processes orthogonal to our experience could be nonlocal at any given
time. No connection need exist between one event and another, as far
as we were concerned, and yet distant phenomena could be tightly
bound by laws beyond our comprehension. This possibility parallels
strongly the possibility of "quantum nonlocality" explored by physicists (Davies 1989).

It is sufficient for our purposes here to note how nature can be seen
as the impenetrable field about which and within which we construct
our hopeful fictions. We can then draw the parallel between natural
and artificial fields. Already we have seen how artificial realities raise
questions about our reality. The artifacts we make within our reality
have equally much to say about what we build in artificial worlds, how,
to what purpose, and for what reasons.

From Poetics to Architecture

We have examined various aspects of cyberspace from a viewpoint that
stresses the power of poetic language over ordinary, reductive language.
Poetic language is language in the process of making and is best studied
by close examination of poetic artifacts, or, better yet, by making poetic
artifacts.

The transition from real space to cyberspace, from prose to poetry,
from fact to fiction, from static to dynamic, from passive to active, from
the fixed in all its forms to the fluid in its everchanging countenance,

is best understood by examining that human effort that combines science and art, the worldly and the spiritual, the contingent and the permanent: architecture.

Even as cyberspace represents the acceptance of the body in the realm of the mind, it attempts to escape the mortal plane by allowing everything to be converted to a common currency of exchange. Architecture, especially visionary architecture, the architecture of the excess of possibility, represents the manifestation of the mind in the realm of the body, but it also attempts to escape the confines of a limiting reality. The story of both these efforts is illuminating, and in both directions. Cyberspace, as a world of our creation, makes us contemplate the possibility that the reality we exist in is already a sort of "cyberspace," and the difficulties we would have in understanding what is real if such were the case. Architecture, in its strategies for dealing with a constraining reality suggests ways in which the limitations of a fictional reality may be surmounted.

Architecture, most fundamentally, is the art of space. There are three fundamental requirements for the perception of space: reference, delimitation, and modulation. If any one is absent, space is indistinguishable from nonspace, being from nothingness. This, of course, is the fundamental observation of categorical relativity. This suggests that cyberspace does not exist until a distance can be perceived between subject and boundary, that is to say, until it is delimited and modulated.

A space modulated so as to allow a subject to observe it but not to inhabit it is usually called sculpture. A space modulated in a way that allows a subject to enter and inhabit it is called architecture. Clearly, these categories overlap a great deal: architecture is sculptural, and sculpture can be inhabited.

We can now draw an association between sculpture and the manner in which we are accustomed to interacting with computers. The interface is a modulated information space that remains external to us, though we may create elaborate spatial visualizations of its inner structure in our minds. Cyberspace, on the other hand, is intrinsically about a space that we enter. To the extent that this space is wholly artificial, even if it occasionally looks "natural," it is a modulated space, an architectural space. But more than asserting that there is architecture within cyberspace, it is more appropriate to say that cyberspace cannot exist without architecture, cyberspace is architecture, albeit of a new kind, itself long dreamed of.

Part Two: Liquid Architecture in Cyberspace

> . . . and we can in our Thought and Imagination contrive perfect Forms of Buildings entirely separate from Matter, by settling and regulating in a certain Order, the Disposition and Conjunction of the Lines and Angles.
>
> —Leon Battista Alberti, *The Ten Books of Architecture*

Visionary Architecture: The Excess of Possibility

Just as poetry differs from prose in its controlled intoxication with meanings to be found beyond the limits of ordinary language, so visionary architecture exceeds ordinary architecture in its search for the conceivable. Visionary architecture, like poetry, seeks an extreme, any extreme: beauty, awe, structure, or the lack of structure, enormous weight, lightness, expense, economy, detail, complexity, universality, uniqueness. In this search for that which is beyond the immediate, it proposes embodiments of ideas that are both powerful and concise. More often than not these proposals are well beyond what can be built. This is not a weakness: in this precisely is to be found the poignancy of vision.

The Space of Art   In imagining how information is to be "spatialized" in cyberspace, it is easy to be overwhelmed by the idea of "entering" the computer in the first place, and to only consider relatively mundane depictions of space: perspectival space, graph space, the space of various simple projection systems. Humanity's library of depictions of space is far richer than that: synchronically and diachronically, across the globe and through time, artists have invented a wealth of spatial systems. What would it be like to be inside a cubist universe? a hieroglyphic universe? a universe of cave drawings or Magritte paintings? Just as alternative renditions of the same reality by different artists, each with a particular style, can bring to our attention otherwise invisible aspects of that reality, so too can different modes of cyberspace provide new ways of interrogating the world.

The development of abstract art by Malevich, Kandinsky, Klee, Mondrian, and other early modern artists prefigures cyberspace in that it is an explicit turning away from representing known nature. Perhaps artists have always invented worlds—one is reminded how varied the representations used in the arts have been—from Chinese watercolors, to Byzantine icons, to the strange, conflicting backgrounds of Leonardo—but those worlds usually made reference to some familiar reality. Even when that reality was of a cosmic or mystical nature, we find an assumption of similarity to the everyday world. Modern artists took on the task of inventing entire worlds without explicit reference to external reality. Malevich is pertinent: his *architectones* are architectural studies imagined to exist in a world beyond gravity, *against gravity*. Created as an architecture without functional program or physical constraint, they are also studies for an *absolute architecture*, in the same sense that we speak of absolute music, architecture for the sake of architecture. Paul Klee, in his dairies, speaks about being a "god" in a universe of his own making.

The paintings of Max Ernst or, even more so, Hieronymous Bosch, come to mind, for their ability to create mysterious new worlds. In the works of these artists we not only see worlds fashioned out of unlikely combinations of a code consisting of familiar elements, but also meaningful crossings of expected conceptual and categorical boundaries.

As with cyberspace, the space of art *is* achitecture, *has* an architecture and *contains* architecture. It *is* architecture in its ability to create a finely controlled create sense of depth, even within depictions that are inherently two-dimensional; it *has* an architecture in its compositional structure; and, by depiction, it *contains* architecture. It can serve as a bridge between cyberspace and architecture.

Theoretical Projects   To the body of work that constitutes the built architecture we know, there is, of course, a counterpart that is unbuilt. Throughout history we find examples of architectural projects of such vast ambition that they simply could not be built using the resources of their time.

Piranesi's series of etchings entitled *Carceri*, or Prisons, marks the beginning of an architectural discourse of the *purposefully unbuildable*. Against the increasing constriction of architectural practice, Piranesi drew an imagined world of complex, evocative architecture. His title recalls a phrase by Georges Bataille: "Man will escape his head as a convict escapes his prison" (quoted in Hollier 1989). Man's head, prime shelter of reason, is both home and dungeon for the imagination. Piranesi's gesture asserts unequivocally that it is more important to escape the prison of the merely reasonable by creating an artificial reality ample enough to contain what can be conceived than it is to apply a Procrustean amputation to all that exceeds the contingencies of the present.

Ledoux, Lequeu, Boullée, each contributed to this struggle: as architectural practice was made more and more prosaic by the encroachment of utility, they responded by inventing a bolder and bolder imaginary counterpart. When the implicit metaphysical underpinnings of the past became obsolete and new explanations were needed to fend off the impoverishment of mere usefulness, Ledoux emphasized *architecture parlante*; an architecture that spoke, architecture and poetry. Lequeu invented a hypereclectic style of what on the surface appears to be an almost indiscriminate blending of manners and images. As with the paintings of Bosch, one is struck both by the heterogeneity of his sources as by the literalness with which they are used. As strange as his assemblages may have looked in his own time, in ours they appear to be more the case than the exception. A short walk in any major city will reveal juxtapositions more strange, literal, and indiscriminate. Boullée, searching for a way to express the sublime potential of architecture, composed architectural forms that aspired to be mountains, even planets, interiors that could suggest the universe. Of his *Cenotaph for Newton*, he wrote : "O Newton, as by the extent of your wisdom and the sublimity of your genius you determined the shape of the earth; I have conceived the idea of enveloping you in your own discovery" (quoted in Kostoff 1985).

The production of visionary architecture continues to the present. It is instructive to scan the manifestos for premonitions of an architecture of cyberspace. Many have contributed to this effort, becoming the world's front line of imagination, building in words and images what we can not yet convince the physical world to bear. Here is a short sampling of their voices, showing an awareness far ahead of their time:[3]

Sant'Elia and Marinetti, 1914: The tremendous antithesis between the ancient and the modern world is the outcome of all those things that exist now and did not exist then. Elements have entered our life whose very possibility the ancients did not even dream. Material possibilities and attitudes of mind have come into being that have had a thousand repercussions, first and foremost of which is the creation of a new ideal of beauty, still obscure and embryonic, but whose fascination is already being felt even by the masses. We have lost the sense of the monumental, of the heavy, of the static; we have enriched our sensibility by a 'taste for the light, the practical, the ephemeral and the swift.' We feel that we are no longer the men of the cathedrals, the palaces, the assembly halls; but of big hotels, railway stations, immense roads, colossal ports, covered markets, brilliantly lit galleries, freeways, demolition and rebuilding schemes.

Gropius: New Ideas on Architecture, 1919: . . . build in the imagination, unconcerned about technical difficulties.

Taut: Down with Seriousism, 1920: In the distance shines our tomorrow. Hurray, three times hurray for our kingdom without force! Hurray for the transparent, the clear! Hurray for purity! Hurray for crystal! Hurray and again hurray for the fluid, the graceful, the angular, the sparkling, the flashing, the light—hurray for everlasting architecture!

Schulze-Fielitz: The Space City, 1960: The space structure is a macro-material capable of modulation, analogous to an intellectual model in physics, according to which the wealth of phenomena can be reduced to a few elementary particles. The physical material is a discontinuum of whole-number units, molecules, atoms, elementary particles. Their combinational capabilities determine the characteristics of the material.

The space city accompanies the profile of the landscape as a crystalline layer; it is itself a landscape, comparable to geological formations with peaks and valleys, ravines, and plateaux, comparable to the leafy area of the forest with its branches. To regenerate existing cities, structures will stretch above their degenerate sections and cause them to fall into disuse.

Constant: New Babylon, 1960: The modern city is dead; it has fallen victim to utility. New Babylon is a project for a city in which it is possible to live. And to live means to be creative.

New Babylon is the object of a mass creativity; it reckons with the activation of the enormous creative potential which, now unused, is present in the masses. It reckons with the disappearance of non-creative work as the result of automation; it reckons with the transformation of morality and thought, it reckons with a new social organization.

[This will be a] . . . netlike pattern . . . [of spaces whose ambience can be varied by] . . . an abundant manipulation of colour, light, sound, climate, by the use of the most varied kinds of technical apparatus, and by psychological procedures. The shaping of the interior at any given moment, the interplay of the various environments takes place in harmony with the experimental life-play of the inhabitants. The city brings about a dynamically active, creative unfolding of life.

One can wander for prolonged periods through the interconnected sectors, entering into the adventure afforded by this unlimited labyrinth. The express traffic on the ground and the helicopters over the terraces cover great distances, making possible a spontaneous change of location.

More recently architects such as Lebbeus Woods have espoused the cause of an "experimental architecture," even an "anti-gravity" architecture, again well ahead of our physical technologies.

Woods: What Does it Mean? 1989: It must be said: experimental architecture may or may not be of redeeming benefit to others, to society-at-large. I believe it is improper to claim such work will benefit anyone at all. This is not its *raison d'etre*. True, a particular work may provide useful innovations, or may even become influential on the thought and work of others (for better or worse), but there is a better chance that most experimental works will, with justice, be consigned to oblivion. This is the risk of the new, the original: it may prove to be merely novel, merely eccentric. Better, most will say, to carry on the grand and proven tradition of architecture than risk the squandering of one's life and talents making nonsense. Perhaps.

It is clear that our ability to imagine architecture far outstrips our ability to build, so far. In most advanced disciplines this marks the difference between applied and pure research, and the value of pure research is undisputed. Architecture has no theoretical laboratory, apart from the studio, and the studio is only open to architects: the world does not share the inventions produced there. Cyberspace can thus be seen as a vast virtual laboratory for the continuing production of new architectural visions, while at the same time returning architecture to a public realm. What then would a cyberspace architecture be like?

Cyberspace Architecture

Architecture has been earthbound, even though its aspirations have not, as the citations collected above show. Buckminster Fuller remarked that he was surprised that, in spite of all the advances made in the technology of building, architecture remained rooted to the ground by the most mundane of its functions, plumbing. Rooted by waste matter, architecture has nevertheless attempted to fly in dreams and projects, follies and cathedrals.

Architecture has never suffered a lack of fertile dreams. Once, however, in times far less advanced technologically, the distance between vision and embodiment was smaller, even though the effort required for that embodiment was often crushing. Most "grand traditions" began with an experimental stage of danger and discovery and did not become fossilized until much later. Hard as it may be for us to fathom, a gothic cathedral was an extended experiment often lasting over a century, at the end of which there was the literal risk of collapse. The dream and the making were one. Curiously, the practice of architecture has become increasingly disengaged from those dreams. Cyberspace permits the schism that has emerged to be bridged once again.

Cyberspace alters the ways in which architecture is conceived and perceived. Beyond computer-aided design (CAD), design computing (DC), or the development of new formal means of describing, generating, and transforming architectural form, encodes architectural knowledge in a way that indicates that our conception of architecture is becoming increasingly musical, that architecture is spatialized music. Computational composition, in turn, combines these new methods with higher-level compositional concepts of overall form subject to local and global constraints to transform an input pattern into a finished work. In principle, and with the proper architectural knowledge, any pattern can be made into a work of architecture, just as any pattern can be made into music. In order for the data pattern to qualify as music or architecture it is passed through compositional "filters," processes that select and massage the data according to the intentions of the architect and the perceptual capacity of the viewer. This "adaptive filtering," to use a neural net term, provides the beginning of the intelligence that constitutes a cyberspace and not a hypergraph. This, of course, means that any information, any data, can become architectonic and habitable, and that cyberspace and cyberspace architecture are one and the same.

A radical transformation of our conception of architecture and the public domain that is implied by cyberspace. The notions of city, square, temple, institution, home, infrastructure are permanently extended. The city, traditionally the continuous city of physical proximity becomes the discontinuous city of cultural and intellectual community. Architecture, normally understood in the context of the first, conventional city, shifts to the structure of relationships, connections and associations that are webbed over and around the simple world of appearances and accommodations of commonplace functions.

I look to my left, and I am in one city; I look to my right, and I am in another. My friends in one can wave to my friends in the other, through my having brought them together.

It is possible to envision architecture nested within architecture. Cyberspace itself is architecture, but it also contains architecture, but now without constraint as to phenomenal size. Cities can exist within chambers as chambers may exist within cities. Since cyberspace signifies the classical object yielding to space and relation, all "landscape" is architecture, and the objects scattered upon the landscape are also architecture. Everything that was once closed now unfolds into a place, and everything invites one to enter the worlds within worlds it contains.

I am in an empty park. I walk around a tree, and I find myself in a crowded chamber. The tree is gone. I call forth a window, and in the distance see the park, leaving.

Liquid Architecture

That is why we can equally well reject the dualism of appearance and essence. The appearance does not hide the essence, it reveals it; it is the essence. The essence of an existent is no longer a property sunk in the cavity of this existent; it is the manifest law which presides over the succession of its appearances, it is the principle of the series.

. . . But essence, as the principle of the series is definitely only the concatenation of appearances; that is, itself an appearance.

. . . The reality of a cup is that it is there and that it is not me. We shall interpret this by saying that the series of its appearances is bound by a principle which does not depend on my whim.

—Jean-Paul Sartre, *Being and Nothingness*

The relationship established between architecture and cyberspace so far is not yet complete. It is not enough to say that there is architecture in cyberspace, nor that *that* architecture is animistic or animated. Cyberspace calls us to consider the difference between animism and animation, and animation and metamorphosis. Animism suggests that entities have a "spirit" that guides their behavior. Animation adds the capability of change in *location*, through time. Metamorphosis is change in *form*, through time *or space*. More broadly, metamorphosis implies changes in one aspect of an entity as a function of other aspects, continuously or discontinuously. I use the term liquid to mean animistic, animated, metamorphic, as well as crossing categorical boundaries, applying the cognitively supercharged operations of poetic thinking.

Cyberspace is liquid. Liquid cyberspace, liquid architecture, liquid cities. Liquid architecture is more than kinetic architecture, robotic architecture, an architecture of fixed parts and variable links. Liquid architecture is an architecture that breathes, pulses, leaps as one form and lands as another. Liquid architecture is an architecture whose form is contingent on the interests of the beholder; it is an architecture that opens to welcome me and closes to defend me; it is an architecture without doors and hallways, where the next room is always where I need it to be and what I need it to be. Liquid architecture makes liquid cities, cities that change at the shift of a value, where visitors with different backgrounds see different landmarks, where neighbor-hoods vary with ideas held in common, and evolve as the ideas mature or dissolve.

The locus of the concept "architecture" in an architecture that fluctuates is drastically shifted: Any particular appearance of the architecture is devalued, and what gains importance is, in Sartre's terms, "the principle of the series." For architecture this is an immense transformation: for the first time in history the architect is called upon to design not the object but the principles by which the object is generated *and varied* in time. For a liquid architecture requires more than just "variations on a theme," it requires the invention of something equivalent to a "grand tradition" of architecture at each step. A work of liquid architecture is no longer a single edifice, but a continuum of edifices, smoothly or rhythmically evolving in both space and time. Judgments of a building's *performance* become akin to the evaluation of dance and theater.

If we described liquid architecture as a symphony in space, this description would still fall short of the promise. A symphony, though it varies within its duration, is still a fixed object and can be repeated. At its fullest expression a liquid architecture is more than that. It is a symphony in space, but a symphony that never repeats and continues to develop. If architecture is an extension of our bodies, shelter and actor for the fragile self, a liquid architecture is that self in the act of becoming its own changing shelter. Like us, it has an identity; but this identity is only revealed fully during the course of its lifetime.

Conclusion

A liquid architecture in cyberspace is clearly a dematerialized architecture. It is an architecture that is no longer satisfied with only space and form and light and all the aspects of the real world. It is an architecture of fluctuating relations between abstract elements. It is an architecture that tends to music.

Music and architecture have followed opposite paths. Music was once the most ephemeral of the arts, surviving only in the memory of the audience and the performers. Architecture was once the most lasting of the arts, reaching as it did into the caverns of the earth, changing only as slowly as the planet itself changes. Symbolic notation, analog recording, and, currently, digital sampling and quantization, and computational composition, have enabled music to become, arguably, the most permanent of the arts. By contrast, the life span of architecture is decreasing rapidly. In many ways architecture has become the least durable of the arts. The dematerialized, dancing, difficult architecture of cyberspace, fluctuating, ethereal, temperamental, transmissible to all parts of the world simultaneously but only indirectly tangible, may also become the most enduring architecture ever conceived.

I am in a familiar place. Have I been here before? I feel I know this place, yet even as I turn something appears to have changed. It is still the same place, but not quite identical to what it was just a moment ago. Like a new performance of an old symphony, its intonation is different, and in the difference between its present and past incarnations something new has been said in a language too subtle for words. Objects and situations that were once thought to have a fixed identity, a generic "self," now possess personality, flaw and flavor. All permanent categories are defeated as the richness of the particular impresses upon me that in this landscape, if I am to benefit fully, attention is both required and rewarding. Those of us who have felt the difference nod to each other in silent acknowledgment, knowing that at the end of specificity lies silence, and what is made speaks for itself, not in words, but in presences, ever changing, liquid.

A Cyberspace Portfolio

The color plates 19–30 constitute a portfolio of algorithmically composed architectural designs that form the beginnings of a body of liquid architecture. This architecture varies according to the structure of its internal and external relations. Several "architectures," derived from one underlying architectural genotype generator, are viewed differently at different times during the course of their evolution, according to the views set forth in the first section of this chapter. A separation is made between the compositional structure of each architectural work, the coordinate system it is mapped to, and the set of primitives that it assembles. The structure itself, the coordinate systems, and the primitives are all seen as dynamically transformable under their own intelligence or in response to external control (by cyberspace/time, in response to other objects, or to users). This fundamental transformability suggests a liquidity in what was once the most concrete of the arts and requires the use of the plural: Architectures.

Notes

1. Bruce Sterling has recently collected five: Gibsonian cyberspace, Barlovian cyberspace, Virtual Reality, Simulation, and Telepresence. The definition above is a concatenation of all five plus a sixth, set forth in "Making Reality a Cyberspace" by Wendy Kellog and others in this volume.

2. Thomas Kuhn discussed the notion of world lines with respect to scientific concepts in a talk on untranslatability between different intellectual frameworks (Kuhn 1990).

3. Several sources of collected architectural and artistic manifestos exist. See, for example, Conrads 1970.

References

Alberti, L. B. [1404–1472] 1986. *The Ten Books of Architecture*. New York: Dover Publications.

Anderson, J. A., and Rosenthal, E. (eds.) 1988. *Neurocomputing: Foundations of Research*. Cambridge, Massachusetts: MIT Press.

Benedikt, M. (ed.) 1990. *Cyberspace: Collected Abstracts*. University of Texas at Austin.

Berman, M. 1989. *Coming to Our Senses*. New York: Simon and Schuster.

Bonner, J. T. 1988. *The Evolution of Complexity by Means of Natural Selection*. Princeton, New Jersey: Princeton University Press.

Brisson, D. W. (ed.) 1978. *Hypergraphics: Visualizing Complex Relationships in Art, Science and Technology*. Boulder, Colorado: Westview.

Conrads, U. 1970. *Programs and Manifestoes on 20th-Century Architecture*. Cambridge, Massachussetts: MIT Press.

Coyne, R. D. 1990. "Tools for Exploring Associative Reasoning in Design." In McCullough, M., Mitchell, W. J. and Purcell, P. (eds.) 1990. *The Electronic Design Studio*. Cambridge, Massachussetts: MIT Press.

Davies. P. (ed.). 1989. *The New Physics*. Cambridge: Cambridge University Press.

Dawkins, R. 1986. *The Blind Watchmaker*.

Feyerabend, P. 1988. *Against Method*. London: Verso.

Gregory, B. 1988. *Inventing Reality: Physics as Language*. New York: Wiley Science Editions.

Helsel, S. K., and Roth, J. P. (eds.) 1991. *Virtual Realities: Theory, Practice and Promise*. London: Meckler.

Hillier, B., and Hanson, J. 1984. *The Social Logic of Space*. Cambridge: Cambridge University Press.

Hollier, D. 1989. *Against Architecture: The Writings of George Bataille*. Cambridge, Massachussetts: MIT Press.

Kostoff, S. 1985. *A History of Architecture*. Oxford: Oxford University Press.

Kuhn, T. 1990. Untranslatabilty, lecture notes, UCLA.

Küppers, B. O. 1990. *Information and the Origin of Life*. Cambridge, Massachussetts: MIT Press.

Langton, C. G. 1989. *Artificial Life: Proceedings of an Interdisciplinary Workshop on the Synthesis and Simulation of Living Systems Held September 1987*. Reading, Massachusetts: Addison-Wesley.

Lorca, F. G. 1989. *Poet in New York*. New York: Noonday Press.

McClelland, J. L. and Rumelhart, D. E. (eds.). 1986. *Parallel Distributed Processing: Exploration in the Microstructure of Cognition*, Vols. 1 and 2. Cambridge, Massachussetts: MIT Press.

Negroponte, N. 1975. *Soft Architecture Machines*. Cambridge, Massachussetts: MIT Press.

Pérez-Gómez, A. 1983. *Architecture and the Crisis of Modern Science*. Cambridge, Massachussetts: MIT Press.

Pickover, C. 1990. *Computers, Pattern, Chaos and Beauty: Graphics from an Unseen World*. New York: St. Martin's Press.

Sartre, J. P. 1956. *Being and Nothingness*. New York: Washington Square Press.

Scarry, E. 1985. *The Body in Pain*. New York: Oxford University Press.

Stirling, B. 1990. "Cyberspace (TM)," *Interzone*, November 1990.

Tzonis, A. 1986. *Classical Architecture: The Poetics of Order*. Cambridge, Massachussetts: MIT Press.

Vidler, A. 1990. "The Building in Pain." *AA Files*, London: AA Press, Spring 1990.

Woods, L. 1989. *OneFiveFour*. New York: Princeton Architectural Press.

9 *Giving Meaning to Place: Semantic Spaces*

Alan Wexelblat

Motivation

Cyberspaces, or virtual realities, provide us with a number of powerful tools. Chief among these is the ability to create and directly interact with objects not available in the everyday world. Cyberspaces are especially powerful as visualization tools.

One of the most important features of any visualization system is the placement, or location, and arrangement of the represented objects. A well-structured view can make things obvious to the viewer and can empower interaction. Similarly, a badly constructed view can obfuscate and impede. A well-structured view has internal consistency and logic, and can be easily understood. In addition, the structure can convey an underlying mental model and can indicate possibilities for interaction.

In cases where the represented objects correspond to physical or potentially physical objects, placement is easy: it corresponds to the placement of the object in the everyday world. The logic and common sense of physical laws with which we are all familiar provide the structure for us to interpret the view. For example, an architect using a cyberspace to explore a not-yet-built building would place the building's components (wall, doors, window, and so forth) in the way that she expected the physical building to look. We understand this view because of our shared conventions about buildings; a room with no exits would appear odd to most viewers.

However, someone who must deal with more abstract information does not have this easy guideline from which to work. The user who wants to view information unrelated to physical location must deal with issues of how to lay out the information in a logical, consistent, and understandable fashion. For example, a user who wants to view a complex system of software modules and design documents has no external referent to aid layout.

In this chapter, I propose a fundamental organization technology to assist in this problem, called *semantic spaces*. Semantic spaces are a general mechanism that can be used to help solve problems of placement and composition. In addition, they give users considerable power in navigating and manipulating data objects in virtual spaces, such as cyberspace. The goal is to construct a theory that can be implemented in many forms, depending on the capabilities of the target system, and that subsumes current practice in a broader, simpler, and more flexible framework. Because of the limits of the print medium, I shall be using primarily two-dimensional examples in the following sections, but the principles I discuss are easily extensible to three or more dimensions.

Semantic Dimensions

The elementary components of semantic spaces are semantic dimensions. Semantic dimensions, as the name implies, are a way of correlating elements of meaning—semantics—with arrangement in a mathematical or virtual space—dimension. The elements of meaning that we will use in constructing our views are derived from properties of the objects themselves, from their interactions with each other, and from the interests expressed by the users.

Staying with our software development example, each module has properties such as *source language*, *lines of code*, *date of last modification*, *author*, and so on. By associating each of these properties with a dimension, we create an N-dimensional information space. Objects are located at positions (or possibly not located) according to the values of the properties on the objects.

To view the spaces thus produced, the property dimensions are mapped to the familiar X, Y, and Z dimensions that we can show on our displays. This allows us to produce displays that are meaningful because of the arrangement of their elements. In this way, we give meaning to the visualization just as the architect gives meaning by locating doors and window.

Semantics, traditionally, consists in the way representations—icons, signs, words—are meaningfully translated into values. The purpose of semantic dimensions is to give representational significance to arrangement and location, for arbitrarily abstract sets of information. In effect, we create a logic of placement in accordance with the structure of the objects.

This logic holds even though the information may have no intuitive semantics. The view is itself abstract and may require learning. The advantage of semantic dimensions is that, because the placement reflects information about the objects, the resulting view is well-structured and can make matters more obvious to the viewer.

There are two *kinds* and five *types* of semantic dimensions, producing the ten combinations shown in Table 9.1. These are described in detail in the following sections.

Kinds of Dimensions

The two major kinds of semantic dimensions are *relative* and *absolute*.

Absolute spaces are the most familiar; they correspond to unary properties of objects such as the *size* and *author* properties mentioned above. An absolute dimension is one that can be described based on the examination of each object in isolation.

It may be argued that in a cyberspace there are no unary properties. Those things we ordinarily think of as unary properties are really relationships between an object and an observer. This becomes important when we allow users to move through the cyberspace, possibly at quite a rapid rate. Even when adopting this relativistic view, though, we are relating to only one object at a time.

An example of a two-dimensional view generated from absolute dimensions is shown in Figure 9.1 where the *size* and *author* properties have been mapped to the X and Y dimensions. In this diagram, each dot represents an object (a code module in this case). This view allows us to understand quickly where each author's energies have been spent.

Table 9.1
Combination of dimension types and kinds

	Relative	Absolute
Linear		
Ray		
Quantum		
Nominal		
Ordinal		
Functional		

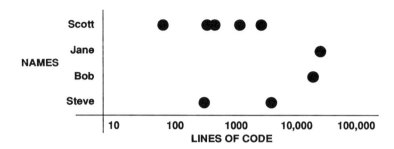

Figure 9.1
Example of absolute dimensions.

Relative dimensions, on the other hand, correspond to properties that can be determined only by examining relationships among the objects or relationships between the objects and the users. These kinds of dimensions are common in knowledge representation and artificial intelligence where relationship properties such as *parent-of* and *is-a* are frequently used. An illustration of this in our example software development domain is shown in Figure 9.2, which maps the calling graph of a set of modules.

Figure 9.2 shows a view in which arrows have been used to show the *is-called-by* property. Because the relationship is not one-to-one, it cannot be collapsed to a single dimension without introducing unacceptable ambiguity.

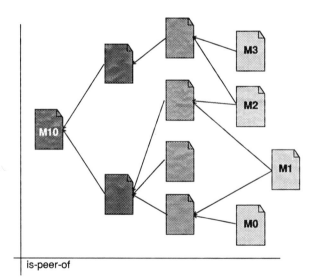

Figure 9.2
Example of relative dimensions.

The X dimension shows the *is-peer-of* property. The Y dimension is used to avoid visual overlap. Note that even though module M10 has no peer in this view, the *is-peer-of* property still requires pairwise examination of M10 and the other modules. Note also that the subtle shifting of module M1 to the right enables us to determine immediately that it is not a peer of M0, M2, and M3, even though all four are called by peer modules.

An important property of absolute dimensions is that location may have precise meaning even without the presence of an object at that location. The meaning is determined by reference to the dimension and determining one's distance from the (possibly imagined) origin. For example, in Figure 9.1, an object at the extreme right-hand edge would have 100,000 or more lines of code.

In a relative dimension, the meaning of a location cannot be determined precisely without reference to other objects in the space. Each object is, in effect, its own origin. For example, in Figure 9.2, an object at the extreme right-hand edge might be called by one or more of M0, M1, M2, or M3. But this cannot be known without reference to the objects in question. Nevertheless, if we were interested in finding modules that were called by M0, this would be the logical place to look.

Types of Dimensions

Within both the major kinds, there are five types. The types roughly correspond to the mathematics of the property being examined.

Linear dimensions map those properties that can be described on a conventional line. The values of these properties roughly correspond to the real numbers in that there are a potentially infinite number of them. For any two items along a linear dimension, one can imagine placing a new object between the two existing ones by refining the accuracy of measurement. For example, software modules M1 and M2 might appear next to each other on a dimension that mapped their times of creation to 5/25/90 and 5/26/90. However, module M3 would appear between them if it were created at a later hour of the day on 5/25 than M1. This refinement of measurement from days to hours is like the refinement used to place 1.5 between 1 and 2.

A *ray dimension* is like a linear dimension except that it is bounded on one end by some fact about the property. Ray dimensions roughly correspond to the positive real numbers. For example, the size dimension in Figure 9.1 is a ray dimension because it makes no sense to talk about software modules that have fewer than zero lines of code in them. This fact about the real world is part of the extra-object information that must be used to present an intelligible view. Just as the room with no exits causes confusion by violating our commonsense assumptions, so too would a module with negative lines of code.

Quantum dimensions are so named because the elements in them, like quanta of energy, fall into "buckets" where the property is either one value or another, but cannot be between. Quantum dimensions roughly correspond to the integers. Quantum dimensions can be described by formal systems or equations, such as the Boolean system or a formula that produces all acceptable English names. For example, the ASCII character set produces a quantum dimension because an ASCII character can be any member of the set, but cannot fall between or outside that range.

Interestingly, some nonobvious properties produce quantum dimensions because of the way they are treated. Variations that may seem significant are ignored and widely varying values are treated the same. For example, the programming language LISP supports special values T and NIL, which roughly correspond to the Boolean values TRUE and FALSE. However, the LISP language definition treats any non-NIL value the same as T. Thus, a seemingly infinite variation collapses down to a quantum dimension with two values: NIL and not-NIL.

Nominal dimensions are similar to quantum dimensions in that they specify values of which the property can assume one or another value, but not between them or outside. Nominal dimensions roughly correspond to our naive notion of sets. The difference between nominal and quantum dimensions is that unlike the formal, abstract description available for quantum dimensions, nominal dimensions can only be described by reference to (naming) some extra-object knowledge. For example, in Figure 9.1 the *author* dimension is a nominal dimension. It consists of the set of names of the possible authors of modules in the system being visualized (thus, author is really shorthand for *software-authors-on-team*). There is no way to specify this dimension without reference to the development team, which is knowledge external to the modules.

Ordinal dimensions are similar to nominal in that they contain an enumerated set of values and in that they are often described by reference to some extra-object knowledge. However, an ordinal dimension additionally imposes an ordering on the members of its set (thus the name). For example, persons standing in a queue form an ordinal dimension—the queue imposes an order on what would otherwise be a nominal dimension. Ordinal dimensions have the characteristic that one can talk about their members in terms such as "first," "second," "third," and so on.

Functional dimensions are those where the value of the property is not simple, but is itself a function or other computational formula. This may seem counterintuitive to noncomputer users, but it is often critical to be able to refer to the *process* by which something is done as well as the *product* of doing it. In addition, it is sometimes impossible to know the value of a property without actually examining it. For example, an object may have a property that corresponds to the time recorded by the National Atomic Clock in Colorado. The value of this property will be different each time it is examined. Therefore, for visualization and understanding purposes, it is easier to represent the function that computes the value, rather than the actual value.

Ordering

In addition to the so-named ordinal dimensions discussed above, some types of dimensions imply an ordering. Others, such as a hue/intensity cone, imply sequentiality but no ordering in that one can start anywhere on the surface of the cone and go in any direction, encountering the colors in no particular order.

Linear and ray dimensions imply an ordering that corresponds to our commonsense understanding of numbers: monotonically increasing or decreasing. To arrange them otherwise would be confusing. This does not mean that every value must be enumerated. As in Figure 9.1, it is often enough to list landmark values.

Other orders are suggested by standard or convention. For example, the ASCII character set is conventionally represented in the order of its characters' numeric code values (space = 32, 'A' = 65, and so on). However, many dimensions suggest no default ordering. In particular, ordering spaces by functional dimensions can be quite difficult or obscure for a nonmathematician.

In some cases, this can be resolved by establishing a convention: *authors' names* could appear in alphabetic order or in order of their seniority on the project. If the convention is well known or obvious, this can be an advantage because it adds information without additionally complicating the view.

Effectively, dimensions can be *preordered* or *postordered*. That is, there may be some property of the dimension that preordains an order for the elements of that dimension. Alternatively, the ordering may be derived after the fact, either by the necessity of creating a sequence to display, or by observation.

Semantic Spaces

To complete the characterization of cyberspace as a rational organizing medium, we can extend the idea of semantic dimensions to semantic spaces. Conventional spaces are defined by coordinate dimensions such as latitude and longitude or by a position such as X, Y, and Z relative to a known origin.

A semantic space is an N-dimensional space where each dimension is a semantic dimension as discussed in the previous sections and N is the number of properties expressed on the objects in the space. The expression of properties is a way of abstracting interesting elements from the infinite set of possible properties computable on an object. Regrettably, a discussion of the way this is done, such as by degree-of-interest functions, is beyond the scope of this chapter. Interested readers should see Furnas 1986 and Fairchild, Meredith, and Wexelblat 1989a.

Thus, a location in a semantic space is an N-tuple $\{P_0, P_1, \ldots, P_N\}$.

Strictly speaking, this characterization is not complete. Frequently, one or more properties of interest will fail to exist on an object we wish to locate. This can happen for any number of reasons: the property may be inappropriate for that object, such as asking for the *author-of* property on the object representing Alan Wexelblat, or the property may not be available in the information space being examined, such as querying this chapter for the text of items listed in the References.

To handle this situation, we extend our notation to include a NOT-FOUND value, represented by α. Thus any of P_N can have the value α.

The existence of this tuple allows us to completely characterize the position of each object and solves the first problem mentioned in the

Motivation section. We can now assign reliable, meaningful locations in cyberspace to an arbitrary set of objects, even when these objects are abstract and fail to present an obvious physical layout.

We can now easily achieve one of our stated goals: that of subsuming current practice. Conventional displays are laid out according to predefined dimensions such as the semantic dimensions discussed above, or by user preference. For the latter case, we can attach properties such as *display-X-position* and *display-Y-position*. By mapping these properties to the X and Y dimensions of a semantic space, we can reproduce any conventional computer display as a semantic space.

It may be argued that this characterization is insufficient in that two or more objects, such as newly created instances of a class, may map to precisely the same location in a semantic space. This objection does not count against the theory; rather, it points out that—for the purposes expressed in the semantic cyberspace—the objects are indistinguishable. This mirrors a frequently seen property of objects in the real world. If my secretary is making two copies of a design specification, I do not care which of the two copies I get.

If it is necessary to make an absolute and unmistakable differentiation among objects (for example, for purposes of reliable reference), then some form of globally unique object identifier can be attached to the object. This property can then be mapped to a dimension wherein all objects are differentiated.

Visualizing Semantic Spaces

For the purposes of visualization, cyberspace locations can be mapped into X, Y, Z locations relative to the user's point of view or to an arbitrary origin. These mappings can be augmented by visual guides such as the arrows of Figure 9.2 and the logarithmic scale of Figure 9.1. Alternatively, the augmentations can themselves express properties, particularly relational ones.

The notations used to augment the basic mappings are based on expressiveness and effectiveness criteria such as visual impact or the ability to be quickly interpreted. For example, Figure 9.3 (adapted from Mack 1987) shows a view of CIS courses using X and Y axes to show in what quarter a course will be taught and by whom, augmented with arrows expressing the *prerequisite-of* property.

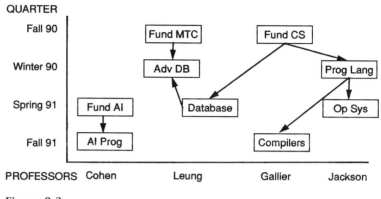

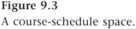

Figure 9.3
A course-schedule space.

The augmentation is effective in that it allows us to detect quickly that the Advanced Databases course is scheduled to be taught before the Database course that is its prerequisite. Cleveland and McGill (1984) deal in detail with issues of effectiveness in graphical displays.

This theory of mappings and augmentations allows us to construct displays of an abstract cyberspace terrain. Although this framework does not, by itself, provide all the answers to the problems of visualization, it does give a reliable, predictable, consistent base on which to build our maps, much as the system of latitude and longitude does for conventional mapmakers.

In addition to solving our layout problems and providing clear views of our information, semantic spaces offer us new power in our interaction with information. Navigation and manipulation take on meaning not possible before. We will now examine some of these possibilities.

Navigational Operations

The ability to traverse areas is a fundamental human ability: it is one of the first things we look for in a child's development; it is the first thing we take away from someone who has violated society's laws. In a cyberspace that corresponds to a physical terrain, navigation, or movement, has meaning by virtue of its analogy with physical action. Movement from one room to another within an imaginary building is meaningful by virtue of the fact that we have been doing it since childhood and have learned where to go at what times.

In an ordinary computer system, movement does not necessarily have the same societally constructed connotations. In most such systems, movement is merely a means of stopping interaction with one object and beginning interaction with another. Movement is necessary because objects are at some distance from one another, but the act of movement itself has no meaning.

However, this loss of meaning need not happen in a semantic cyberspace. An effect of constructing cyberspace along semantic dimensions is to render the actions of motion meaningful in and of themselves. This is so because, as discussed above, the space itself has meaning, possibly even when no objects are present.

For example, consider the space of design documents shown in Figure 9.4. This figure shows a two-dimensional space of documents where each major document is assigned a number when it is first created. As new versions are made of original documents, a decimal number is assigned. Thus, "Doc 2.1" is the first revision of the second original design document.

If my focus of attention in the system is on Doc 2.1 and I move to the right, I have not only left one object, I have moved into a region where I can anticipate finding documents like Doc 3.1, Doc 4.1, and so on. The fact that Doc 3.1 is not present in this view is made readily apparent by this motion.

This allows us to perform visual searches quickly and helps us find objects that may more or less closely match our desires. We can move the focus of our attention (our "cyber-cursor") to the point in cyberspace at which the properties of interest have the correct values. Navigation

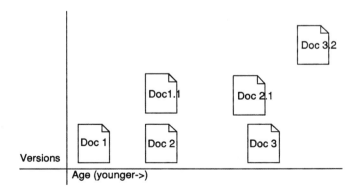

Figure 9.4
A space of design documents.

is accomplished by moving along dimensions until the values of our location match the desired values. This is similar to navigating in Manhattan, New York, where we have a grid of numbered avenues and streets to guide our walks. In effect, what we are doing is creating a description of an object; an actual item matching that description may or may not exist.

In actuality this kind of navigation—which is analogous to walking or flying—turns out to be both inefficient and undesirable, as shown in Shook's evaluation of SemNet (Shook 1986; see also Fairchild and Poltrock 1987, Poltrock et al. 1986). However, we are not limited to this kind of navigation. Cyberspace allows us to take advantage of movement modes, such as teleportation and movement through higher-order dimensions, that are simply not available in the physical world.

If there is an object at that location, then we have achieved our objective. However, if there is no object, we can simply "look around" the X, Y, Z space around us and find objects that more or less match our criteria.

This sort of approximate matching is important in many areas of software development, such as reuse (see Lubars 1988 for discussion and Fairchild, Meredith, and Wexelblat 1989a for examples). Though difficult for a computer, humans make these approximations constantly. Having objects arranged according to semantic dimensions allows us to take advantage of physical proximity to aid our searching.

Manipulation Operations

In the previous section, I mentioned that we were creating a description of an object that might not exist. It is possible, of course, to manipulate the description of an object that *does* exist. We can move objects around in semantic spaces, just as we do in physical spaces. Like navigation, manipulation in semantic spaces has meaning.

Movement in physical spaces, like the other things examined in this chapter, is also often meaningful. From "crossing the Rubicon" to entering forbidden airspaces, we are familiar with the possibility that movement can change meaning. Moving an object between contexts can also be meaningful in the physical world: for example, Egyptian artifacts take on new meanings when moved from ancient tombs to modern museum display halls.

In a semantic space, we change the meaning of objects by moving them along one or more dimensions. For example, in Figure 9.3, we could correct the name of "Doc 3.2" to "Doc 3.1" by moving the object down one row. In effect, what we are doing by moving an object through a semantic space is that we are editing the N-tuple that represents the location of the object in cyberspace. Because the N dimensions are not arbitrarily assigned values, but instead represent information about the object, we are in fact making a semantic change—editing the object by moving it.

Although theoretically simple, manipulation in actual object spaces can become quite complex. Objects may fail to have some or all of the properties we want. Thus the N-tuple representing the object may be sparse. In the discussion of functional dimensions, we mentioned an object that had the property of returning the National Atomic Clock time value each time it was accessed. In real systems, it is unlikely that one would have multiple objects with this property. Therefore, most objects will fail to appear in this functional dimension.

Additionally, some properties are *recalcitrant*. That is, they resist manipulation because of their nature. For example, objects may have a property such as *creator*, which shows the name of the person who originated the object. While the value of such a property may have been originally recorded in error, and thus require editing, the *creator* property should generally be resistant to casual change.

To make modification work fully, the cyberspace system must be reified. That is, the means for modifying the system must be contained within the system. For example, if we wish to add or delete properties from objects, this should be accomplished by the same mechanisms we use for more "ordinary" manipulations. Also, we should be able to switch between visualizing different sets of properties using these same mechanisms. Although reification is well understood in theory, practical implementations of reified systems are hard to come by. Further, it may not be simple to present a reified cyberspace in a way that is as natural to understand as what we have been discussing.

We should be able to construct a "flight rules" view in which we can select dimensional mappings and control how movement affects objects. Flight rules would, for example, allow us to control the recalcitrance of properties and the ordering conventions discussed above.

Phase Spaces

Up to now, we have considered cyberspace to be a user-inhabited place, a space in which users visualize, navigate among, and manipulate information from within. However, it is also possible to imagine viewing the system from a vantage point that appears to be outside the cyberspace, treating it as a phase space.

I say "appears to be" because this is the view afforded to the user. In terms of the semantic space theory, what happens is that the user moves along an orthogonal (or higher-order) dimension. It is only the limits of our technology and physiology which require us to visualize four or more dimensions in three.

Such a view might appear similar to Figure 9.5, where the stars are data objects or, if seen from far enough away, groups of objects. The notion of clustering objects based on distance from the viewer brings to mind the ideas of fish-eye views. In a fish-eye view, objects near to the viewer and objects with high importance are made visually more prominent. Less important objects are deemphasized and aggregated into clusters. The entire cluster is represented as a single object until the viewer moves closer or otherwise indicates a higher level of interest in the aggregate object. This has the effect of simplifying complex displays.

This kind of view enables us to perform cluster analyses and similar static measurements of the set of objects in the cyberspace. We can also rotate the space as a whole to get new perspectives on it.

It is important to remember, though, that a cyberspace is not a static place. Static measurements, while important, do not tell the whole

Figure 9.5
An "outside" view.

story. The space, and especially the objects in the space, are changing as work is done to add, delete, and modify the system represented in the space.

One way to look at this evolution is to set the objects in motion while retaining a motionless view. Starting at a specified time zero, the history of changes to objects is replayed, with each time interval (day, hour, etc.) representing one "frame" of the "animation." With this technique the history of the system can be seen in the movement, appearance, and disappearance of objects. The objects move (change their position) in response to changes in the values of their properties, just as before. However, we can now get a feel for how the system as a whole acts over time.

Another way this can be used, as done by Meredith (1991), is to use complexity measures to transform nonlinear spaces into linear visualizations. The effect of this is to collapse dependent dimensions into one linear variable (represented below as K1, K2, and K3). This turns a large number of interrelated properties into two or three orthogonal values that easily map to X, Y, and Z. The composite change over time of the system or object to be studied can then be plotted, resulting in figures similar to Figure 9.6, which shows a system climbing a slow curve until it reaches a point of exponential growth.

In the study of chaotic systems this kind of view led to the discovery of strange attractors (Gleick 1987). In a software environment, we can use this to get a feel for how the system under development or in use grows and changes. We can also examine factors that previously were hard to visualize, such as the dynamic shift of responsibility and work burden over the lifetime of a project.

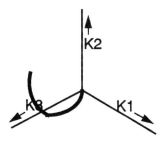

Figure 9.6
A composed phase view.

Conclusion

By setting up on a foundation of semantic dimensions and semantic spaces, cyberspaces can provide a vehicle for building and testing visualizations of large amounts of complex data no matter how abstract. This is achieved by virtue of the multiplicity of meaningful dimensions available as properties of the objects being represented. Because the dimensions convey information, both movement and locations in a semantic cyberspace have meaning. Navigation and visual search are available in ways not possible before.

Just as each point along a dimension describes a value or relationship, the N-tuple of a complete location in semantic cyberspace describes the value of the set of properties and relationships of interest about an object. Just as movement along a dimension corresponds to change in a property or relationship, movement through N-dimensional cyberspace corresponds to continual redefinition of an object. Thus we give meaning to place, and operations like "up two and over one" are transformed from arbitrary movements into meaningful semantic operations.

Acknowledgments

I would like to thank my colleagues who helped me develop these ideas: Kim Fairchild, now at the National University of Singapore and Greg Meredith at MCC. I would also like to thank the organizers of and participants in the First Cyberspace Conference. William Bricken made insightful and helpful comments on an earlier version of this chapter.

References

Cleveland, W. S., and McGill, R. "Graphical Perception: Theory, experimentation and application to the development of graphical methods," *J. Am. Stat. Assoc.* 79, 387, September 1984.

Fairchild, K., and Poltrock, S. *Soaring Through Knowledge Space: SemNet 2.1* (videotape). Presented at CHI'87, Toronto, 1987.

Fairchild, K., Meredith, G., and Wexelblat, A. "The Tourist Artificial Reality," *CHI'89 Conference Proceedings*, Austin, TX, May 1989a.

Fairchild, K., Meredith, G., and Wexelblat, A. "A Formal Structure for Automatic Icons," *Interacting with Computers*, Vol. 1, #2, August 1989b.

Furnas, G. W. "Generalized Fisheye Views," *CHI'86 Conference Proceedings,* Boston, MA, May 1986.

Gleick, James. *Chaos: Making a New Science,* Viking Penguin Inc., New York, 1987.

Lubars, Mitch. *Environmental Support for Reuse,* MCC Technical Report STP-120-88, April 1988.

Mackinlay, Jock. "Automating the Design of Graphical Presentations of Relational Information," *ACM Transactions on Graphics,* Vol. 5, #2, April 1986.

Meredith, Greg. *Software EEG,* MCC Technical Report STP-BC-142-91, April 1991.

Poltrock, S., Shook, R., Fairchild, K., Lofgren, J., Tarlton, P., Tarton, M., Hauser, M. *Three-dimensional interfaces: The promise and the problems,* MCC Technical Report HI-291-86, 1986.

Shook, Robert E. *SemNet: A Conceptual and Interface Evaluation,* MCC Technical Report HI-320-86, 1986.

10 The Lessons of Lucasfilm's Habitat*

Chip Morningstar
F. Randall Farmer

Introduction

Lucasfilm's Habitat project was one of the first attempts to create a very large-scale, commercial, many-user, graphical virtual environment. A far cry from many laboratory research efforts based on sophisticated interface hardware and tens of thousands of dollars per user of dedicated computing power, Habitat is built on top of an ordinary commercial on-line service and uses an inexpensive—some would say "toy"— home computer to support user interaction. In spite of these somewhat plebeian underpinnings, Habitat is ambitious in its scope. The system we developed can support a population of thousands of users in a single shared cyberspace. Habitat presents its users with a real-time animated view into an on-line simulated world in which users can communicate, play games, go on adventures, fall in love, get married, get divorced, start businesses, found religions, wage wars, protest against them, and experiment with self-government.

The Habitat project proved to be a rich source of insights into the nitty-gritty reality of actually implementing a serious, commercially viable cyberspace environment. Our experiences developing the Habitat system, and managing the virtual world that resulted, offer a number of interesting and important lessons for prospective cyberspace architects. The purpose of this chapter is to discuss some of these lessons. Our hope is that the next generation of builders of virtual worlds can benefit from our experiences and (especially) from our mistakes.

---

* Lucasfilm's Habitat was created by Lucasfilm Games, a division of LucasArts Entertainment Company, in association with Quantum Computer Services, Inc.

Due to space limitations, we will not be able to go into as much technical detail as we might like; this will have to be left to a future publication. Similarly, we will only be able to touch briefly upon some of the history of the project as a business venture, which is a fascinating subject of its own. Although we will conclude with a brief discussion of some of the future directions for this technology, a more detailed exposition on this topic will also have to wait for a future occasion.

The essential lesson that we have abstracted from our experiences with Habitat is that a cyberspace is defined more by the interactions among the actors within it than by the technology with which it is implemented. While we find much of the work presently being done on elaborate interface technologies—DataGloves, head-mounted displays, special-purpose rendering engines, and so on—both exciting and promising, the almost mystical euphoria that currently seems to surround all this hardware is, in our opinion, both excessive and somewhat misplaced. We can't help having a nagging sense that it's all a bit of a distraction from the really pressing issues. At the core of *our* vision is the idea that cyberspace is necessarily a *many-participant environment*. It seems to us that the things that are important to the inhabitants of such an environment are the capabilities available to them, the characteristics of the other people they encounter there, and the ways these various participants can affect one another. Beyond a foundation set of communications capabilities, the details of the technology used to present this environment to its participants, while sexy and interesting, are of relatively peripheral concern.

What Is Habitat?

Habitat is a "many-player online virtual environment" (its purpose is to be an entertainment medium; consequently, the users are called "players"). Each player uses his or her home computer as a frontend, communicating over a commercial packet-switching data network to a centralized backend system. The frontend provides the user interface, generating a real-time animated display of what is going on and translating input from the player into requests to the backend. The backend maintains the world model, enforcing the rules and keeping each player's frontend informed about the constantly changing state of the universe. The backend enables the players to interact not only with the world but with each other.

Habitat was inspired by a long tradition of "computer hacker science fiction," notably Vernor Vinge's story, "True Names" (1981), as well as many fond childhood memories of games of make-believe, more recent memories of role-playing games and the like, and numerous other influences too thoroughly blended to pinpoint. To this we added a dash of silliness, a touch of cyberpunk (Gibson 1984, Sterling 1986), and a predilection for object-oriented programming (Abelson and Sussman 1985).

The initial incarnation of Habitat uses a Commodore 64 for the frontend. Figure 10.1 is a typical screen from this version of the system.[1] The largest part of the screen is devoted to the graphics display. This is an animated view of the player's current location in the Habitat world. The scene consists of various objects arrayed on the screen, such as the houses and tree. The players are represented by animated figures that we call "Avatars." Avatars are usually, though not exclusively, humanoid in appearance. In Figure 10.1 you can see two of them, carrying on a conversation.

Avatars can move around, pick up, put down, and manipulate objects, talk to each other, and gesture, each under the control of an individual player. Control is through the joystick, which enables the player to point at things and issue commands. Talking is accomplished by typing on the keyboard. The text that a player types is displayed over his or her Avatar's head in a cartoon-style "word balloon."

Figure 10.1
A typical Habitat scene (© 1986 LucasArts Entertainment Company).

A Habitat world is made up of a large number of discrete locations that we call "regions." In its prime, the prototype Habitat world consisted of around 20,000 of them. Each region can adjoin up to four other regions, which can be reached simply by walking your Avatar to one or another edge of the screen. Doorways and other passages can connect to additional regions. Each region contains a set of objects that define the things that an Avatar can do there and the scene that the player sees on the computer screen.

Some of the objects are structural, such as the ground or the sky. Many are just scenic, such as the tree or the mailbox. Most objects, however, have some function that they perform. For example, doors transport Avatars from one region to another and may be opened, closed, locked, and unlocked. ATMs (Automatic Token Machines) enable access to an Avatar's bank account.[2] Vending machines dispense useful goods in exchange for Habitat money. Many objects are portable and may be carried around in an Avatar's hands or pockets. These include various kinds of containers, money, weapons, tools, and exotic magical implements. Table 10.1 lists some of the most important types of objects and their functions. The complete list of object types numbers in the hundreds.

Implementation

The following, along with several programmer-years of tedious and expensive detail that we won't cover here, is how the system works:

At the heart of the Habitat implementation is an object-oriented model of the universe.

The frontend consists of a system kernel and a collection of objects. The kernel handles memory management, display generation, disk I/O, telecommunications, and other "operating system" functions. The objects implement the semantics of the world itself. Each type of Habitat object has a definition consisting of a set of resources, including animation cells to drive the display, audio data, and executable code. An object's executable code implements a series of standard behaviors, each of which is invoked by a different player command or system event. The model is similar to that found in an object-oriented programming system such as Smalltalk (Goldberg and Robson 1983), with its classes, methods, and messages. These resources consume significant

Table 10.1
Some important objects

Object Class	Function
ATM	Automatic Token Machine; Access to an Avatar's Bank Account
Avatar	Represents the player in the Habitat world
Bag, Box	Containers in which things may be carried
Book	Document for Avatars to read (e.g., the daily newspaper)
Bureaucrat-in-a-box	Communication with system operators
Change-o-matic	Device to change Avatar gender
Chest, Safe	Containers in which things may be stored
Club, Gun, Knife,	Various weapons
Compass	Points direction to West Pole
Door	Passage from one region to another; can be locked
Drugs	Various types; changes Avatar body state, e.g., cure wounds
Elevator	Transportation from one floor of a tall building to another
Flashlight	Provides light in dark places
Fountain	Scenic highlight; provides communication to system designers
Game piece	Enable various board games: backgammon, checkers, chess, etc.
Garbage can	Disposes of unwanted objects
Glue	System building tool; attached objects together
Ground, Sky	The underpinnings of the world
Head	An Avatar's head; comes in many styles; for customization
Key	Unlocks doors and other containers
Knick-Knack	Generic inert object; for decorative purposes
Magic Wand	Various types, can do almost anything
Plant, Rock, Tree	Generic scenic objects
Region	The foundation of reality
Sensor	Various types, detects otherwise invisible conditions in the world
Sign	Allows attachment of text to other objects
Stun gun	Nonlethal weapon
Teleport booth	Means of quick long-distance transport; analogous to phone booth
Tokens	Habitat money
Vendroid	Vending machine; sells things

amounts of scarce frontend memory, so we can't keep them all in core at the same time. Fortunately, their definitions are invariant, so we simply swap them in from disk as we need them, discarding less recently used resources to make room.

When an object is instantiated, we allocate a block of memory to contain the object's state. The first several bytes of an object's state information take the same form in all objects, and include such things as the object's screen location and display attributes. This standard information is interpreted by the system kernel as it generates the display and manages the run-time environment. The remainder of the state information varies with the object type and is accessed only by the object's behavior code.

Object behaviors are invoked by the kernel in response to player input. Each object responds to a set of standard verbs that map directly onto the commands available to the player. Each behavior is simply a subroutine that executes the indicated action; to do this it may invoke the behaviors of other objects or send request messages to the backend. Besides the standard verb behaviors, objects may have additional behaviors that are invoked by messages that arrive asynchronously from the backend.

The backend also maintains an object-oriented representation of the world. As in the frontend, objects on the backend possess executable behaviors and in-memory state information. In addition, since the backend maintains a persistent global state for the entire Habitat world, the objects are also represented by database records that may be stored on disk when not "in use." Backend object behaviors are invoked by messages from the frontend. Each of these backend behaviors works in roughly the same way: a message is received from a player's frontend requesting some action; the action is taken and some state changes to the world result; the backend behavior sends a response message back to the frontend informing it of the results of its request and notification messages to the frontends of any other players who are in the same region, informing *them* of what has taken place.

The Lessons

In order to say as much as we can in a limited space, we will describe what we think we learned through a series of principles or assertions surrounded by supporting reasoning and illustrative anecdotes. As cyberspace develops, a more formal and thorough exposition may be called for.

We mentioned our primary principle earlier:

• *The idea of a many-user environment is central to cyberspace.*

It is our deeply held conviction that one of the defining characteristics of a cyberspace system is that it represents a many-user environment. This stems from the fact that what (in our opinion) people seek in a virtual world is richness, complexity, and depth. With our best science and technology we do not possess the ability to produce an automaton that even approaches the complexity of a real human being, let alone a society. Our approach, then, was and is not even to attempt this, but instead to use the computational *medium* to augment the communications channels between real people.

If what we are constructing is a many-user environment, it naturally follows that some sort of communications capability must be fundamental to our system. However, we must take into account an observation that is the second of our principles:

• *Communications bandwidth is a scarce resource.*

This point was driven home to us by one of Habitat's nastier, externally imposed, design constraints, namely, that it provide a satisfactory experience to the player over a 300-baud serial telephone connection (one routed, moreover, through commercial packet-switching networks that impose an additional, uncontrollable latency of 100 to 5000 milliseconds on each packet transmitted).

Even in a more technically advanced network, however, bandwidth remains scarce in the sense that economists use the term: available carrying capacity is not unlimited. The law of supply and demand suggests that no matter how much capacity is available, you always want more. When communications technology advances to the point where we all have multigigabaud fiber-optic connections into our homes, computational technology will have advanced to match. Our processors' expanding appetite for data will mean that the search for ever more sophisticated data compression techniques will *still* be a hot research area (though what we are compressing may at that point be high-resolution volumetric time-series or something even more esoteric) (Drexler 1986).

Computer scientists tend to be reductionists who like to organize systems in terms of primitive elements that can be easily manipulated within the context of a simple formal model. Typically, you adopt a small variety of very simple primitives, which are then used in large numbers. For a graphics-oriented cyberspace system, the temptation is to build upon bit-mapped images or polygons or some other *graphic* primitive. These sorts of representations, however, are invitations to disaster. They arise from an inappropriate fixation on display technology, rather than on the underlying purpose of the system.

However, the most significant part of what *we* wish to be communicating are human behaviors. These, fortunately, can be represented quite compactly, provided we adopt a relatively abstract, high-level description that deals with behavioral concepts directly. This leads to our third principle:

- *An object-oriented data representation is essential.*

Taken at its face value, this assertion is unlikely to be controversial, as object-oriented programming is currently the methodology of choice among the software engineering cognoscenti. However, what we mean here is not only that you should adopt an object-oriented approach, but that the basic objects from which you build the system should correspond more or less to the objects in the user's conceptual model of the virtual world, that is, people, places, and artifacts. You could, of course, use object-oriented programming techniques to build a system based on, say, polygons, but that would not help to cope with the fundamental problem.

The goal is to enable the communications between machines to take place primarily at the behavioral level (what people and things are doing) rather than at the presentation level (how the scene is changing). The description of a place in a virtual world should be in terms of what is there rather than what it looks like. Interactions between objects should be described by functional models rather than by physical ones. The computation necessary to translate between these higher-level representations and the lower-level representations required for direct user interaction is an essentially local function. At the local processor, display-rendering techniques may be arbitrarily elaborate and physical models arbitrarily sophisticated. The data channel capacities required for such computations, however, need not and should not be squeezed into the limited bandwidth available between the local processor and remote ones. Attempting to do so just leads to disasters such as NAPLPS (ANSI 1983, Alber 1985), which couples dreadful performance with a display model firmly anchored in the technology of the 1970s.

Once we began working at the conceptual rather than the presentation level, we were struck by the following observation:

- *The implementation platform is relatively unimportant.*

The presentation level and the conceptual level cannot (and should not) be *totally* isolated from each other. However, defining a cyberspace in terms of the configuration and behavior of objects, rather than their presentation, enables us to span a vast range of computational and display capabilities among the participants in a system. This range extends both upward and downward. As an extreme example, a typical scenic object, such as a tree, can be represented by a handful of parameter values. At the lowest conceivable end of things might be an ancient Altair 8800 with a 300-baud ASCII dumb terminal, where the interface is reduced to fragments of text and the user sees the humble string so familiar to the players of text adventure games: "There is a tree here." At the high end, you might have a powerful processor that generates the image of the tree by growing a fractal model and rendering it three dimensions at high resolution, the finest details ray-traced in real time, complete with branches waving in the breeze and the sound of wind in the leaves coming through your headphones in high-fidelity digital stereo. And these two users might be looking at the same tree in the same place in the same world and talking to each other as they do so. Both of these scenarios are implausible at the moment, the first because nobody would suffer with such a crude interface when better ones are so readily available, the second because the computational hardware does not yet exist. The point, however, is that this approach covers the ground between systems already obsolete and ones that are as yet gleams in their designers' eyes. Two consequences of this are significant. The first is that we can build effective cyberspace systems today. Habitat exists as ample proof of this principle. The second is that it is conceivable that with a modicum of cleverness and foresight you could start building a system with today's technology that could evolve smoothly as tomorrow's technology develops. The availability of pathways for growth is important in the real world, especially if cyberspace is to become a significant communications medium (as we obviously think it should).

Given that we see cyberspace as fundamentally a communications medium rather than simply a user interface model, and given the style of object-oriented approach that we advocate, another point becomes clear:

• *Data communications standards are vital.*

However, our concerns about cyberspace data communications standards center less upon data transport protocols than upon the definition of the data being transported. The mechanisms required for reliably getting bits from point A to point B are not terribly interesting to us. This is not because these mechanisms are not essential (they obviously are) nor because they do not pose significant research and engineering challenges (they clearly do). It is because we were focused on the unique communications needs of an object-based cyberspace. We were concerned with the protocols for sending messages between objects, that is, for communicating behavior rather than presentation, and for communicating object definitions from one system to another.

Communicating object definitions seems to us to be an especially important problem, and one that we really did not have an opportunity to address in Habitat. It *will* be necessary to address this problem if we are to have a dynamic system in the future. Once the size of the system's user base has grown modestly large, it becomes impractical to distribute a new release of the system software every time one wants to add a new class of object. However, we feel the ability to add new classes of objects over time is crucial if the system is to be able to evolve.

While we are on the subject of communications standards, we would like to make some remarks about the ISO Reference Model of Open System Interconnection (ISO 1986). This seven-layer model has become a centerpiece of most discussions about data communications standards today. It is so firmly established in the data communications standards community that it is virtually impossible to find a serious contemporary publication on the subject that does not begin with some variation on Figure 10.2. Unfortunately, while the bottom four or five layers of this model provide a more or less sound framework for considering data transport issues, we believe that the model's Presen-

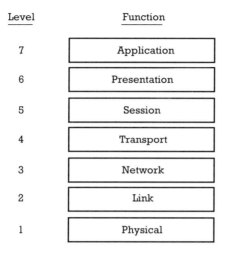

Level	Function
7	Application
6	Presentation
5	Session
4	Transport
3	Network
2	Link
1	Physical

Figure 10.2
The 7-layer ISO reference model of open system interconnection.

tation and Application layers are not very helpful when considering cyberspace data communications.

We have two main quarrels with the ISO model: first, it partitions the general data communications problem in a way that is a poor match for the needs of a cyberspace system; second, and more important, we think that the model itself is an active source of confusion because it focuses the attention of system designers on the wrong set of issues and thus leads them to spend their time solving the wrong set of problems. We know because this happened to us. "Presentation" and "Application" are simply the wrong abstractions for the higher levels of a cyberspace communications protocol. A "Presentation" protocol presumes that at least some characteristics of the display are embedded in the protocol. The discussions above should give some indication why we think that such a presumption is both unnecessary and unwise. Certainly, an "Application" protocol presumes a degree of foreknowledge of the message environment that is incompatible with the sort of dynamically evolving object system we envision.

A better model would be to substitute a different pair of top layers (Figure 10.3): a Message layer, which defines the means by which objects can address one another and standard methods of encapsulating structured data and encoding low-level data types (numbers); and a Definition layer built on top of the Message layer, which defines a

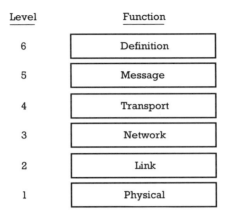

Level	Function
6	Definition
5	Message
4	Transport
3	Network
2	Link
1	Physical

Figure 10.3
A possible alternative protocol model.

standard representation for object definitions so that object classes can migrate from machine to machine. One might argue that these are simply Presentation and Application with different labels. However, the differences are so easily reconciled. In particular, we think the ISO model has, however unintentionally, systematically deflected workers in the field from considering many of the issues that concern us.

World Building

There were two sorts of implementation challenges that Habitat posed. The first was the challenge of creating a working piece of technology—developing the animation engine, the object-oriented virtual memory, the message-passing pseudo operating system, and squeezing them all into the ludicrous Commodore 64 (the backend system also posed interesting technical problems, but its constraints were not as vicious). The second challenge was the creation and management of the Habitat world itself. It is the experiences from the latter exercise that we think will be most relevant to future cyberspace designers.

Initially, we were our own worst enemies in this undertaking, victims of a way of thinking to which all engineers are dangerously susceptible. This way of thinking is characterized by the conceit that all things may be planned in advance and then directly implemented according to the plan's detailed specification. For persons schooled in the design and construction of systems based on simple, well-defined, and well-understood foundation principles, this is a natural attitude to have.

Moreover, it is entirely appropriate when undertaking most engineering projects. It is a frame of mind that is an essential part of a good engineer's conceptual tool kit. Alas, in keeping with Maslow's assertion that "to the person who has only a hammer, all the world looks like a nail," it is a frame of mind that is easy to carry beyond its range of applicability. This happens when a system exceeds the threshold of complexity above which the human mind loses its ability to maintain a complete and coherent model.

One generally hears about systems crossing the complexity threshold when they become very large. For example, the Space Shuttle and the B-2 bomber are both systems above this threshold, necessitating extraordinarily involved, cumbersome, and time-consuming procedures to keep the design under control—procedures that are at once vastly expensive and only partially successful. To a degree, the complexity of a problem can be dissolved by "throwing money" at it: faster computers, more managers, more bureaucratic procedures, and so on. However, such capital-intensive management techniques are a luxury not available to most projects. Furthermore, although these "solutions" to the complexity problem may be out of reach of most projects, alas the complexity threshold itself is not. Smaller systems can suffer from the same sorts of problems. It is possible to push much smaller and less elaborate systems over the complexity threshold simply by introducing chaotic elements that are outside the designers' sphere of control or understanding. The most significant of such chaotic elements are autonomous computational agents (other computers). This is why, for example, debugging even very simple communications protocols often proves surprisingly difficult. Furthermore, a special circle of living hell awaits the implementors of systems involving that most important category of autonomous computational agents of all: groups of interacting human beings. This leads directly to our next (and possibly most controversial) assertion:

• *Detailed central planning is impossible; don't even try.*

The constructivist prejudice that leads engineers into the kinds of problems just mentioned has received more study from economists, philosophers, and sociologists (Popper 1962, 1972; Hayek 1973, 1978, 1989; Sowell 1987) than from researchers in the software engineering community. Game and simulation designers are experienced in creating closed virtual worlds for individuals and small groups. However, they have had no reason to learn to deal with large populations of simultaneous users. Each user or small group is unrelated to the others, and the same world can be used over and over again. If you are playing an adventure game, the fact that thousands of other people elsewhere in the (real) world are playing the same game has no effect on your experience. It is reasonable for the creator of such a world to spend tens or even hundreds of hours crafting the environment for each hour that a user will spend interacting with it, since that user's hour of experience will be duplicated tens of thousands of times by tens of thousands of other individual users.

Builders of today's on-line services and communications networks are experienced in dealing with large user populations, but they do not, in general, create elaborate environments. Furthermore, in a system designed to deliver information or communications services, large numbers of users are simply a load problem rather than a complexity problem. All users get the same information or services; the comments in the previous paragraph regarding duplication of experience apply here as well. It is not necessary to match the size and complexity of the information space to the size of the user population. While it may turn out that the quantity of information available on a service is largely a function of the size of the user population itself, this information can generally be organized into a systematic structure that can still be maintained by a few people. The bulk of this information is produced by the users themselves, rather than the system designers. (This observation, in fact, is the first clue to the solution to our problem.)

Our original, contractual specification for Habitat called for us to create a world capable of supporting a population of 20,000 Avatars, with expansion plans for up to 50,000. By any reckoning this was a large undertaking and complexity problems would certainly be expected. However, in practice we exceeded the complexity threshold very early in development. By the time the population of our on-line community had reached around 50 we were in over our heads (and these 50 were "insiders" who were prepared to be tolerant of holes and rough edges).

Moreover, a virtual world such as Habitat needs to scale with its population. For 20,000 Avatars we needed 20,000 "houses," organized into towns and cities with associated traffic arteries and shopping and recreational areas. We needed wilderness areas between the towns so that everyone would not be jammed together into the same place. Most of all, we needed things for 20,000 people to do. They needed interesting places to visit—and since they can't all be in the same place at the same time, they needed a *lot* of interesting places to visit—and things to do in those places. Each of those houses, towns, roads, shops, forests, theaters, arenas, and other places is a distinct entity that someone needs to design and create. Attempting to play the role of omniscient central planners, we were swamped.

Automated tools may be created to aid the generation of areas that naturally possess a high degree of regularity and structure, such as apartment buildings and road networks. We created a number of such tools, whose spiritual descendents will no doubt be found in the standard bag of tricks of future cyberspace architects. However, the very properties that make some parts of the world amenable to such techniques also make those same parts of the world among the least important. It is really not a problem if every apartment building looks pretty much like every other. It is a big problem if every enchanted forest looks the same. Places whose value lies in their uniqueness, or at least in their differentiation from the places around them, need to be crafted by hand. This is an incredibly labor-intensive and time-consuming process. Furthermore, even very imaginative people are limited in the range of variation that they can produce, especially if they are working in a virgin environment uninfluenced by the works and reactions of other designers.

Running the World

The world design problem might still be tractable, however, if all players had the same goals, interests, motivations, and types of behavior. Real people, however, are all different. For the designer of an ordinary game or simulation, human diversity is not a major problem, since he or she gets to establish the goals and motivations on the participants' behalf, and to specify the activities available to them in order to channel events in the preferred direction. Habitat, however, was deliberately open-ended and pluralistic. The idea behind our world was precisely that it did not come with a fixed set of objectives for its inhabitants, but rather provided a broad palette of possible activities from which the players could choose, driven by their own internal inclinations. It was our intention to provide a variety of possible experiences, ranging from events with established rules and goals (a treasure hunt, for example) to activities propelled by the players' personal motivations (starting a business, running the newspaper) to completely free-form, purely existential activities (hanging out with friends and conversing). Most activities, however, involved some degree of planning and setup on our part. We were to be like the cruise director on a ocean voyage, but it turned out we were still thinking like game designers.

The first goal-directed event planned for Habitat was a rather involved treasure hunt called the "D'nalsi Island Adventure." It took us hours to design, weeks to build (including a 100-region island), and days to coordinate the actors involved. It was designed much like the puzzles in an adventure game. We thought it would occupy our players for days. In fact, the puzzle was solved in about 8 hours by a person who had figured out the critical clue in the first 15 minutes. Many of the players hadn't even had a chance to get into the game. The result was that one person had had a wonderful experience, dozens of others were left bewildered, and a huge investment in design and setup time had been consumed in an eye blink. We expected that there would be a wide range of "adventuring" skills in the Habitat audience. What wasn't so obvious until afterward was that this meant that most people didn't have a very good time, if for no other reason than that they never really got to participate. It would clearly be foolish and impractical for us to do things like this on a regular basis.

Again and again we found that activities based on often unconscious assumptions about player behavior had completely unexpected outcomes (when they were not simply outright failures). It was clear that we were not in control. The more people we involved in something, the less in control we were. We could influence things, we could set up interesting situations, we could provide opportunities for things to happen, but we could not predict or dictate the outcome. Social engineering is, at best, an inexact science, even in protocyberspaces. Or, as some wag once said, "In the most carefully constructed experiment under the most carefully controlled conditions, the organism will do whatever it damn well pleases."

Propelled by these experiences, we shifted into a style of operations in which we let the players themselves drive the direction of the design. This proved far more effective. Instead of trying to push the community in the direction we thought it should go, an exercise rather like herding mice, we tried to observe what people were doing and aid them in it. We became facilitators as much as designers and implementors. This often meant adding new features and new regions to the system at a frantic pace, but almost all of what we added was used and appreciated, since it was well matched to people's needs and desires. As the experts on how the system worked, we could often suggest new activities for people to try or ways of doing things that people might not have thought of. In this way we were able to have considerable influence on the system's development in spite of the fact that we didn't really hold the steering wheel—more influence, in fact, than we had had when we were operating under the delusion that we controlled everything.

Indeed, the challenges posed by large systems in general are prompting some researchers to question the centralized, planning-dominated attitude that we have criticized here, and to propose alternative approaches based on evolutionary and market principles (Miller and Drexler 1988a, 1988b; Drexler and Miller 1988). These principles appear applicable to complex systems of all types, not merely those involving interacting human beings.

The Great Debate

Among the objects we made available to Avatars in Habitat were guns and various other sorts of weapons. We included these because we felt that players should be able to "materially" effect each other in ways that went beyond simply talking, ways that required real moral choices to be made by the participants. We recognized the age-old storyteller's dictum that conflict is the essence of drama. Death in Habitat was, of course, not like death in the real world! When an Avatar is killed, he or she is teleported back home, head in hands (literally), pockets empty, and any object in hand at the time dropped on the ground at the scene of the crime. Any possessions carried at the time are lost. It was more like a setback in a game of "Chutes and Ladders" than real mortality. Nevertheless, the death metaphor had a profound effect on people's perceptions. This potential for murder, assault, and other mayhem in Habitat was, to put it mildly, controversial. The controversy was further fueled by the potential for lesser crimes. For instance, one Avatar could steal something from another Avatar simply by snatching the object out its owner's hands and running off with it.

We had imposed very few rules on the world at the start. There was much debate among the players as to the form that Habitat society should take. At the core of much of the debate was an unresolved philosophical question: Is an Avatar an extension of a human being (thus entitled to be treated as you would treat a real person) or a Pac-Man-like critter destined to die a thousand deaths or something else entirely? Is Habitat murder a crime? Should all weapons be banned? Or is it all "just a game"? To make a point, one of the players took to randomly shooting people as they roamed around. The debate was sufficiently vigorous that we took a systematic poll of the players. The result was ambiguous: 50 percent said that Habitat murder was a crime and shouldn't be a part of the world, while the other 50 percent said it was an important part of the fun.

We compromised by changing the system to allow thievery and gunplay only outside the city limits. The wilderness would be wild and dangerous while civilization would be orderly and safe. This did not resolve the debate, however. One of the outstanding proponents of the antiviolence point of view was motivated to open the first Habitat church, the Order of the Holy Walnut (in real life he was a Greek Orthodox priest). His canons forbid his disciples to carry weapons, steal, or participate in violence of any kind. His church became quite popular and he became a very highly respected member of the Habitat community.

Furthermore, while we had made direct theft impossible, one could still engage in indirect theft by stealing things set on the ground momentarily or otherwise left unattended. And the violence still possible in the outlands continued to bother some players. Many people thought that such crimes ought to be prevented or at least punished somehow, but they had no idea how to do so. They were accustomed to a world in which law and justice were always provided by somebody else. Somebody eventually made the suggestion that there ought to be a Sheriff. We quickly figured out how to create a voting mechanism and rounded up some volunteers to hold an election. A public debate in the town meeting hall was heavily attended, with the three Avatars who had chosen to run making statements and fielding questions. The election was held, and the town of Populopolis acquired a Sheriff.

For weeks the Sheriff was nothing but a figurehead, though he was a respected figure and commanded a certain amount of moral authority. We were stumped about what powers to give him. Should he have the right to shoot anyone anywhere? Give him a more powerful gun? A magic wand to zap people off to jail? What about courts? Laws? Lawyers? Again we surveyed the players, eventually settling on a set of questions that could be answered via a referendum. Unfortunately, we were unable to act on the results before the pilot operations ended and the version of the system in which these events took place was shut down. It was clear, however, that there are two basic camps: anarchists and statists. This division of characters and world views is an issue that will need to be addressed by future cyberspace architects. However, our view remains that a virtual world need not be set up with a "default" government, but can instead evolve one as needed.

A Warning

Given the above exhortation that control should be released to the users, we need to inject a note of caution and present our next assertion:

• *You can't trust* anyone.

This may seem like a contradiction of much of the preceding, but it really is not. Designers and operators of a cyberspace system must inhabit two levels of "virtuality" at once. The first we call the "infrastructure level," the level of implementation, where the laws that govern "reality" have their genesis. The second we call the "experiential level," which is what the users see and interact with. It is important that there not be "leakage" between these two levels. The first level defines the physics of the world. If its integrity is breached, the consequences can range from aesthetic unpleasantness (the audience catches a glimpse of the scaffolding behind the false front) to psychological disruption (somebody does something "impossible," thereby violating users' expectations and damaging their fantasy) to catastrophic failure (somebody crashes the system). When we exhort cyberspace system designers to give control to the users, we mean control at the experiential level. When we say that you can't trust anyone, we mean that you can't trust them with access to the infrastructure level. Some stories from Habitat will illustrate this.

When designing a piece of software, you generally assume that the software is the sole intermediary between the user and the underlying data being manipulated (possibly multiple applications will work with the same data, but the principle remains the same). In general, the user need not be aware of how data are encoded and structured inside the application. Indeed, the very purpose of a good application is to shield the user from the ugly technical details. While it is conceivable that a technically astute person who is willing to invest the time and effort could decipher the internal structure of things, this would be an unusual thing to do as there is rarely much advantage to be gained. The purpose of the application itself is, after all, to make access to and manipulation of the data easier than digging around at the level of bits and bytes. There are exceptions to this, however. For example, most game programs deliberately impose obstacles on their players in order for play to be challenging. By tinkering around with the insides of such a program—dumping the data files and studying them, disassembling the program itself and possibly modifying it—it may be possible to "cheat." However, this sort of cheating has the flavor of cheating at solitaire: the consequences adhere to the cheater alone. There is a difference, in that disassembling a game program is a puzzle-solving exercise in its own right, whereas cheating at solitaire is pointless, but the satisfactions to be gained from either, if any, are entirely personal.

If, however, a computer game involves multiple players, then delving into the program's internals can enable one to truly cheat, in the sense that one gains an unfair advantage over the other players, an advantage moreover of which they may be unaware. Habitat is such a multiplayer game. When we were designing the software, our "prime directive" was, "The backend shall not assume the validity of anything a player computer tells it." This is because we needed to protect ourselves against the possibility that a clever user had hacked around with his copy of the frontend program to add "custom features." For example, we could not implement any of the sort of "skill and action" elements found in traditional video games wherein dexterity with the joystick determines the outcome of, say, armed combat, because we couldn't guard against users modifying their copy of the program to tell the backend that they had "hit," whether they actually had or not. Indeed, our partners at QuantumLink warned us of this very eventuality before we even started—they already had users who did this sort of thing with their regular system. Would anyone actually go to the trouble of disassembling and studying 100K or so of incredibly tight and bizarrely threaded 6502 machine code just to tinker? As it turns out, the answer is yes.

People did. We were not 100 percent rigorous in following our own rule. It turned out that there were a few features whose implementation was greatly eased by breaking the rule in situations where, in our judgment, the consequences would not be material if some people "cheated" by hacking their own systems. Darned if some people didn't hack their systems to cheat in exactly these ways.

Care must be taken in the design of the world as well. One incident that occurred during our pilot test involved a small group of players exploiting a bug in our world database that they interpreted as a feature. First, some background. Avatars were hatched with 2000 Tokens in their bank account, and each day that they logged in the received another 100T. Avatars could acquire additional funds by engaging in business, winning contests, finding buried treasure, and so on. They could spend their Tokens on, among other things, various items for sale in vending machines called Vendroids. There were also Pawn Machines, which would buy objects back (at a discount, of course).

In order to make this automated economy a little more interesting, each Vendroid had its own prices for the items in it. This was so that we could have local price variation (a widget would cost a little less if you bought it at Jack's Place instead of The Emporium). It turned out that in two Vendroids across town from each other were two items for sale whose prices we had inadvertently set lower than what a Pawn Machine would buy them back for: Dolls (for sale at 75T, hock for 100T) and Crystal Balls (for sale at 18,000T, hock at 30,000T!). Naturally, a couple of people discovered this. One night they took all their money, walked to the Doll Vendroid, bought as many Dolls as they could, then took them across town and pawned them. By shuttling back and forth between the Doll Vendroid and the Pawn Shop for *hours*, they amassed sufficient funds to buy a Crystal Ball, whereupon they continued the process with Crystal Balls and a couple orders of magnitude higher cash flow. The final result was at least three Avatars with hundreds of thousands of Tokens each. We only discovered this the next morning when our daily database status report said that the money supply had quintupled overnight.

We assumed that the precipitous increase in "T1" was due to some sort of bug in the software. We were puzzled that no bug report had been submitted. By poking around a bit we discovered that a few people had suddenly acquired enormous bank balances. We sent Habitat mail to the two richest, inquiring as to where they had gotten all that money overnight. Their reply was, "We got it fair and square! And we're not going to tell you how!" After much abject pleading on our part they eventually did tell us, and we fixed the erroneous pricing. Fortunately, the whole scam turned out well, as the nouveau riche Avatars used their bulging bankrolls to underwrite a series of treasure hunt games that they conducted on their own initiative, much to the enjoyment of many other players on the system.

Keeping "Reality" Consistent

The urge to breach the boundary between the infrastructure level and the experiential level is not confined to the players. The system operators are also subject to this temptation, though their motivation is expediency in accomplishing their legitimate purposes rather than gaining illegitimate advantage. However, to the degree to which it is possible, we vigorously endorse the following principle:

• *Work within the system.*

Wherever possible, things that can be done within the framework of the experiential level should be. The result will be smoother operation and greater harmony among the user community. This admonition applies to both the technical and the sociological aspects of the system.

For example, with the players in control, the Habitat world would have grown much larger and more diverse than it did had we ourselves not been a technical bottleneck. All new region generation and feature implementation had to go through us, since there were no means for players to create new parts of the world on their own. Region creation was an esoteric technical specialty, requiring a plethora of obscure tools and a good working knowledge of the treacherous minefield of limitations imposed by the Commodore 64. It also required much behind-the-scenes activity of the sort that would probably spoil the illusion for many. One of the goals of a next generation Habitat-like system ought to be to permit far greater creative involvement by the participants *without* requiring them to ascend to full-fledged guruhood to do so.

A further example of working within the system, this time in a social sense, is illustrated by the following experience:

One of the more popular events in Habitat took place late in the test, the brainchild of one of the more active players who had recently become a QuantumLink employee. It was called the "Dungeon of Death." For weeks, ads appeared in Habitat's newspaper, *The Rant*, announcing that the Duo of Dread, DEATH and THE SHADOW, were challenging all comers to enter their lair. Soon, on the outskirts of town, the entrance to a dungeon appeared. Out front was a sign reading, "Danger! Enter at your own risk!" Two system operators were logged in as DEATH and THE SHADOW, armed with specially concocted guns that could kill in one shot, rather than the usual twelve. These two characters roamed the dungeon blasting away at anyone they encountered. They were also equipped with special magic wands that cured any damage done to them by other Avatars, so that they wouldn't themselves be killed. To make things worse, the place was littered with cul-de-sacs, pathological connections between regions, and various other nasty and usually fatal features. It was clear that any explorer had better be prepared to "die" several times before mastering the dungeon. The rewards were pretty good: 1000 Tokens minimum and access to a special Vendroid that sold magic teleportation wands. Furthermore, given clear notice, players took the precaution of emptying their pockets before entering, so that the actual cost of getting "killed" was minimal.

One evening, one of us was given the chance to play the role of DEATH. When we logged in, we found him in one of the dead ends with four other Avatars who were trapped there. We started shooting, as did they. However, the last operator to run DEATH had not bothered to use his special wand to heal any accumulated damage, so the character of DEATH was suddenly and unexpectedly "killed" in the encounter. As we mentioned earlier, when an Avatar is killed, any object in his hands is dropped on the ground. In this case, said object was the special kill-in-one-shot gun, which was immediately picked up by one of the regular players who then made off with it. This gun was not something that regular players were supposed to have. What should we do?

It turned out that this was not the first time this had happened. During the previous night's mayhem the special gun was similarly absconded with. In this case, the person playing DEATH was one of the regular system operators, who, accustomed to operating the regular Q-Link service, had simply ordered the player to give back the gun. The player considered that he had obtained the weapon as part of the normal course of the game and balked at this, whereupon the operator threatened to cancel the player's account and kick him off the system if he did not comply. The player gave the gun back, but was quite upset about the whole affair, as were many of his friends and associates on the system. Their world model had been painfully violated.

When it happened to us, we played the whole incident within the role of DEATH. We sent a message to the Avatar who had the gun, threatening to come and kill her if she didn't give it back. She replied that all she had to do was stay in town and DEATH couldn't touch her (which was true, if we stayed within the system). OK, we figured, she's smart. We negotiated a deal whereby DEATH would ransom the gun for 10,000 Tokens. An elaborate arrangement was made to meet in the center of town to make the exchange, with a neutral third Avatar acting as an intermediary to ensure that neither party cheated. Of course, word got around and by the time of the exchange there were numerous spectators. We played the role of DEATH to the hilt, with lots of hokey melodramatic touches. The event was a sensation. It was written up in the newspaper the next morning and was the talk of the town for days. The Avatar involved was left with a wonderful story about having cheated DEATH, we got the gun back, and everybody went away happy.

These two very different responses to an ordinary operational problem illustrate our point. Operating within the participants' world model produced a very satisfactory result. On the other hand, taking what seemed like the expedient course, which involved violating the world model, provoked upset and dismay. Working within the system was clearly the preferred course in this case.

Current Status

As of this writing, the North American incarnation of Lucasfilm's Habitat, QuantumLink's Club Caribe, has been operating for almost three years. It uses our original Commodore 64 frontend and a somewhat stripped-down version of our original Stratus backend software. Club Caribe now sustains a population of some 15,000 participants.

A technically more advanced version, called Fujitsu Habitat, has been operating for over a year in Japan, available on NIFtyServe. The initial frontend for this version is the new Fujitsu FM Towns personal computer, though ports to several other popular Japanese machines are planned. This version of the system benefits from the additional computational power and graphics capabilities of a newer platform, as well as the Towns' built-in CD-ROM for object imagery and sounds. However, the virtuality of the system is essentially unchanged and Fujitsu has not made significant alterations to the user interface or to any of the underlying concepts.

Future Directions

There are several directions in which this work can be extended. Most obvious is to implement the system on more advanced hardware, enabling a more sophisticated display. A number of extensions to the user interface also suggest themselves. However, the line of development most interesting to us is to expand on the idea of making the development and expansion of the world itself part of the users' sphere of control. There are two major research areas in this. Unfortunately, we can only touch on them briefly here.

The first area to investigate involves the elimination of the centralized backend. The backend is a communications and processing bottleneck that will not withstand growth above too large a size. While we can support tens of thousands of users with this model, it is not really feasible to support millions. Making the system fully distributed, however, requires solving a number of difficult problems. The most significant of these is the prevention of cheating. Obviously, the owner of the network node that implements some part of the world has an incentive to tilt things in his favor. We think that this problem can be addressed by secure operating system technologies based on public-key cryptographic techniques (Rivest, Shamir, and Adelman 1978; Miller et al. 1987).

The second fertile area of investigation involves user configuration of the world itself. This requires finding ways to represent the design and creation of regions and objects as part of the underlying fantasy. Doing this will require changes to our conception of the world. In particular, we don't think it will be possible to conceal all of the underpinnings to those who work with them. However, all we really need to do is to find abstractions for those underpinnings that fit into the fantasy itself. Though challenging, this is, in our opinion, eminently feasible.

Conclusion

We feel that the defining characteristic of cyberspace is the sharedness of the virtual environment, and not the display technology used to transport users into that environment. Such a cyberspace is feasible today, if you can live without head-mounted displays and other expensive graphics hardware. Habitat serves as an existence proof of this contention.

It seems clear to us that an object-oriented world model is a key ingredient in any cyberspace implementation. We feel that we have gained some insight into the data representation and communications needs of such a system. While we think that it may be premature to start establishing detailed technical standards for these things, it is time to begin the discussions that will lead to such standards in the future.

Finally, we have come to believe that the most significant challenge for cyberspace developers is to come to grips with the problems of world creation and management. While we have only made the first inroads into these problems, a few things have become clear. The most important of these is that managing a cyberspace world is not like managing the world inside a single-user application or even a conventional on-line service. Instead, it is more like governing an actual nation. Cyberspace architects will benefit from study of the principles of sociology and economics as much as from the principles of computer science. We advocate an agoric, evolutionary approach to world building rather than a centralized, socialistic one.

We would like to conclude with a final, if ironical, admonition, one that we hope will not be seen as overly contentious:

• *Get real.*

In a discussion of cyberspace on Usenet, one worker in the field dismissed Club Caribe (Habitat's current incarnation) as uninteresting, with a comment to the effect that most of the activity consisted of inane and trivial conversation. Indeed, the observation was largely correct. However, we hope some of the anecdotes recounted above will give some indication that more is going on than those inane and trivial conversations might indicate. Further, to dismiss the system on this basis is to dismiss the users themselves. They are paying money for this service. *They* don't view what they do as inane and trivial, or they wouldn't do it. To insist this presumes that one knows better than they what they should be doing. Such presumption is another manifestation of the omniscient central planner who dictates all that happens, a role that this entire chapter is trying to deflect you from seeking. In a real system that is going to be used by real people, it is a mistake to assume that the users will all undertake the sorts of noble and sublime activities that you created the system to enable. Most of them will not. Cyberspace may indeed change humanity, but only if it begins with humanity as it really is.

Acknowledgments

We would like to acknowledge the contributions of some of the many people who helped make Habitat possible. At Lucasfilm, Aric Wilmunder wrote much of the Commodore 64 frontend software; Ron Gilbert, Charlie Kelner, and Noah Falstein also provided invaluable programming and design support; Gary Winnick and Ken Macklin were responsible for all the artwork; Chris Grigg did the sound; Steve Arnold provided outstanding management support; and George Lucas gave us the freedom to undertake a project that for all he knew was both impossible and insane. At Quantum, Janet Hunter wrote the guts of the backend; Ken Huntsman and Mike Ficco provided valuable assistance with communications protocols. Kazuo Fukuda and his crew at Fujitsu have carried our vision of Habitat to Japan and made it their own. Phil Salin, our boss at AMIX, let us steal the time to write this chapter and even paid for us to attend the First Conference on Cyberspace, even though its immediate relevance to our present business may have seemed a bit obscure at the time. We'd also like to thank Michael Benedikt, Don Fussell, and their cohorts for organizing the conference and thereby prompting us to start putting our thoughts and experiences into writing.

Notes

1. One of the questions we are asked most frequently is, "Why the Commodore 64?" Many people somehow get the impression that this was a technical decision, but the real explanation has to do with business, not technology. Habitat was initially developed by Lucasfilm as a commercial product for QuantumLink, an on-line service (then) exclusively for owners of the Commodore 64. At the time we started (1985), the Commodore 64 was the mainstay of the recreational computing market.

Since then it has declined dramatically in both its commercial and technical significance. However, when we began the project, we didn't get a choice of platforms. The nature of the deal was such that both the Commodore 64 for the frontend and the existing QuantumLink host system (a brace of Stratus fault-tolerant minicomputers) for the backend were givens.

2. Habitat contains its own fully-fledged economy, with money, banks, and so on. Habitat's unit of currency is the Token, reflecting the fact that it is a token economy and to acknowledge the long and honorable association between tokens and video games. Incidently, the Habitat Token is a 23-sided plastic coin slightly larger than an American quarter, with a portrait of Vernor Vinge and the motto "Fiat Lucre" on its face, and the text "Good for one fare" on the back; these details are difficult to make out on the Commodore 64 screen.

References

Abelson, Harold, and Sussman, Gerald Jay, *Structure and Interpretation of Computer Programs* (MIT Press, Cambridge, Massachusetts, 1985).

Alber, Antone F., *Videotex/Teletext: Principles and Practices* (McGraw-Hill, New York, 1985).

American National Standards Institute, *Videotex/Teletext Presentation Level Protocol Syntax*, North American PLPS (ANSI, December 1983).

Drexler, K. Eric, *Engines of Creation* (Anchor Press, Doubleday, New York, 1986).

Drexler, K. Eric, and Miller, Mark S., "Incentive Engineering for Computational Resource Management." In Huberman, B. A., ed., *The Ecology of Computation* (Elsevier Science Publishing, Amsterdam, 1988).

Gibson, William, *Neuromancer* (Ace Books, New York, 1984).

Goldberg, Adele, and Robson, David, *Smalltalk-80: The Language and Its Implementation* (Addison-Wesley, Reading, Massachusetts, 1983).

Hayek, Friedrich A., *Law Legislation and Liberty*. Vol. 1, *Rules and Order* (University of Chicago Press, Chicago, 1973).

Hayek, Friedrich A., *New Studies in Philosophy, Politics, Economics, and the History of Ideas* (University of Chicago Press, Chicago, 1978).

Hayek, Friedrich A., *The Fatal Conceit* (University of Chicago Press, Chicago, 1989).

International Standards Organization, *Information Processing Systems—Open System Interconnection—Transport Service Definition*, International Standard number 8072, (ISO, Switzerland, June 1986).

Miller, Mark S., Bobrow, Daniel G., Tribble, Eric Dean, and Levy, David Jacob, "Logical Secrets." In Shapiro, Ehud Y., ed., *Concurrent Prolog: Collected Papers*, 2 vols. (MIT Press, Cambridge, Massachusetts 1987).

Miller, Mark S., and Drexler, K. Eric, "Comparative Ecology: A Computational Perspective." In Huberman, B. A., ed., *The Ecology of Computation* (Elsevier Science Publishing, Amsterdam, 1988a).

Miller, Mark S., and Drexler, K. Eric, "Markets and Computation: Agoric Open Systems," In Huberman, B. A., ed., *The Ecology of Computation* (Elsevier Science Publishing, Amsterdam, 1988b).

Popper, Karl R., *The Open Society and Its Enemies,* 5th ed. (Princeton University Press, Princeton, New Jersey, 1962).

Popper, Karl R., *Objective Knowledge: An Evolutionary Approach* (Oxford University Press, Oxford, 1972).

Rivest, R., Shamir, A., and Adelman, L., "A Method for Obtaining Digital Signatures and Public-Key Cryptosystems." *Communications of the ACM* 21, no. 2 (February 1978).

Sowell, Thomas, *A Conflict of Visions* (William Morrow, New York, 1987).

Sterling, Bruce, ed., *Mirrorshades: The Cyberpunk Anthology* (Arbor House, New York, 1986).

Vinge, Vernor, "True Names," *Binary Star #5* (Dell Publishing Co., New York, 1981).

11 Collaborative Engines for Multiparticipant Cyberspaces

Carl Tollander

Introduction

Overview

A virtual world requires infrastructure; a substrate in which the integrity of its appearances and behaviors can be played out and extended within the context of individual expectations, needs, and desires. When the theater for this interaction is the computer, the limitations of any given machine become strong mediating influences for the collaborative efforts of the author-participants. The mechanisms used must therefore *simulate* the intentions of the participants when those participants are separated by distance, time, the focus of their attentions, and the differences between the computing resources available to them locally.

A *cyberspace* provides this infrastructure. It is a self-sustaining simulation environment in which the behavior of objects is determined within a context selected by the activity and point of view of a participant in real time. This differs from other kinds of simulations where the user attempts to be a detached observer and the point of view is primarily part of an interface to a set of rendering capabilities. Rather than having "users," cyberspaces have *participants*, each of which can be a human agent or computer program.

In this chapter, a *collaborative cyberspace* is one in which the sum of the actions of one or more participants and elements of the space select among possible categorizations of object behavior. This determines not only the range of possible behaviors of the object, but also how its attributes are modeled. A *collaborative engine* that recombines these categorizations to produce revised expectations of what a participant may experience will be described. These expectations are used to produce quasi-stable models, which can then be rendered.

In a collaborative cyberspace, directed perception (gaze) is an action in the space as much as navigation and manipulation are, and is treated by the engine in a similar manner. This has the effect of making the mechanisms used to author the space and the mechanisms used to inhabit and navigate within it identical. Participants and the collaborative engine mutually influence both situations and the semantics of and between objects. Creation, modification, and semantic linkage of objects occur in real time.

Within a collaborative cyberspace, we can capture one of the most salient aspects of "everyday reality"—namely, the closer you look, the more there is. The engine does not spend valuable system resources attempting to maintain parts of the space that participants don't care about (aside perhaps from some background physics). Resources are mobilized to expand on those portions where the participants are directing their attention. Unless explicitly saved, unattended objects simply fade away or lose detail, since the participants no longer "expect" them to be there. These objects can be recreated at a later time according to resident expectations at each participant's "deck." Since the focus of attention is different for each participant, each one may experience the world slightly differently.

Obstacles for Multiparticipant Cyberspaces

Efforts to construct multiple participant cyberspaces have had to confront a number of computational obstacles. These are difficult to overcome, and the usual global approaches are to give up real-time responsiveness or concentrate on fast I/O while limiting world complexity. In more detail, these obstacles are:

• Maintaining global consistency of centralized database is $O(n^2)$ in participants and objects. This computing power requirement limits the size of the overall cyberspace as well as the number of participants. The global approach does not "scale up" well.

• Even single-participant, noncollaborative approaches are subject to an explosion in the computational resources required for constraint satisfaction and maintenance between objects for any but the simplest of "toy" spaces. While a number of programming tricks and perceptual illusions have been used in the past to address highly specialized problems, general purpose solutions are elusive.

• Issues such as database divergence and real-time update problems for participants distributed over large geographic areas are limited by network performance. Availability of bandwidth in multiparticipant systems is difficult to predict very far in advance. Despite the increasing availabilty of advanced networks such as ISDN, it will still be hard to provide enough bandwidth to meet the needs of an increasing number of users and complex cyberspaces.

• The expected wide variety in rendering hardware, clothing, and modeling systems software limits interoperability in common spaces and interchangability of cyberspace artifacts.

• Existing models assume "mission-critical" applications, where all communications and states are held to a strong model of precision and logical consistency. Cyberspace is likely to be different. Most participants will seldom be concerned with criticality, any more than they are in "normal" conversation. We should not assume that the audience will be the same one that uses today's personal computers and software.

• Existing approaches do not adapt well to the scale of interaction. For example, it is difficult to have a consistent space that keeps you from walking through walls and allows you to fix watches at the same time. An engine is required that can alter the dynamics of the space and activities available as the participants' context of scale and modes of interaction change with time.

These problems raise many issues similar to those found in parallel and distributed systems. It is a central contention of this chapter that these problems are *not* merely implementation issues, but must be confronted early in the design stage of collaborative systems. While it is too early in the development of the collaborative cyberspace approach to say with confidence that we can solve all these problems, it is intended that the collaborative engine described here will at least allow us to begin to address them in a coherent manner.

Goals for Collaborative Engines

At this point, I will describe reasonable design goals for a collaborative cyberspace engine. Some of these goals are selected to help address the multiparticipant space problems listed above. Others are intended to facilitate development of the engine or make it available to a wider audience.

• The space should gain detail as you interact with it and focus attention on it.

• The level of detail in items in the space should be related to the scale of participation. For example, things in the "foreground" should have more detail than things in the "background," regardless of the sensory modalities involved.

• There should be a means to tune the ratio between responsiveness and resolution.

• The entry level cost should be low. There should be nothing in the design that prejudices it toward implementation on expensive hardware. The potential audience is much larger than the current computing community that might expect to pay a premium.

• Accordingly, increased capability should be added incrementally by merging existing modules. Users should be able to *compose* more capable decks by temporarily adding together existing decks.

• It is assumed that most participants will be mainly concerned with correlated models (dynamically and additively built and modified) rather than mission-critical models, where all object semantics of the world are known beforehand. Mission-critical models should be a special case of correlated models.

• Intermediate steps in the development of the system should have stand-alone uses and provide early payoffs. We should not have to wait for the full design to be realized before we can begin using its parts.

• Privileged roles in the space should be minimized. The collaborative space should be self-regulating; unification of interaction and creation should leave little need for the role of "dungeon master."

• Simple abstract interaction semantics are required for both human and machine participants. The number and complexity of operations should be minimized.

• Exchange of data between "decks" should be self-adapting to varying availability of bandwidth.

• Manipulations of a space should occur in real time with respect to the mechanisms used to render them.

Note in particular that we have avoided including interaction technologies such as helmets, gloves, or wands in any detail. These, along with any specific rendering technologies, are assumed to be under continuous development in parallel with the collaborative engine, but are outside the scope of this chapter. I attempt to make only the most minimal assumptions about them. While interaction technologies are extremely important to the eventual realization of cyberspaces, the primary emphasis at this stage should be to design a new collaborative media rather than to describe interaction and interface technologies.

In this chapter, I will outline a basic design for a collaborative cyberspace engine based on a theory of natural selection. Following that will be a discussion of the implications of such an approach. Some potential applications will be considered. I will conclude with a description of some of the major issues involved in the research and development of a collaborative engine. In the process, it will be suggested that one of the most commonly accepted features of cyberspace, that of a maintainable, shared, globally consistent and consensual environment faces severe difficulties and may have to be abandoned. I hope to show that a new vision of collaborative cyberspace is attainable and will make possible both a more interesting and a wider variety of applications.

Collaborative Cyberspace as a Selective System

Selective Systems

The literature of evolving systems is large and complex, giving rise to a variety of (sometimes conflicting) views of "natural" selection. The interested reader is referred to Allen and Starr 1982, Bobick 1987, Farmer et al. 1986, Dawkins 1987, Langton 1989, Ridley 1986, Weber et al. 1988 for a sampling of the diverse approaches to the issues involved. For the purposes of this chapter, a *selective system* is any multidimensional environment in which the behavior of entities and the relationships between them evolve according to situations in which those entities find themselves "nearby" each other. The meaning of "nearby," and indeed the dimensionality of the space itself is a function of meaningful categorizations (taxonomies) of the entities (further entities themselves) in the environment. These categorizations arise in concert with certain selective influences that will be discussed in more detail shortly. Where the environment is conceptual, the primary selective influence is that of one or more participants.

One of the salient features of our perception of everyday reality is the mutability of our categorizations. In this variant of the constructivist view, we create our realities "on the fly" by the activity of constantly categorizing and recategorizing low-level perceptions and sensory input. Standing in a forest clearing, one realizes that no single categorization of the surroundings can be adequate for any but the most minimal set of navigational paths or activities. Indeed, our ability to function in any environment is largely dependent on our ability to create new behaviors by recombining our responses and perspectives within a rapidly changing context. Our categorizations must *evolve* over time. We may say that given categorizations of our worlds are *selected for* from other potential sets of categories. Nonetheless, these sets exist only as general directions to be elaborated on when selected. There is no globally optimum set of categorizations at any particular instant, since the problem space is in constant flux.

In a cyberspace system based on selection, then, the representation of a given object is not created statically and stored for future use, but is *reconstructed* in real time by combining and recombining categorizations under the guidance of selective influences. These include recent context (existing semantics), the continued success of the prevailing categorizations (expectations) about the object's appearance and behavior against competing categories, and the interest of the participants as manifested through their directed attention to some subset of the existing possible expectations. Under these influences, objects in the cyberspace will evolve at varying rates. This will be explored in more detail later in this section, but first I will explore some of the theoretical underpinnings for what will come later.

Edelman's Theory of Neuronal Group Selection

Considerable work investigating selective mechanisms in the brain has been done by Gerald Edelman. Much of the work in this chapter owes a great deal to his theory of Neuronal Group Selection (NGS). At the core of his theory is the notion that many brain processes operate by natural selection in a kind of *neural darwinism*, and that this process is an extension of the same processes governing its formation and growth Edelman 1987. Specialization of cell function in the developing individual is determined by the characteristics of the neighborhood of other cells in which it finds itself. Cells in the developing nervous system tend to migrate to areas in the brain favorable to their further specialization.

In Edelman's theory, the unit of selection is the *neuronal group*. The brain contains large numbers of these groups of nerve cells. Each such group of neurons responds more or less specifically to a particular pattern of input. Groups can share input patterns, although their responses, even to identical stimuli, will always differ slightly due to differences in their structure and the current context of their connections to other groups. Neuronal groups that are strongly correlated in their responses within a given context form *repertoires* that respond more or less specifically to categories of input. The rates of change of *primary repertoires* slow down and their configurations become relatively fixed shortly after birth. *Secondary repertoires* are formed by changing the strength of connections between, within, and among primary repertoires due to the situational context of the organism, and are accordingly in constant flux. The formation of secondary repertoires, that is, the constant categorization and recategorization of phenomena in accord with changing contexts, is an important foundation of perception and memory.

The content of a secondary repertoire makes it respond more or less specifically in a particular context. The lack of specificity is called the *degeneracy* of the repertoire, and figures strongly in the timing of recombination. In a completely degenerate repertoire, all groups respond to all inputs and the repertoire fails to distinguish between signals. In a completely nondegenerate repertoire, the groups are so specific that there would need to be an infinite number of them to respond to all possible nearby contexts (see Figure 11.1). A stable situation occurs when the degeneracy is somewhere in between these extremes. In this case, the repertoire's categorization of the situation tends to be more successful.

A distinguishing characteristic of stable repertoires in general is their tendency to grow by reorganizing neural groups from other, less stable repertoires, thus producing new categorizations that may further success. A kind of competition ensues, with more or less stable repertoires raiding less stable repertoires for neural groups. A growing repertoire is frequently stimulated, and its intergroup connections become stronger as a result. A repertoire that cannot successfully organize other

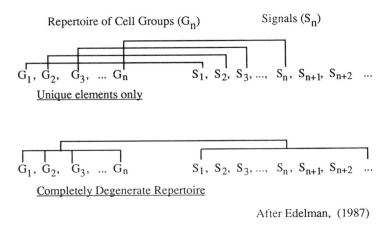

Figure 11.1
Specificity and degeneracy of repertoires.

groups is stimulated less frequently and its cohesiveness decreases. Its member groups can then be colonized by other repertoires.

Repertoires are self-pruning; they do not grow without limit. If one becomes too degenerate or too specific, it will be less likely to attract new groups and may "evaporate." A completely specific repertoire has no common basis for correlation. It will "shed" groups until a stronger correlation between those remaining can occur. By contrast, a completely degenerate repertoire does nothing "interesting," and so is less likely to form correlations with other groups or otherwise be selected. The cohesiveness of secondary repertoires decays with time, so if it is not reinforced, a given categorization tends to fade away. In this way, the very opportunism that enables that categorization to form in the first place creates a limit on its extent.

While secondary repertoires arise and fade away, it is proper to think of the space of repertoires as evolving under the guidance of a set of selective influences. This is possible due to the very large number of neuronal groups involved; evolution is to a large extent a population effect. Evolving dynamic systems are often said to be *self-organizing*. Although there are many other factors involved, in general the probability that such systems will produce islands of stability (reasonable categorizations) increases with the size and variability of the system. In short, they scale up well.

The introduction of variation into an evolving system based on Edelman's model becomes important to keep the set of repertoires from the extremes of specificity and degeneracy. The importance of variation in self-organizing systems was first pointed out by Ashby (1956). More recently, Mpitsos et al. (1989) has suggested that neural group variability may be central to the ability of even simple animals to adapt to complex and unpredictable environments. Variation thus has an important consequence with respect to self-organization in NGS theory. It provides for the long-term stability of behaviors as adaptive entities by keeping them from becoming too specific. This stability of more basic behaviors increases the probability that a proven constructed adaptation can be quickly reconstructed again in a similar manner in similar circumstances. Proven behaviors need not be stored in detail, since they are not, strictly speaking, *retrieved*.

Implications of Edelman's NGS Theory for Collaborative Cyberspaces

Recall that a cyberspace is basically a simulation, where the model that is being simulated is highly dynamic and controlled in real time by a participant. The model is composed of and created by participants' *expectations*. A participant expects an object to behave in a certain way, or to have a certain appearance, and further that these expectations relate in some way to those about other objects in the space. Expectations extend to whether an object will possess a certain quality or attribute. They are formed by the interplay between the relative scale of objects, the prior intentions of the participants, and the resolution at which the objects are rendered. From a data representation standpoint, these in turn are functions of what characteristics of particular objects are *more or less* present. An object, then, can be represented as a more or less persistent *fuzzy set* (Zadeh 1973, Tong 1985) of expectations (see Figures 11.2a, 11.2b), which can have varying degrees of membership in other objects. If the degree of membership is below a certain point, it is less likely to be modeled in the space if time or computing resources are at a premium. This might affect the depth and kind of simulation used—Monte Carlo, event-based, actor-based, or others.

Even though Edelman's work is intended to explain neural processes of memory and perception, much of his theory can be applied to collaborative cyberspaces. Neuronal groups and repertoires implement the objects of a cyberspace. A major divergence of this work from the NGS theory is that the expectations making up an object or a space can

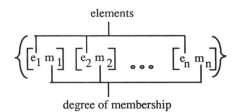

Figure 11.2a
A fuzzy set.

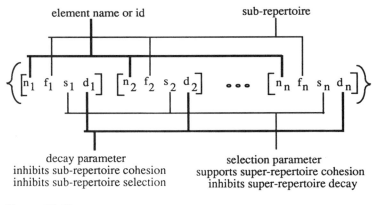

Figure 11.2b
A fuzzy set representation for a repertoire.

be nested to arbitrary levels. In a collaborative cyberspace, the notions of primary and secondary repertoires refer to relative measures of the degree of persistence of an expectation over time. Objects can operate as both primary and secondary repertoires as the simulation progresses. The dynamics of object creation and internal manipulation are identical to the space in which it is embedded, thus *every object in a cyberspace is itself a cyberspace.* (In the remainder of the chapter, I will use the terms *object, repertoire,* and *cyberspace* more or less interchangably, depending on context). We will see later that this deep mutability at all levels of simulation will greatly simplify the semantics of participant interaction.

Since each repertoire contains a number of entities (neuronal groups or other repertoires) that respond more or less specifically to a set of signals, each repertoire can be said to be a *classifier* of those signals (here a signal is the detection of some activity within some other repertoire).

Action at the level of the secondary repertoire is to categorize and recategorize the classifiers of the primary repertoire. As we have identified repertoires with objects/cyberspaces, the usual distinctions between perception and description are relaxed. To perceive an object in a collaborative cyberspace is in large part to *construct* it, and vice versa. As we saw in the section above describing NGS, it is possible (at least in biological systems) for variation to contribute to the stability of a category of possible objects, such that they can be reconstructed as needed, rather than stored. Objects are not necessarily constructed the same way twice. Edelman (1990) might say that in the brain, we do not *remember* anything the same way twice.

A study of NGS theory thus has several implications for the design of cyberspace systems. First, the notion of repertoires helps bridge the traditional gap between perceiver and perceived in simulation systems by stressing the perceiver's role in producing objects as categories of phenomena. Second, it suggests a means for computer representation of objects within a cyberspace. Third, it points the way to a means by which databases of objects can remain stable over time while under the influence of a variety of diverse participant actions.

The Collaborative Engine

This brings us to the notion of the *collaborative engine*. While the engine does not exist as a centralized mechanism, it is associated with collections of repertoires called *consoles*. A console is simply a boundary drawn around a group of repertoires by a participant that isolates that group from other such groups. Communication between consoles occurs via a propagation mechanism described below. A *deck* is a locus of physical interaction with human participants and can be associated with any number of consoles. Decks contain rendering mechanisms that convert expectations into renderable models, as well as the hardware to actually perform interaction with human users. The precise mechanisms for multisensory rendering are a separate problem outside the scope of this chapter. It is assumed that interaction technologies, such as advanced forms of eye phones and data gloves, will develop separately.

Note that the agent of attention can either be a human user or a computer program. A participant may attend to some subset of consoles associated with his deck. Any console with an agent of attention is called an *attended console*. As we shall see, consoles will, to some extent, be able to evolve without a participant's direct intervention. Not all consoles will necessarily be attended.

Four major activities are performed by the collaborative engine. They will be examined in turn.

Mechanisms of Selection

The first activity is *selection*. Primary and secondary repertoires are held together by the strength of the membership of their components (fuzzy sets). A repertoire also has an additional multiplier parameter that controls its ability to cohere and form new alliances with other repertoires. The collaborative engine increases this parameter when the repertoire is selected and introduces a slow decay in its value when it fails to be selected over some time interval. For natural selection to occur, there must be some environmental influences promoting this selection parameter in some neural groups and/or possibly inhibiting it in others. In the cyberspace concept proposed here, the space of objects evolves under the collaborative guidance of a number of these influences.

One influence is the focus of a participant's attention. Such a focus can occur either by fixing the gaze or other sense on some part of the space or by manipulating it in some way. If one stares at a corner of a box, the corner is selected. If one opens the box (even without looking at it), the box is selected. Focus of attention can be multisensory. All that is necessary is for a participant to somehow "mark" the space, indicating that it is at least momentarily of importance. Navigation is a kind of attentional mechanism. Walking a path selects it. Most of the participant's activity in the space is interpreted by the collaborative engine as the manipulation of attentional focus.

Unlike a point-and-click menu interface, where there is only one focus, in the collaborative engine there can be a number of foci. Choosing a new one does not invalidate the old one (though they do decay, see below). A participant can choose a number of foci, thereby indicating to the engine that those elements of the space should "go together" in some way. This ability to juxtapose elements is one of the primary means of authoring new cyberspaces, or directing the evolution of existing ones.

Initially, human participants may find this difficult, since they may have no special knowledge of the underlying structure of the space and thus may have difficulty telling what parts of the space they are attending to. Further, the consequences of their attention may not be immediately evident. Accordingly, they may wish to have an auxiliary *fish-eye* display (similar in function and appearance to a fish-eye lens), in which repertoires and consoles are represented iconically. In such a display, the participant can see the connections between repertoires and their relationships to his focus of attention. Spaces near the center of the display are nearer to the center of attention; spaces further away are less important to the focus. By manipulating the icons in the display, he can make minor corrections to his intentions. The fish-eye display can act as a kind of "training wheels" for the novice, or as an authoring aid for the more experienced.

A second influence of selection is the recombination of repertoires. A repertoire that is successful at recombining with other repertoires is given greater chance of being successful in the future. The collaborative engine boosts the selection parameter of a repertoire that continues to grow and inhibits it for one that does not. The natural tendency of repertoires to be self-limiting in the absence of new variety (see below) prevents runaway positive feedback loops from causing one cyberspace to take over a console completely. Repertoires can, to a large extent, be self-selecting in the absence of the constant attention of a participant.

The last selective influence we will look at is the propagation of repertoires across console boundaries. The actual mechanisms are discussed in more detail below, but for now it should be noted that consoles do not actually send messages between themselves. When a participant creates a console, he causes that console to *listen* to some number of other consoles. When a repertoire in another console relates strongly to one within the first console, a shallow copy of the weaker repertoire is made and brought across the console boundary to be integrated into the stronger repertoire. As with the normal recombination of repertoires mentioned above, this causes the collaborative engine to select the new combined repertoire.

Mechanisms of Repertoire Combination and Amplification

The second major activity of the collaborative engine is the combination and amplification of repertoires/cyberspaces. Here, combination refers to how linkages are created and maintained. Amplification refers to how repertoires grow and shrink, using the mechanisms of combination as a base.

A secondary repertoire is a *metric space*, that is, a space with a distance measure such that some objects in the space are nearer to each other than others. Since the objects (secondary repertoires) in the space are themselves metric spaces, this distance metric can be defined in terms of *degree of membership* in the "parent" space. In addition, the objects in the space are merely sets of linked, built-up concepts, so it is appropriate to refer to the metrics involved as *conceptual distance metrics*. The application of these measures forms the criteria for creating new semantic linkages between repertoires and accentuating or attenuating existing ones. While the number of means of prescribing conceptual distance is extremely large, it is taken as a given that the metrics should, to a large degree, evolve with the space. A collaborative cyberspace with a distance measure is also *content addressable,* since any other repertoire that matches more or less closely will be deemed "nearby" by the engine. The subcyberspaces in a cyberspace collaboratively determine a distance metric against which other spaces are compared; a repertoire is its own metric.

The collaborative engine creates linkages between repertoires in proportion to their mutual conceptual distance and their local selection paramenters described above. Given that computational or temporal resources are limited, the engine will be most likely to direct those resources toward repertoires with larger selection parameters. All linkages have variable strength on the unit interval.

A given cyberspace is opportunistic, in the sense that its conceptual distance metric will attract the most "attention" from other cyberspaces if it is more general. Its selection parameters will be increased if it is successful in growing. Thus the spaces that survive the best over time are those that contain the most variety. While a space contains no particular intentional mechanisms, it does, in a sense, compete for the attention of the participants having access to its particular console.

Variation is a vital ingredient of adaptability in animal behavior. Far from being detached observers, participant selection activities play a vital part in the ability of the cyberspaces within a console to maintain and increase their level of variation. First, individual participants themselves are a source of variation, in that selective acts may be performed according to whim. Even if produced according to some predefined plan, such actions would rarely be precisely synchronized with the state of a console, and would thus be likely to produce varied results even if executed on identical copies of some initial console state. A second source of participant variation is the interaction of the multiple participants across consoles on similar repertoires. When the first source of variation is factored in, this produces a highly nonlinear and chaotic system that produces a great deal of variation in the system disguised as noise. Rather than something to be avoided, this "noise" is in fact of great value to the collaborative engine in keeping its constituent cyberspaces "healthy."

Since the membership of a given repertoire may change with time, the conceptual distance metric holding it within a parent space will change as well. Rather than refiguring all linkages constantly, each linkage is allowed to *decay* with time unless its parent or constituents are periodically reinforced by some selection activity. Spaces not selected will tend to lose cohesion and fade away. Thus, a repertoire that is more parsimonious to a participant than another will be able to "steal" constituents of the more hapless repertoire. Recalling Edelman's theory, parsimony is related to the notion of specificity. A repertoire that is too specific or not specific enough is less likely to be selected. It will cease to grow and thus fade away. By controlling the flow of variety between consoles, the collaborative engine maintains a control on specificity, which keeps any given cyberspace from runaway growth while promoting diversity.

Participants are provided with a means to tune the *decay rate* of any given repertoire. This affects the rate of growth of that repertoire. By alternating decay and selection, a kind of optimization goes on, in which a repertoire can migrate to the place in the console where it best fits. This is illustrated in Figure 11.2b, where the selection parameter strengthens cohesion in supersets and the decay parameter weakens cohesion in subsets. A repertoire can also *tunnel* to another area of the console by being propagated to another console and then back again, (see Figure 11.3).

To compare conceptual distances, the collaborative engine requires a *sampling algorithm* to sample a space of repertoires and determine

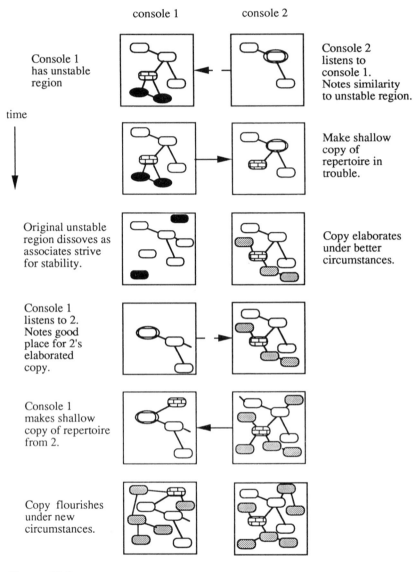

Figure 11.3
Repertoires can "tunnel" to a new place in the network by moving between consoles.

which repertoires will be compared next. Consoles and repertoires are read-only. Candidates for comparison are other repertoires within a console and other repertoires in consoles to which the enclosing console is listening. The latter are sampled at a lower rate and/or larger scale, with the "root" of each repertoire for sampling purposes located at the portion of the repertoire with the largest selection parameter. The engine compares conceptual distances based on the depth the sampling can reach within the time available. Sampling and comparison thus can take place in real time since results at some scale can be guaranteed for some arbitrarily small time interval (O'Reilly and Cromarty 1985). This mechanism also allows the engine to adapt to the available interconsole and interdeck bandwidth.

Propagation of Repertoires

The third major activity of the collaborative engine is the propagation of repertoires across console boundaries. This occurs for two reasons. First, to help maintain consistency between similar repertoires in different consoles. Second, as a means of importing variation from other consoles.

As was pointed out previously, a repertoire may exist in more than one console as a shallow copy (only the most salient features are copied). Two repertoires related in this way may come to resemble each other only superficially after several rounds of recombination. However, if their consoles are set up to "listen" to each other, interconsole sampling will tend to import their differences, which will be recombined and integrated locally.

From an evolutionary standpoint, the primary value of one console to another is as a source of variation; that is, they provide separate environments where diverse selection mechanisms can produce the variability required to keep the console-local repertoires healthy. From the participants' standpoints, separate consoles provide a means to separate and control selective influences and thereby facilitate the ongoing "sculpting" of the cyberspace.

The notion of consoles as sources of variation and control leads us to think of consoles as information "channels" (see Figure 11.4). This provides us with a simple model of communications between decks. The consoles in one deck listen to (sample) a set of representative consoles from another deck. Implementations of these "representative"

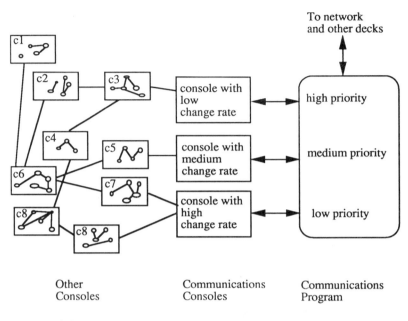

Figure 11.4
Consoles as interdeck communication channels.

consoles may be slightly different from others, with special connections to networking hardware, but to the collaborative engine, they operate the same as any other. Together, the representative consoles comprise a virtual multichannel communication scheme, in which different consoles represent differing scales of variation.

Modeling Repertoires for Rendering
The fourth and last major activity of the collaborative engine is the production of models of the underlying secondary repertoires for eventual sensory rendering by the deck. The precise methods of rendering are outside the scope of this chapter. It is assumed that one can render from a relatively stable model if such can be produced. Stable models are needed to smooth out the numerous small fluctuations that can arise in the operation of the collaborative engine. Such fluctuations could confound a rendering engine where the space changes radically before a frame can be fully rendered. If an even looser definition of rendering is used, that is, any interface effector to a human or nonhuman participant, then nonmonotonic issues (Ginsberg 1987) arise if the participant incorporates heuristics to any significant degree.

For our purposes, a model is a set of consoles that samples another set of consoles. The sampled set is the one spoken about thus far in the chapter. The consoles in the rendering model are characterized by low decay rates and an attenuating multiplier on selection parameters. As a result, consoles in rendering models change more slowly than other consoles, giving the rendering engines (or other computer programs) the sufficient illusion of a more stable space. Models can have varying decay and attenuation parameters, depending on the requirements of the programs and hardware with which they interface.

Rendering can be seen as an even more general phenomena within the collaborative engine. The consoles in a deck that listen to consoles in another deck are in fact models of the second deck, albeit rapidly changing ones. However, the *scale* of sampling is different in modeling since sampling is related to the rate of change of the console. By contrast, in recombination a secondary repertoire is sampled relative to the recent amount of selection it has undergone. Note the similarities between a description of a primary repertoire and models. Their structure and dynamics are very similar, though they are maintained for different reasons. It is assumed to be no more difficult to render a primary repertoire than a model. In some sense, a primary repertoire *is* a model. The consequences of this are discussed in more detail below.

While outside the scope of this paper, there are some rendering and encoding mechanisms that roughly parallel the modeling techniques described here; for example, the engine's sampling of repertoires relative to selection is similar in many respects to Barnsley's Collage Theorem used in image representation (Barnsley 1988).

Discussion

Semantics of Participation

The semantics of participation are very simple. All participants, whether human or computer program, interact with the collaborative engine via four basic activities.

1. The participant may *establish a console* at any time. This can be done in two ways. Either a boundary can be declared around some set of repertoires, or a shallow copy can be created of the same set. A shallow copy is a time or scale-related *precis* of an existing repertoire.

2. A participant may *establish connectivity between consoles* by specifying which consoles "listen" to which other consoles. Repertoires enclosed in a boundary are isolated from interaction with repertoires outside the boundary unless a participant explicitly allows them to eavesdrop on the activity of another console.

3. A participant can *direct the focus of attention* for some set of consoles within a deck. Navigation and creation are folded together in focusing actions to determine the moment-to-moment evolution of objects in the cyberspace.

4. Participants may *control console mutability* by slowing down or speeding up the potential rate of change. This is done by controlling the balance between selection and decay parameters for a given set of repertoires inside a console.

Of these activities, the one to be used most often in real time is the third. Those wishing to author new spaces for others to work with will make the most use of the fourth.

Establishment of Primary Repertoires

All collaborative cyberspaces will begin from a few primary repertoires, which are simply repertoires slowed down by participant authors controlling the mutability of consoles. Secondary repertoires arise by recombining the primary repertoires. These repertoires/models/cyberspaces are the "seed objects" around which the collaborative engine grows the larger space. They are the artifacts of cyberspace that absorb and structure the participants' experience.

Chaos and Entropy

Some might find disturbing the notion of all participants experiencing the same virtual world slightly differently. Common sense seems to dictate that their experiences would diverge. Indeed, if we take Edelman's theory seriously, and mental phenomena is indeed governed by selective processes, then the very mechanisms we use to perceive and understand an intensely nonlinear and chaotic physical world are themselves nonlinear and chaotic. Nonetheless, people manage to function and even thrive in a physical world characterized by wide variations in personal perspective and idiosyncratic mental models.

Indeed, according to Miller and Drexler (1988) such systems (which they call *agoric open systems*) are plentiful in markets and law as well as biology. If seen as games, then they change the playing field as they are played. Huberman has noted that in open, cooperative, computing systems, chaotic effects are unavoidable (Huberman and Hogg 1988). The collaborative systems described herein, it appears, must operate by similar population effects (in this case natural selection), rather than logical propagation methods.

Even the simplest of nonlinear systems are often characterized by unexpected complexity and deterministic chaos. Entropy is a slippery term, but could be defined as the tendency of a closed dynamic system towards a chaotic state (and thus loss of information and structure) for certain parameter domains. The selective systems maintained by the collaborative engine are particularly susceptible to this kind of entropic effect. In fact, operation of the engine is in large part built upon the expectation that the entropy and chaos that the engine will encounter can be tamed through the *management of variation*. There is more reason for optimism on this matter than there would have been six years ago, due to a large and excellent body of work in the area. (For example, see Thompson and Stewart 1986, Huberman and Hogg 1988.)

Ownership and Privacy

Ownership and privacy present special problems and opportunities in a collaborative space. Due to the high mutability of the secondary repertoires within a console, there will be great difficulty in establishing what a repertoire or cyberspace *is* at any given instant, not to mention who (or what) is responsible for it. Accordingly, ownership will probably be more an attribute of decks and consoles. For now, I will bypass the problem of participant ownership of deck hardware and concentrate on consoles. It should be understood that the legal definition of ownership is problematic in collaborative cyberspace. It is discussed here to suggest guidelines for the development of participant etiquette.

A console must be attended (at least periodically) to establish ownership. Of course, if a human participant "owns" a computer participant, the latter's attended consoles belong to the human user.

Ownership is more directly determined by the sources of selection actions for the console. Each of these sources introduce some interesting issues that must await the creation of the first collaborative engine to be resolved. The source of focus selection is the participant. How are conflicts between human and computer participants resolved? Selection by recombination implies that the console and the spaces within are self-owning. Selection by propagation places the locus of ownership on the console that was being listened to when propagation occurred. There is no way short of creating a mutual eavesdropping between two consoles to notify a console of its ownership, and that could negate the reason for having a boundary between them in the first place.

Privacy does not exist in the active sense since there is no protection on a console once a listening connection is established. There is no way for a console to tell who or what is listening. Nonetheless, a kind of virtual privacy prevails, since the listener cannot precisely duplicate the context in which the "listenee" functions, and is not necessarily granted knowledge of what consoles the listenee is listening to. So even though a console is wide open, for a noninvited eavesdropper the context in which a particular configuration of spaces is information is difficult if not impossible to reproduce. Thus the casual eavesdropper would have great difficulty making use of stray signals between consoles in any intentional way. Since consoles are read-only, they cannot be corrupted unless their sources of selection are manipulated. Participants control those and participant ownership rights have been defined above.

The Deck: What It Might Be Like

The appearance and attributes of a deck for use with the collaborative engine must be speculative at this point (see Figure 11.5), but there are some hallmarks that might differentiate it from decks intended for use with less collaborative spaces.

1. As mentioned previously, an iconic (fish-eye) interface will be available to monitor and fine tune the activity of chosen consoles. This device provides a training mechanism for the novice if needed as well as an authoring tool for the more experienced user.

2. While the deck will probably include instrumented clothing or other interface devices, they will interface with a primary repertoire representing a *body image*. One's body image will be its own cyberspace, which maps the "real" body image into another cyberspace.

3. The modeling mechanisms described above make possible a variety of interchangeable rendering engines of varying costs and abilities that can work within the same collaborative space.

4. The abstract capabilities of the collaborative engine can be associated with consoles of arbitrarily small size. A deck is essentially made up of consoles and clothing. The hardware to support one console looks much the same as any other. Therefore, decks are *composable*. Participants can pool their decks together to get more capable decks.

These last two points greatly affect the economics of participation. Users can start out fairly cheaply, pooling their resources to create more capable decks, all the while interacting with far more expensive machinery maintained by those that can afford it.

Audience and Involvement

Many investigators have emphasized totality of involvement and suspension of disbelief as a hallmark of cyberspace. While the collaborative engine certainly makes that possible, it also opens up the possibilities for cyberspaces with varying degrees of participation.

The domain of the cyberspace depends entirely on the seed objects one brings to it and one's own skills. A given human participant might participate in a number of task-specific cyberspaces simultaneously as a means of controlling and monitoring those tasks.

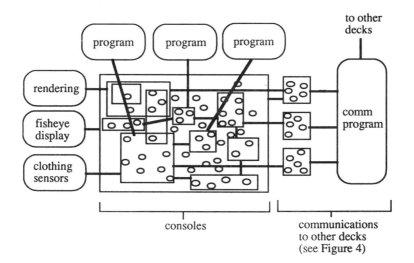

Figure 11.5
Possible console layout.

Potential Applications

Many commercial opportunities will arise surrounding the collaborative engine. A cyberspace system that assumes participants distributed geographically will require specialized network support services. Experts in nonlinear dynamics may be needed to configure, tune, and troubleshoot large systems. Services will arise to support interfacing computer participants to primary repertoires and education of new users. As with other, noncollaborative cyberspace systems, demands will arise for many different kinds and combinations of deck hardware, linking existing senses together in new ways.

As the technology of collaborative engines matures, many new applications of cyberspace will become possible. A few are listed here.

1. By far the largest market will be in the creation and trade of cyberspace seed objects and primary repertoires. These can be created at all scales and levels of complexity. It can be indulged in by any participant, by simply "freezing," or lowering the mutability of some space. It might be expected that those with the greatest understanding of the way primary repertories inspire secondary repertoires would have the greatest commercial success as "conceptual sculptors."

2. In traditionally mission-critical applications (such as telerobotics, where low bandwidth and time delays are a problem) collaborative spaces could be valuable in merging directives for action with what the robot on the scene deems possible. The result might be much greater scientific return for exploratory missions.

3. "Oral" traditions might make a comeback as storytellers use the collaborative engine to create new fictional universes in real-time response to the attention of participants. Those participating in the story might be in different locales or belong to diverse cultures. It might also be a new way to preserve and extend cultural diversity, which will also create new sources of variation for the collaborative spaces to structure.

4. The fledgling area of Computer-Assisted Collaborative Negotiation (Samarasan 1988, Campbell 1989) will acquire a new tool. Much of the process of negotiation depends on parties to the negotiation deciding on what is being negotiated about. Much like the storyteller idea, the collaborative engine can highlight areas of agreement and clarify issues.

5. Hypertext and hypermedia authoring are largely a matter of linking together diverse documents in a task-related manner. One of the strengths of the engine will be as a research tool, finding new and perhaps nonobvious links in large hypertext databases such as Nelson's Project Xanadu (Nelson 1990).

6. The relatively simple semantics of interaction allows participants to achieve fine control over the kinaesthetics, appearance, and mutability of their body images in cyberspace, since a body image is itself a cyberspace. This opens up vast possibilities for cyberspace theater (Walser 1990, Laurel 1989).

Research Issues: How Do We Get There?

Many of the following enabling technologies for the collaborative engine already exist or are well underway.

1. Edelman and his group have produced several computer demonstrations illustrating his theory. Their Darwin III system has acquired the fine motor skills required to track, reach for, and touch a moving object with no external prompting. The Darwin II system uses multiple classifiers to learn to identify handwritten characters at varying scales and rotations (Edelman 1987). While the collaborative engine will deal with somewhat larger grain concepts than the simulation of artificial neurons, there will still be enough going on for population effects to occur.

2. The activity of providing variable value links between repertoires requires a mechanism for partial set membership. The technology of fuzzy sets (Tong et al. 1985) and databases can provide this and is sufficiently mature to be realized in several existing commercial software packages. There are a number of possible fuzzy logics available to assist in the combination of repertoires and it remains to be seen how these operate within an open system such as the engine.

3. At the heart of the collaborative engine are Content Addressable Memories (CAMS), which allow retrieval of patterns based on their context rather than specific keys (Kohonen 1987). A very large body of work in neural networks has been accomplished in the last five years, of which the a significant part has been in the area of CAM development (ICNN 1987, ICNN 1988, Anderson 1988, DARPA 1988).

4. Objects in the collaborative engine are distributed and mutable in real time. Distributed databases and distributed object schemas (Carriero, Gelernter, and Leichter 1986; Leler 1990) are becoming better understood daily and form the basis for an emerging industry.

5. Commercial storage and distribution mechanisms for cyberspace object components seem destined to occur within hypermedia systems. Rick Mascitti and the author are currently investigating the mapping of cyberspace components into a large, typed link hypermedia environment.

6. Operations in the engine take place asynchronously and in parallel. Medium- to fine-grained parallel processors are in the market now. A major challenge is find low-cost processors that can be mapped to the computational requirements of collaborative engines. Recent work in analog VLSI (Mead 1989) shows promise in this regard.

7. Networking between decks is essential. Telecommunications networking is a constant factor of everyday life. The engine has been designed so that reasonable interactions can take place over networks where bandwidth is variable or at a premium. The quality of interaction rises as more bandwidth becomes available, but ISDN should not be necessary for entry-level functionality. The major challenge here is to create an expandable multichannel protocol on top of existing standards.

8. Rendering from 3-D models is an established technology. Representations that can model a smooth range of resolutions will eventually be required. One of the most promising of these representations at the current time are Barnsley fractals (Barnsley 1988), which have the added advantage of having some similarities to the representation discussed here for repertoires. At the time of this writing, however, little work has been done to develop fractal representations for nonvisual sensory modalities.

9. The interaction between repertoires and consoles is intensively nonlinear and will be marked by deterministic chaos. Several questions in this area are relevant to the collaborative engine.

  • Selective systems are dynamic, nonlinear, and often exhibit deterministic chaos. What are the appropriate formalisms to allow us to take advantage of existing work in these areas?

• How does variation operate and how can it be used to ensure divergence and consistency within and between consoles?

• How do changes in variation manifest within the experience of the participants? One example might be how fractal dimensionality (texture) changes with variation.

• How do changes in rate of console and repertoire mutability affect changes in participant experience? How does the engine control mutability changes?

Fortunately, the fields of nonlinear dynamics and chaos theory are subjects of intensive study around the world. Chaos has been studied in everything from neural networks to planetary rings. Considerable insight into the problems outlined above in the section on chaos has already occurred.

I have admittedly concentrated here on technical research issues with which I am somewhat familiar. Numerous other fields of study, particularly in the social sciences and philosophy, will be involved in the development of collaborative engines.

Potential "Gotchas"

There are, of course, some potential obstacles that might doom the enterprise or restrict its scope. While much is known about nonlinear dynamics, there may be undiscovered aspects of the engine that make the computations involved intractable. Some aspects of the ideas here have been deliberately left vague to allow for "workarounds" in later designs should this occur.

Another obstacle arises when there is a significant mismatch between what a user expects and what the simulation provides. So-called *simulator sickness* arises in some flight simulators when the inertial behavior of the simulator does not match the expectations provided by the visual behavior. If this is not simply an eye-vestibular coordination problem, it could pose a problem for a cyberspace in which the participant is to some extent in a constant state of low-grade surprise. Mechanisms would have to be developed to constantly adapt the ratio of resolution to control to maintain a "comfort zone." This is an exercise in control theory, and not an intractable problem. Finding the proper zones for each individual user might be more challenging.

Collaborative engines scale up nicely; in fact, they depend on a lot of traffic within and between consoles to operate properly. This makes it difficult to do simple experiments or provide examples until the first engines are built. This inhibits design. There is no workaround for this; we will simply have to find new ways to describe systems of this type. A partial solution can be found by pacing development. As a first step, the intention is to develop fuzzy, mutable CAMs to a point where we can produce relatively simple primary repertoires within a single console. This will give us some experience with the formation of secondary repertoires and we can go on from there. It is expected that the development of cyberspace "clothing" and other input/output devices will proceed in parallel with efforts to build the first simple engines. It is possible that simple "conceptual sculpture" devices could become available within a year or two.

Conclusion

In this chapter I have identified an approach to collaborative cyberspace based on natural selection. A cyberspace is simply a metric space in which other cyberspaces have variable-strength membership. Objects and the dynamics of their interaction in the space are cyberspaces; juxtapositions of attributes and values are each in turn represented as a cyberspace. Collaboration here is used to delineate a set of operations over a set of cyberspaces, in which a group of selective influences under the control of the participants direct the evolution of the space. The collaborative engine suggested herein can provide an *arena* where these distributed, decentralized influences can operate in concert to produce and maintain shared spaces.

Variation in participant interactions and in the representations of shared worlds plays a major role in the operation of the collaborative engine. Such variation assists in maintaining approximate consistency between worlds on different decks without the concern of more traditional and strict database update mechanisms.

The hallmark of this approach to collaborative cyberspace is *participation*. No user can inhabit a collaborative cyberspace for long without becoming involved in its moment-to-moment evolution. Even "passively" watching has a selective effect. Participants can take a more active role by establishing boundaries between sets of spaces, called *consoles*, that cause a designated set of spaces to evolve separately with less interference from spaces outside a boundary.

The approach is inspired by Edelman's theory of Neuronal Group Selection. It is understood that mixing metaphors about computers and brains can sometimes lead to misleading conclusions about both. It should be kept in mind that in talking about this approach that I am simply pointing out correspondences between some aspects of a promising though unproven theory of brain function and the goals put forward above for a collaborative engine. If the NGS theory is somehow shown to be incorrect, it should not necessarily affect work on the engine.

Collaborative cyberspace is intended to be an extension to the kinds of exploration we do everyday; a constantly developing vehicle for the exploration of relationships between scale, intention, resolution, and control. Cyberspace, after all, comes from the Greek word *kybernetes*, meaning "helmsman." Whether computer- or biologically based, a cyberspace is simply a steerable space. Collaborative engines will open new channels of communication between our private, internal cyberspaces and those we share.

Acknowledgments

The author would like to thank Sonia Lyris, Ellen Campbell, Dan Shapiro, Randy Walser, William Bricken, Meredith Bricken, and Jacque Scace for many helpful discussions and for their support.

References

Allen, T. F. H., and Thomas B. Starr, *Hierarchy: Perspectives for Ecological Complexity*, University of Chicago Press, Chicago, 1982.

Anderson, Dana Z., ed., *Neural Information Processing Systems*, American Institute of Physics, New York, 1988.

Ashby, W. Ross, *Variety, Constraint, and the Law of Requisite Variety*, in *An Introduction to Cybernetics*, Chapman and Hall, London, 1956.

Barnsley, Michael F., *Fractals Everywhere*, Academic Press, London, 1988.

Bobick, Aaron F., *Natural Object Categorization*, Ph.D. diss., Massachusetts Institute of Technology, technical report AI-TR1001, July 1987.

Campbell, Ellen C., "Computer-Assisted Negotiations: A Systems Study and Analysis," Master's thesis, San Jose State University, August 1989.

Carrriero, Nicholas, David Gelernter and Jerry Leichter, "Distributed Data Structures in Linda," in *Proceedings of the ACM Symposium on Principles of Programming Languages*, St. Petersburg, Fla., Jan. 13–15, 1986.

DARPA, *DARPA Neural Network Study*, Armed Forces Communications and Electronics Association International Press, Fairfax, Virginia, 1988.

Dawkins, Richard, *The Blind Watchmaker*, W. W. Norton and Company, New York, 1987.

Edelman, Gerald M., *Neural Darwinism: The Theory of Neuronal Group Selection*, Basic Books, New York, 1987.

Edelman, Gerald M., *The Remembered Present: A Biological Theory of Conciousness*, Basic Books, New York, 1990.

Farmer, Doyne, Alan Lapedes, Norman Packard, and Burton Wendroff, eds, *Evolution, Games, and Learning: Proceedings of the Fifth Annual International Conference of the Center for Nonlinear Studies*, North-Holland, Amsterdam, 1986.

Ginsberg, Matthew L., ed., *Readings in Nonmonotonic Reasoning*, Morgan Kaufmann Publishers, Inc., Los Altos, Ca., 1987.

Huberman, Bernardo A., and Tad Hogg, "The Behavior of Computational Ecologies," in *The Ecology of Computation*, B. A. Huberman, ed., Elsevier Science Publishing B. V., Amsterdam, 1988, pp. 77–115.

ICNN, *Proceedings of the IEEE International Conference on Neural Networks* (San Diego, Ca., July 24–27, 1988).

IEEE, *Proceedings of the IEEE First Conference on Neural Networks*, (San Diego, Ca., June 21–24, 1987).

Kohonen, Teuvo, *Content-Addressable Memories*, Springer-Verlag, Berlin, 1987.

Langton, Christopher G., ed., *Artificial Life*, Addison-Wesley, Reading, Mass., 1989.

Laurel, B., "On Dramatic Interaction," *Verbum*, Vol. 3, Number 3, 1989.

Leler, W., "Linda Meets Unix," *Computer*, IEEE Computer Society, Vol. 23, Number 2, February 1990.

Mead, Carver, *Analog VLSI and Neural Systems*, Addison-Wesley, Reading, Mass., 1989.

Miller, Mark, and K. Eric Drexler, "Comparative Ecology: A Computational Perspective," in *The Ecology of Computation*, B. A. Huberman, ed., Elsevier Science Publishing B. V., Amsterdam, 1988, pp. 51–76.

Mpitsos, G. J., H. C. Creech, C. S. Cohan, and M. Mendelson, "Variability and Chaos: Neurointegrative Principles in Self-Organization of Motor Patterns," in *Dynamic Patterns in Complex Systems*, J. A. S. Kalso, A. J. Mandell, and M. F. Shesinger, eds., World Scientific Publishing, 1988, pp. 162–190.

Nelson, Theodor, *Literary Machines*, Mindful Press, Sausalito, Ca., 1990.

O'Reilly, Cindy, and Andrew Cromarty, "'Fast' Is Not 'Real-Time': Designing Effective Real-Time AI systems," in *Proceedings of the SPIE: Applications of AI*, J. F. Gilmore, ed., Vol. 548, April 1985.

Ridley, Mark, *Evolution and Classification*, Longman Group Limited, London, 1986.

Samarasan, Dhanesh K., *Collaborative Modelling and Negotiation*, in *COIS88: Conference on Office Information Systems*, Palo Alto, Ca., March 23–26, 1988, Association for Computing Machinery, 1988.

Thompson, J. M. T., and H. B. Stewart, *Nonlinear Dynamics and Chaos: Geometrical Methods for Engineers and Scientists*, John Wiley & Sons Ltd., 1986.

Tong, Richard, Victor Askman, James Cunningham, and Carl Tollander, *Rubric: An Environment for Full Text Information Retrieval*, *Proceedings of the 8th International ACM SIGIR Conference on R&D in Information Retrieval*, Montreal, 1985.

Walser, Randy, *Elements of a Cyberspace Playhouse*, in *Proceedings of the National Computer Graphics Association '90*, Anaheim, Ca., March 19–22, 1990.

Weber, Bruce H., David J. Depew, and James Smith, eds., *Entropy, Information, and Evolution*, MIT Press, Cambridge, Massachusetts, 1988.

Zadeh, L. A., "Outline of a New Approach to the Analysis of Complex Systems and Decision Processes," *IEEE Transactions of Systems, Man and Cybernetics*, SMC-3:28-44, 1973.

12 Notes on the Structure of Cyberspace and the Ballistic Actors Model

Tim McFadden

Introduction

This chapter is motivated by a remarkable series of science fiction novels that pioneered many concepts of cyberspace and by the current research in virtual realities. The novels are *Neuromancer* (Gibson 1984), *Count Zero* (Gibson 1987), *Mona Lisa Overdrive* (Gibson 1989), *Software* (Rucker 1982), and *Wetware* (Rucker 1988).

We refer to these as *the books*.

Cyberspace. A consensual hallucination experienced daily by billions of legitimate operators, in every nation, by children being taught mathematical concepts . . . A graphic representation of data abstracted from the banks of every computer in the human system. Unthinkable complexity. Lines of light ranged in the nonspace of the mind, clusters and constellations of data. Like city lights, receding . . ." (Gibson 1984, 51)

This chapter describes a plausible, more formal model of such a cyberspace and studies the problem, How can we know anything in cyberspace? In outline, the chapter

1. clarifies the working assumptions and casts the formal model in terms of information theory, the Actors model, and the ABCL/R model. This draws upon the current work in Distributed Artificial Intelligence (DAI).

2. studies the basic epistemological problems with the formal model. What is in cyberspace? How are agents' points of view constructed? How are scenes constructed? How do agents cooperate, and how are their actions arbitrated?

3. presents a cyberdeck design using the Ballistic Actors Model (BAM).

4. discusses how such a huge system as cyberspace may be understood or controlled at all, from the point of view of its physical and computational complexity.

In general, using our current intellectual tools in a formal manner makes clear two major problems in building a cyberspace:

1. Mathematics is not now adapted to large systems in which each of the subcomponents follows slightly different rules or in which the slightly different point of view of each subcomponent must be taken into account. Living systems are an obvious example, to which cyberspace has some similarities.

2. Cooperation and arbitration between intelligent agents is still an unsolved problem in DAI (and everywhere else).

These are of course in addition to the well-known problems of building livable, independent virtual realities and of living with software bugs.

The cyberspace of *the books* implicitly assumes the success of three paradigms:

• The Program of Reductionism. Reductionism argues that scientific theories that explain phenomena on one level can be reduced to explanations at a simpler level. For example, the cognitive properties of animal intelligence can be reduced to neurophysiology, which can be reduced to cell biology, and so forth, to quarks and leptons. This is necessary if the medical and wetware advances of *the books* are to be made.

• Strong Artificial Intelligence (AI). Strong AI argues that human intelligence and consciousness is explainable algorithmically, with some large number of rules and procedures. This is necessary if AIs are to become conscious, as in *the books*.

• The Information Science Model. This model is often implicit and is derived from appeals to reductionism and strong AI. It argues that we can represent the world correctly as information. This is necessary if human experience and consciousness can be supplied through the hardware of *the books*. Can the raw material for cyberspace be intellectually disassembled from our world and fed into cyberspace over an optical fiber cable?

A cyberspace is not only just a hugely complex information network, but it is also a system that gives humans (or other intelligent agents) points of view and scenes of interaction. Most current computer networks are passive switches that allow agents outside the network to communicate. This might be called the skin of cyberspace, but a full cyberspace is an scene for societies of interacting agents, which also permits humans to have points of view and roles in these societies. A point of view (POV) abstracts and organizes the action in cyberspace so that it is comprehensible for a human being.

Cyberspace is an exercise in the outer boundaries of our understanding of the world as information and is yet another way of tracking our old concerns about the "human use of human beings" or the "computer's use of human beings." We may have (1) cyber-sweatshops, in which workers can only act within a severely restricted and sensually impoverished world; (2) a world in which many new social niches have been added, a more complicated, interdependent civilization with hundreds of different city architectures; (3) new worlds in which ordinary world experience and cyberspace experience are intermixed in many ways; and (4) new worlds in which a human bodymind is allowed to play out thousands of times the possibilities that can be presented in the everyday world. In fact cyberspace may be used for a systematic exploration of the boundaries of human experience.

In the above sense, cyberspace makes us confront in a more technological manner the problems of human experience. If human connection to cyberspace with everydaylike qualities of experience becomes widespread, then the interface that provides the experience—say, a cyberdeck—becomes a quantifiable metric of human experience as well as a commodity. *Experience will become a substance and a commodity.* An industry standard for current virtual reality systems is a good example and could well appear in this decade.

Human neurophysiology soon becomes an issue. To make this more concrete, take the example of clouds. Most people like to watch clouds and it causes the lateral geniculate nucleus (LGN) of the brain to be reprogrammed, if nothing else. Current virtual reality systems do not have the resolution to display moving clouds accurately. Call this level of display the *cloud barrier*. Until our virtual realities pass the cloud barrier, they will not be engaging the human nervous system to the degree that the everyday world does. However, displays with far lower resolution can engage human interest and participation, with surprising ease.

If we take cyberspace seriously as future engineering, as is apparently possible within our current scientific models, then *the books* raise some interesting possibilities that might enable us (1) to completely describe the mind, or as we consider it here, the bodymind, as information and algorithms (that is, life is a machine; in fact, this is becoming a pervasive belief among scientists); (2) to provide bodyminds with artificial input that provides the same experience as the ordinary world; (3) to "run" minds on computers, creating "consciousness" equivalent to that of ordinary world bodyminds; and (4) to have artificial bodyminds behave in the world and experience it much as humans do, that is, as "boppers."

Is all this possible? Perhaps not. Human experience may be basically tied to some chaotic, molecular, or quantum mechanical phenomena, which does not scale up.

The books take two old themes to new extremes—computer networks (cyberspace) and autonomous, conscious robots (*boppers*, as in Rucker 1982. Boppers are not usually discussed in the context of cyberspace, but this chapter will argue that most of the technology that will make cyberspace possible is directly applicable to boppers.)

These questions are part of a current, serious debate. Roger Penrose, a physicist, has written an influential book, *The Emperor's New Mind* (1989), which argues that the mind is fundamentally not algorithmic. Rudy Rucker, a mathematician and author of two of *the books*, also wrote *Mind Tools: The Five Levels of Mathematical Reality* (1987), which argues that everything, including mind, is information (in a different sense than is this chapter, as we shall see) and therefore, presumably, a matter of algorithmic complexity.

Davis and Hersh, in *The Mathematical Experience* (1981) and *Descartes' Dream* (1986), discuss the encroachment of mathematical, algorithmic thought and the information processing model upon all of science and much of everyday life. The central question of these two books, and of many, many others like it in the cognitive sciences, is: Can human experience be reduced to the information model?

Personally, and without answers to such questions, I have tried to make this chapter the one I would like to have read after reading *Neuromancer* and *Software*.

Assumptions

As much of our inspiration comes from science fiction, we have to make our working assumptions clear. Here *world* means our universe, from the lowest scaffolding of physics and chemistry to human consciousness and whatever is in it:

Cyberspace happens to run in our world.

In the absence of any critical experiments that can differentiate mind from body, we assume only that we are familiar with *bodymind*. We know that the bodymind can compute, eat, sleep, communicate, survive, etc., but we do not know how to separate "mind" from the endorphins, neuropeptides, etc., which make up the biochemical operation of the bodymind, from the electrical signals in neurons, from the brain's neural network itself, and so on.

Consciousness is an important issue in cyberspace. We know that our bodyminds are made up of world "building blocks" of quarks, leptons, fields, and whatever makes it all "run" together.

As no crucial experiments or agreed-upon theories exist to relate the experience of consciousness to these building blocks, in this chapter, we consider consciousness to be a "delusion."

A Definition of Cyberspace

We want to arrive at a more formal version of Gibson's cyberspace-is-a-consensual-hallucination definition. The above delusional definition of consciousness makes this a lot easier!

We use the word *delusion* rather than the word *hallucination* because the world can meaningfully and usefully delude us into experiencing something, like color vision, that is not an hallucination. We use the word *information* in the sense of Shannon-type information theory.

An *information source* produces a message or sequence of messages. A *transmitter* operates on the message to produce a signal for transmission over the *channel* (usually noisy) to the *receiver,* which usually just performs the inverse operation of the transmitter. The *destination* is the end point of the message.

Shannon and Weaver point out that "the significant aspect is that the actual message is one selected from a set of possible messages" (1963, 3). Information theory does not determine how the information source chooses the messages, but it does have a great deal to say about how the transmitter produces signals.

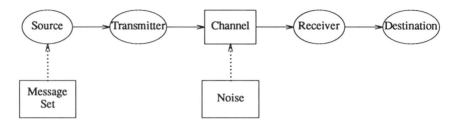

Figure 12.1
These are the components for information transmission in a Shannon-type information theory.

Definition: Acquired information is information that has been generated by just such a procedure and that is physically preserved in its transmitted form.

For example, a recorded message of an human voice is acquired information, but if it is processed by a receiver and not otherwise preserved, if it is broadcast by a loudspeaker, for instance, it is soon lost into the thermal noise level.

So our world is *not* made up of information, in the sense of Shannon's definition. There is no list of possible messages for the universe and no independent mechanism waiting around to choose them. However, from experience, we know that we can acquire information about it by asking questions.

Because quantum mechanics specifically forbids simultaneous measurement of position and momenta beneath a certain accuracy, not all questions can be answered: complete information about the state of a system cannot be acquired.

Definition: An *information space* is an abstract space of acquired information, which has an information source and an information receiver.

For example, all the tiles in the frescoes of the Mediterranean are part of an information space, as are all books, computer memories, disks, CPUs, etc, whether on earth or not, if there is someone ready to read them (as a receiver). The receiver is required by definition. A rock and a geologist are not an information space; there is no message source. A book written in a forgotten language is not an information space, because no one can read it.

An information space might not be connected in any way. One's calculator is not usually connected to the telephone company's local switching computer, nor is the latter connected to a space probe. The gathered information from surveys of the river Nile's flood plane of four thousand years ago was an important information space. Just as were the accounting systems of two thousand years ago, where mathematics began to develop with commerce.

In all these cases, the local entropy is drastically lowered, as the entropy of the surrounding system is raised.

Definition: A *cyberspace* is an information space with the following properties:

1. It is *connected* by a network of information channels, such that, if any part of the information is available to a receiver, then all of it may be.

2. There is a set of *agents* that may change the information and well-known *protocols* for exchanging information between agents. This is the "consensual" part of the original definition. Agents may also be part of the information space and hence be objects of change. These agents may interact with the world to gather information to/from cyberspace, and they may be in fact be boppers.

3. There are *agents* that can transform, abstract, and represent the information in the cyberspace so that a human can experience it as humans experience the space and "everyday" objects of the world. Humans may be "in" cyberspace as they are in space. This is the "delusional" part of the original definition.

A cyberspace with only properties (1), (2), and a primitive version of (3), I will call a *pre-cyberspace*. No one knows how to make a cyberspace, but the technology of pre-cyberspaces is evolving quickly. The first implementations will obviously be on computer networks, which will connect their memories and databases and play host to the agents above. The current Internet (see Comer 1988) is a good example of a pre-cyberspace. A cyberspace will allow humans to see, move, navigate, etc., much like they move about a room or fly an airplane. Cyberspace may also introduce totally novel experiences.

It is just these properties that allow for boppers to be part of a cyberspace, navigating through our world just as primitive robots do now. Bopper actors in cyberspace may contain sophisticated models of our world. These models may of course be available to other actors in cyberspace as raw material for scene construction.

Bond and Gasser give an overview of distributed artificial intelligence (DAI) that might be taken as a partial description of cyberspace.

Open systems are systems that have the following features: (a) they are composed of independently developed parts in continuous evolution, (b) they are concurrent and asynchronous, and they have decentralized control based on debate and negotiation, (c) they exhibit many local inconsistencies, (d) they consist of agents with boundaries visible to the agents constituting the system. (1988, 8)

These definitions try to emphasize that a cyberspace is encoded in an information space that in turn must be maintained by separate agents. *This means that a cyberspace may be maintained by a computer network but is not identical with it.* Computers may play the role of agents within the information space and thus be objects of change.

Definition: A *bopper* is any autonomous, mobile part of cyberspace that is also in direct physical interaction with the world. A bopper is intelligent, mobile, autonomous, etc.—that is, a very smart, classical robot. Boppers may be agents of other cyberspace actors and may be part of a cooperative system or society.

Definition: A *cyberdeck* is the particular agent that transforms and represents the information of a cyberspace into the "delusional" space in the human sense above. The cyberdeck contains the hardware and software that allow a human to experience cyberspace as a space in which to navigate.

We see boppers and cyberdecks as *dual* to each other in many senses. They are symmetric complements. Both boppers and cyberdecks must construct some sort of model of the world. In the case of a cyberdeck, the model is the target of the translation of symbolic actions in cyberspace from the informational to the human. In the case of the bopper, the model is an informational one of the basically non-informational, nonalgorithmic world.

A Model Pre-Cyberspace

Models of Distributed Processing
No single model gives quite every point of view that is useful to us. The following, which are quite recent, come out of the field of distributed

systems (see Bond and Gasser 1988; Lampson, Paul, and Siegert 1981, for example). There is not sufficient room here to do justice to the field or to the work of Carl Hewitt, Gul Agha, Akinori Yonezawa, and others. We give a few formal definitions and some of its flavor. In fact, one motivation for choosing them is that the authors spend many pages *justifying* the models.

The Actors Model   Agha and Hewitt point out that "the actor abstraction has been developed to exploit message-passing as a basis for concurrent computation" (1988, 398). The actor formalism models concurrent computation in distributed systems, by assuming reliable queued asynchronous message-passing among computational agents (actors), which can assume new behaviors as well as create new actors.

Communications in the form of tasks are sent to other actors with known addresses. An actor may only send messages to those actors on his list of acquaintances. To get the feel of this see Figure 12.2, an actor event diagram. An actor is created, receives messages, sends out tasks (new messages), creates other actors, and assumes a new behavior. See Agha 1986 for more concrete examples. Figure 12.3 shows how overdraft protection might operate in the actors model.

There are special actors, called *receptionists,* which are the only actors free to receive communications from the outside world.

In the *universe of actors model,* even what we might consider constants are actors; the constant *identifier 3* is bound to the mail address of the *actor 3.* It behaves like the constant 3. Thus in an expression "x + 3," instead of a constant "3" appearing in the code for an actor, the "3" will be used as a mail address and the actor "3" will be asked for its value, which will be 3.

It must be emphasized here how different Actors is from most programming languages. Commands in actor language are not serialized—that is, the separate statements may be executed concurrently. The metaphor that comes to mind is that of a hungry, massively parallel system that comes upon millions of lines of ordinary serial code and chews them up until the smallest unit separable process has been assigned a processor.

There is no notion of *copying* actors in the actor model.

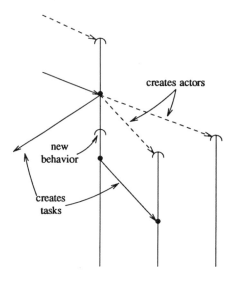

Figure 12.2
Sample actor behavior. Actors can create new actors, send messages to create new tasks, and assume new behaviors.

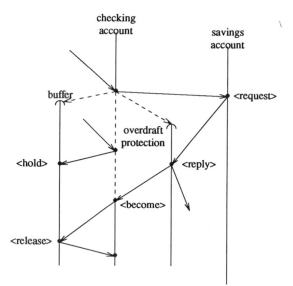

Figure 12.3
An actor system that implements overdraft protection. Buffer and overdraft protection actors are created to hold messages and to receive replies from savings (Agha 1986).

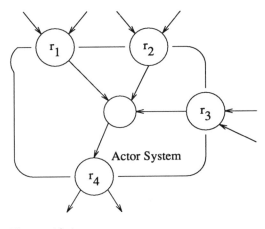

Figure 12.4
Actor system with receptionists. Other actors may only interact through the set of receptionists (Agha 1986).

The ABCL Model   ABCL is a very general model of distributed processing that emphasizes parallelism (for computing power) and object-orientation (for modeling power), or object-oriented concurrency.

Like Actors, it relies on message passing for communication between its elements. The ABCL group recognizes the Actors' formalism as the origin of its approach.

The ABCL system has several distinct components and languages. We will give a general overview and describe ABCL/R below.

There exists an ABCL compiler, and it goes beyond being a theoretical model since systems have been implemented using it. For example, there is an ABCL version of the operating system XINU. XINU was developed by Douglas Comer and others.

When a message is received, an object executes a sequence of the following four basic actions: (1) message passing, (2) creation of objects, (3) referencing and updating of the contents of local memory, and (4) various operations (such as arithmetic operations and list processing) on values that are stored in its local memory and passed around in messages.

There is a specialization of ABCL, ABCL/R, which allows objects to be self-reflective. According to Yonezawa, "Reflection is the process of (an actor) reasoning about and acting upon itself" (1990, 45).

The ABCL/R model is designed for causally connected self representation. This means that objects have metaobjects that model the

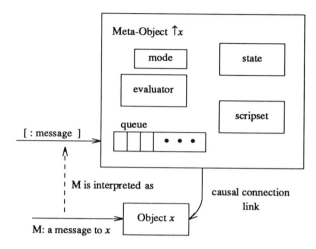

Figure 12.5
An ABCL/R object with its metaobject. The metaobject runs the lower object according to a script (Yonezawa 1990).

structural and computational aspects of the object. The metaobject of an object x is signified ↑x. x is the *denotation* of ↑x. Yonezawa writes, "The values of scriptset, state, evaluator are objects which represent the set of scripts . . . the state of the memory, and the evaluator. The value of mode indicates the current mode of the denotation object, which is either :dormant or :active" (1990, 52).

Using this, we can model the case in which an object does not have the script to run a message. It can, however, inherit (acquire) the script for the message from another object. This is called *dynamic acquisition (inheritance)* of scripts.

Figure 12.6 shows how ABCL/R can be used to change the scripts of agents to accomplish different jobs.

Comparing the Use of Actors and ABCL   Actors and ABCL make similar assumptions about message passing, except that in Actors, the arrival order of communications sent is both arbitrary and entirely unknown.

Actors is more general than ABCL, and we wish to keep its powerful asynchronicity. We also want to be able to express some particular actors' behavior in a succinct, general form.

The significance to us is that we want to populate cyberspace with actors but to specify much of the mechanics in ABCL/R.

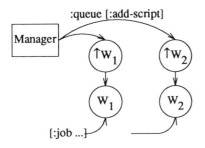

Figure 12.6
Script downloading to metaobjects. A manager agent downloads new scripts to metaobjects, reprogramming the lower objects (Yonezawa 1990).

The Indra's Net Model of Cyberspace

Our goal here is to imagine a system that will model, in a definite sense, however abstractly, many of the aspects of a cyberspace. It abstracts away communications problems, so that the problems of cooperation and navigation can be made plainer.

The Holon Concept   The *holon* concept has been ubiquitous in the study of AI, cybernetics, and living systems. It appears by other names in Minsky 1985 or as "domule" in Tenney and Sandell 1988, for example. The word *holon* itself was coined by Arthur Koestler in 1967. (See also Koestler 1971, 192.)

Wholes and *parts* do not exist in any absolute sense in the domain of living organisms or of social organizations. This "Janus" effect is a fundamental characteristic of subwholes in all types of hierarchies.

Holons have *cohesion* and *separateness*. The two conceptual poles of the holon are *the self-assertive tendency* and *the integrative tendency*.

As stated above, the concept is still ambiguous, which means that it can be a practical description of the world. The holon concept is used here to summarize cooperative behavior that we do not yet understand. How agents cooperate is an unsolved problem, but they do cooperate and cyberspace cannot exist unless they do. Figure 12.7 shows a schematic hierarchy.

Holons reflect organizational behavior. According to Werner, "Hidden in a protocol specification is a global architectural specification of the distributed system. This global architecture is what we call a social group" (1989, 30).

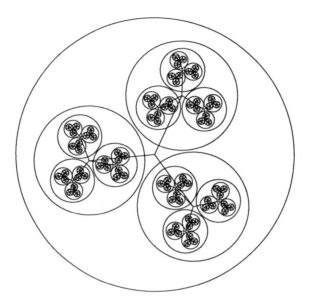

Figure 12.7
If each circle is viewed as an holon, then the figure portrays the relationships:
above (integrative) and below (assertive).

To paraphrase Bond and Gasser 1988, holons deal in announcements,
bidding, awarding, linkage, result sharing, "cause-to-want," "convince,"
etc.

We represent holons as pairs of sets of actor behaviors (in BNF
notation):

\<holon> ::= (\<integrative_behavior>, \<assertive_behavior>)

where \<integrative_behavior> is the set of behaviors that directly act
on messages received from other actors, that involve the above bidding,
awarding, committing, linkage, etc., and \<assertive_behavior> is the
set of behaviors that sends orders, creates other actors, or sends
representations of the actor's point of view, etc.

Most actors participate in the following:

common holon = ({uses common message protocols,

     moves in the consensual space,

     uses common data encoding,

     uses the standard reference frame, ..},

     {sends identifying messages,

     creates other actors,

               ...})

These behaviors will be clarified in the following sections.

The Definition of the Model   *Definition:* An *Indra's Net Model of Cyberspace* *(INMC)* is a pre-cyberspace that implements a large system of actors.

We express the underlying mechanics of this system of actors as the operation of ABCL/R objects. The implementation may be *lazy* in important aspects; that is, not all of the implied interconnectivity as described below may be used in a particular instance, but the properties will be used to run particular thought experiments. Specifically:

• Each actor is the *denotation* of some ABCL/R object.

• Each actor may have any other actor or set of actors as an acquaintance. This implies that messages may be effectively *broadcast* to all the other actors in INMC. In actual instances, we will *multicast,* whereby messages will at least reach a set of actors that will accept them. Hence, actors may distribute copies of their models of INMC, including pictures of their state.

• Each actor may maintain an arbitrarily detailed model of INMC (the Indra's Net property, see below).

• Any actor may communicate with the world outside of INMC.

• Each actor may participate in the common holon.

The major conceptual differences between INMC and most computer networks are the higher possible interconnection of all nodes and the possible existence of a model of the network at each node. These appear because we do not want to worry about how messages get between actors (eliminates topological dependencies), and we do not want to limit in any way what may appear at a node. In particular, all actor behaviors may already be present at each node.

Although the nodes are distinguishable, we do not want to make them look different a priori.

How does INMC relate to the cyberspace of *the books*?

In *Neuromancer,* jacking into a cyberdeck immediately presents one with a view of cyberspace from a certain position. One can then move about from grid point to grid point in a perceptual 3-space. Some parts of cyberspace may only be entered by using some authority or by "ice-breaking," hacking through security. But what agent presents this view of cyberspace? If two agents provided such a view, which one should be acted on? Security screens, "ice," prevent some objects from being changed by unauthorized agents, but what prevents agents from changing the large public structure of cyberspace? If two agents want a cube in a public area to be two different colors, which decides what color it is?

In INMC, these problems are apparent immediately. Any actor can send any other actor a view of INMC or scenes in INMC. There is no mechanism for deciding on which to act. Movement of an actor may be accomplished by sending other actors successive pictures or behaviors that show the movement or by having a designated actor track the movement for a particular scene. Much of INMC may not be visible to a given actor because it does not know the mail addresses of the actors involved.

INMC captures these problems of cyberspace in a really bare form.

The Indra's Net Property   As we know from *the books*, a *grid point*, an idea from *Neuromancer,* may contain a data base of arbitrary complexity (including an arbitrary detailed model of INMC). A grid point could be university library.

This is a self-referential, holographic property that does not have easy mathematical analogs. Indeed, the nearest analog is in the distributed properties of neural nets. The mathematical description of neural nets (and hence the brain) is still in its infancy.

The analogy of Indra's Net is very old, from Hindu mythology, whereby the universe is seen as a great net with a jewel at each intersection that reflects every other jewel in the net.

So by *the books*, INMC must have the Indra's Net property: any node may contain a model of INMC and each node "reflects" every other node, that is, has every other node as an acquaintance.

Where Is Cyberspace?

The purpose of this section is to underline the fact that there is no obvious "natural space" analog for motion or physical interaction in cyberspace. Cyberspace is a consensual delusion; there is only a consensual, arbitrary space (or spaces), in the mathematical sense of "space."

When actors send scenes to each other, the scenes will be regions in the consensual space. If we want humans to experience cyberspace as some sort of space, then that space will have to be arbitrary. Holons are more basic to cyberspace than points and lines.

We can express the above by saying that cyberspace is naturally *top-down* and *symbolic* rather than bottom-up and spatial.

Our world is "bottom-up:" large numbers of atomic things get together, somehow cooperate, and we get our everyday world. There is no known set of axioms or rules for the world.

For the opposite reasons, cyberspace is "top-down." Its lowermost objects may have features only by *declaration*, "this is a blue cube, of duration 5 minutes," leaving the cyberdeck to fill in the details.

It is also top-down especially because we may have to impose high-level, global rules for the overall interaction of agents in cyberspace. As of today, there is no set of rules, however complicated, that is likely to generate a working cyberspace bottom-up.

To be more concrete, we will need to choose some arbitrary space for cyberspace. So let us make up the *standard reference frame (SRF)*—most objects in cyberspace will be given coordinates in a Euclidean 3-space coordinate system centered at MIT, for example. Major landmarks will be located approximately as they are geographically in the world. The Crays at NASA/Ames will be southwest of the Crays at Lawrence Livermore Labs. This gets us the grid points from *the books*.

Part of the standard reference frame is that all actors have access to clocks synchronized with world time. However, the real distances will not do, so we introduce an arbitrary unit of length in cyberspace, called the *Gibson* (G). Humans are visualized as having feet about one Gibson long in cyberspace. We define the *Rucker* (R), to be the reciprocal of the Gibson. Boppers have world feet one Rucker long.

Participating in cyberspace gives boppers a distinct advantage over isolated agents. At any given time, they can be programmed by different agents and can use the resources of cyberspace to analyze situations. Boppers are agents in our world and can play a more direct role in our history. Cyberspace agents may, for example, use automated teller machines to bribe human designers or legislators. *The books* are wonderful exercises in this vein of thought.

Design for a Cyberdeck: The Ballistic Actors Model

This model is one answer to the question, What is in cyberspace? Where is it? How does one get to it? Here Actors and ABCL/R are specialized to the *Ballistic Actors Model* (BAM).

We relate actors to the standard reference frame (SRF) by associating points in the frame with the mail addresses of receptionists:

<spot> ::= (<coordinates>, <mail_address_of_receptionist>)

Definition: A *spot* is the mail address of some receptionist of an actor system associated with the coordinates of a point in SRF. This emphasizes the arbitrary nature of SRF. This relation is many-to-one; spots can be piled up on top of each other at the same coordinates. However, mail addresses are unique.

A *scene* is any collection of spots. Spots and scenes may be broadcast like any other message in INMC. They may be used, for example, to broadcast the position of a receptionist in INMC. Actors may join scenes by sending and receiving traffic like this:

.

.

.

send <my_coords, my_mail_address> to broadcast

.

.

.

This *scene gas* traffic plays the role of light in INMC. Unless an actor sends out this information there is no other way to know where it is, except by pre-assignment. Of course this gives no hint as to how actors cooperate to maintain changing scenes or agree upon utilities like long-lived barriers, data bases, trees, etc. In BAM, the cyberdeck supplies the perceptual picture of cyberspace or uses one sent by another actor.

An actor can move its spots around by broadcasting their different positions, one after another. Similarly, different scenes, perhaps large regions of cyberspace, can be broadcast. In order to participate in a scene, an actor must send messages to the mail address of the spot to see what happens. This is an example of "what is in cyberspace," advertisements for spots, and it motivates the word *ballistic* in BAM. The diametrically opposed alternative of having a single agent run the action in cyberspace is impractical.

In our world, macroscopic objects supply the props for human scenes and their inertia, volume, density, hardness, etc., provide for some stability. Our consciousness participates in this and gives us a world view for a few years. However, in cyberspace, stability of objects is a matter of cooperation between agents, which means that the results can be arbitrary. If there is disagreement between actors about what is in a scene, how does an actor decide which scene to act? That is, what "happened" in cyberspace; scene A or scene B? Either can be broadcast. As Winograd and Flores (1986) point out, ". . . the essence of language as a human activity lies not in its ability to reflect the world, but in its characteristic of creating commitment."

The common holon summarizes the sort of behavior necessary for a stable cyberspacee. *To participate in cyberspace, each actor gives up some autonomy.*

BAM is specialized further; actors give up some autonomy of action and some autonomy of navigation in order to make INMC look something like a cyberspace. This is done by having scenes displayed and arbitrated by a specialized actor system.

Definition: A map spot is a specialized spot that advertizes the position of a *mapper agent* (some receptionist of an actor system), which will negotiate with a set of independent actors to display a scene and arbitrate disputes over proposed changes. The mapper may also have rules that automatically changes the scene with time.

<map_spot> ::= (<coordinate>, <mail_address_map_coord>)

where <mail_address_map_coord> is a special public mail address, which may be calculated from the <coordinate>. For example, f(<coordinate>, <corporation_security_code>), <coordinate> + <utility_display>, <coordinate> + <summary_view_of_cyberspace_from_ here_for_$10/hr.>, etc. If mail addresses are just bit strings, then all the security problems in networks today apply to them.

To actually see in cyberspace necessitates such a system of spots and a map-type protocol related to the SRF. In this manner, there could be many maps of cyberspace. A cyberdeck would see the maps according to its authority, with all the problems and excitement as detailed in *the books*. In order to see into a region, a deck would send a special message—call it a *ping*—to a map_spot of <mail_address_map_coord> and receive a description of the region, perhaps a gift batch of icons and resources describing what's there or "ice." Ice is a term from *the books*, meaning an active security screen.

Of course, if all the actors in a scene in fact follow some set of protocols, then an arbitrator is not necessary. In BAM there is no global, INMC-wide arbitrator, and there may be scenes that are free from arbitration, whatever the behavior of the players. Hence, there could exist situations in which the global structure of INMC was quite different, depending on which spots were pinged. There is no escape from the problem. *All general cyberspaces may diversify.*

The Cyberdeck in the Ballistic Actors Model

How does a cyberdeck fit into the Ballistic Actors Model?

A *point of view* (POV), is an interpretation or abstraction of a cyberspace scene into semantic terms for some agent.

A cyberdeck constructs a POV for a human being. The replies to messages by other actors may not be suitable for human understanding or for allowing some agent to make a decision. The raw traffic of cyberspace may be almost unintelligible to humans and most agents. For example, humans like to have scenes displayed in three dimensions and in color. A far more limited agent may be able to deal with only certain well-bounded situations.

This interpretation may be as simple as translating the responses from spots; for example, the translation may be "I am a blue cube." The semantics may only be relevant to the behavior of a particular actor system or they may be the human semantics necessary to represent a scene in cyberspace to the user of a cyberdeck.

There may be many different POVs of the same scene. *Points of view imply delusions.*

Figure 12.8 shows the components of a cyberdeck design in BAM. The components are summarized below:

• *Transducer (T)*. Basically an input/output device, it is a specialized receptionist. Unlike standard actors, in addition to being able to exchange messages with other actors, *transducers* sense events in the world, encode them into information, and can create messages from them. They can conversely, on receipt of a message, drive physical actuators to produce effects in the world. They are at the edge of cyberspace. They might be CRTs, TV cameras, robot arms, etc.

By the Shannon information formalism, transducers correspond to the transmitter of figure 12.1. They can only generate messages from a given message set. Put another way, they can ask only a certain set of questions of the world unless reprogrammed.

• *Translator*. Translates to/from transducer messages to/from the scene languages of the POV. It is necessary to translate from the voltage signals of a joystick, tracking ball, TV camera, EEG trace, etc., to messages that control scenes.

A joystick may provide voltage signals, but this still must be translated into coordinate movement, for example.

• *POV (point of view)*. A mostly passive component that contains an representation of what the agent experiences or sees of cyberspace.

• *POV constructor*. Assembles a POV from mapper messages. How this is to be done is of course an unsolved problem; it must clearly show what its interpretation of the consensual delusion is. It constructs an abstraction of the information given from the mappers. Here the cyberdeck can supply its own wallpaper of cyberspace.

• *Scene manager*. Manages the current scene by negotiating with mappers. The scene manager will send changes to a scene to the mapper to see what happens. It will move the cyberdeck's spot around in a scene. The scene manager/POV pair might be compared to the knowledge

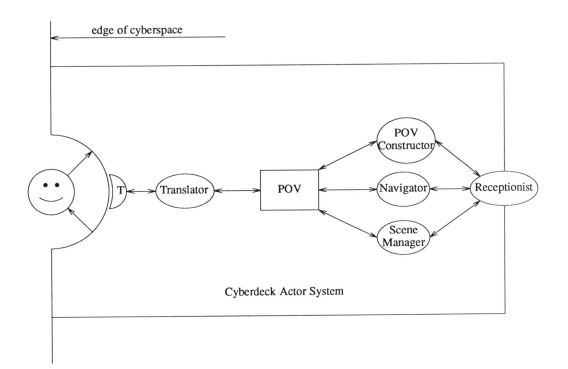

Figure 12.8
Design for a cyberdeck.

source/blackboard combination in Hearsay II (see Fennell and Lesser 1988, 106).

• *Navigator.* Chooses between mappers. It takes care of long journeys and scene changes.

• *Receptionist.* Communicates with the other actors of INMC. Most of their traffic will be scenes.

• *The edge of cyberspace.* Any actor that is a transducer is at the edge of cyberspace.

The significance of this design is that any cyberdeck will have to be at least as functionally complicated. Only major relationships are shown; each component may communicate with any other.

Here's a sample cyberdeck interaction. The user indicates a movement and this is translated into the POV. The POV constructor may just adjust the user's picture slightly if the change is really very local. The scene manager will move the deck's spot around. If the change is not very local, the navigator may interact with the distant mapper to see if a new view of the scene is necessary. New views are processed by the POV constructor.

Real Networks and Actor Identity

We will now discuss what happens when more realistic assumptions are made about cyberspace than are used for INMC, calling the more practical system the *network*.

The following assumptions make the picture a lot more realistic:

1. Limits to CPU speed

2. Limits on the number and availability of CPUs

3. Transmission and queueing delays in message passing

4. Differences in the sizes and speeds of CPUs available to different actors

5. Broadcasting that becomes impractical for most cyberdecks

In INMC there were no "natural forces," but on the network, transmission and queueing delays can dampen the response to movement, for example, with a joystick. Any system with feedback will work slower.

These constraints make something like a mapper necessary anyway, that is, a cyberdeck will have to rely upon distant agents.

These limitations motivate the use of ABCL/R's dynamic acquisition of scripts. Some mappers may support a sort of actor farm, which will allow distant cyberdecks to run scripts on the actors that the mapper "owns," an example of actors giving up autonomy. Let the mapper support an ABCL/R metaobject, ↑runme, which will accept a cyberdeck's script; the actor runme will run the script's behavior. Whole actor systems can be copied around the network, making viruses and worms natural to cyberspace. Junk mail will also be a huge problem. The object ↑↑runme is of course capable of modifying the new script on the spot. The *identity* of an actor soon becomes a nebulous concept.

From the current popularity of *core wars*-style virus competitions, in which competing processes try to gain control of some bits in memory, the question, Who owns a CPU? also has no easy answer.

A given cyberdeck must of course have the proper *authority* with the mapper it's trying to reprogram. (This sounds more like the Vudun and Loas of *the books* at every step!) As a practical matter, a cyberdeck would ping any agent that can "get things done."

Disparities in system capacity can exclude smaller systems from fully participating in script downloading.

If CPU ownership is nebulous, then so is bopper ownership.

The Physical and Algorithmic Complexity of Cyberspace

The Physical Complexity of Cyberspace

Cyberspace will be a large, physically complex system; however it will not be a cooperative system in the sense of physical complexity theory, (see Nicolis and Prigogine 1989, Prigogine 1980). This is because current computer components are supposed to be isolated from each other and only connected through logic gates. Ideally, each bit of cyberspace will be physically independent of any other. This will make it a hugely low entropy system.

In cyberspace, we can say in general that we lose the effects of the law of large numbers, the central limit theorem, and the effects of entropy. These are the laws that explain why ordinary objects, made up of many smaller particles, are pretty stable most of the time. All the hardworking laws that keep lots of particles in line will be gone.

In the world, most objects are in states in phase space surrounded by many other states that are indistinguishable macroscopically, as in thermodynamic equilibrium. However, in cyberspace, there is no such situation. There is no guarantee that states very near a current state may not be catastrophic, leading to massive disorganization and destruction of information.

There is no natural direction of time in cyberspace. Anyone who is paying the rent can run time anyway they want.

Our universe had an initial low entropy big bang, but cyberspace might be viewed as steady-state; cyberdecks, information links, and perhaps AIs inject low entropy messages constantly. By analogy, animals maintain their low entropy state by eating other low entropy living things. (See Schrödinger 1944.) In cyberspace, any stability will have to be algorithmic.

The Algorithmic Complexity of Cyberspace

Rudy Rucker (1987) puts forth the hypothesis that implies that without using a program on the same order of size as cyberspace, we cannot make statements about its stability or evolution. Indeed, how can we model cyberspace when it itself may embody so much of our mathematical knowledge?

Here, Chaitin's 1965 theorem is to the point:

If T is a theory of mathematics and T is
i. finitely given, and
ii. consistent, then
iii. there is a number t such that T cannot prove that any specific bit string has complexity greater than t. (quoted in Rucker 1987, 287)

Definition: The *complexity* of a bit string M is the smallest number n such that there is an n-symbol program P that makes a universal Turing machine U print out M and stop.

Looked at in one way, Chaitin's Theorem says that I can't stand on top of my own head. I cannot be smarter than I am. There is a fixed, finite limit to the complexity of the objects about which my logical discourse can say anything significant. (Rucker 1987, 287; see also Chaitin 1987)

This is similar to Stephan Wolfram's observation that, in many cases, we cannot predict how a given cellular automaton will evolve; we can only watch it evolve.

Especially if cyberspace has many different agents with many different points of view, strategies, and rules, it will be immensely complicated because of the difficulty of abstracting away these different points of view and rule systems. Our current mathematics is not adapted to dealing with thousands of different rule systems side by side. It is adapted to taking some particular rule system, a set of axioms, and developing it. *How, then, can we comprehend cyberspace well enough to describe or maintain it?*

Our universe has a large-scale geometric structure, but the space of cyberspace is very arbitrary. How can we describe it as an object? Cyberspace does have a Gödel number, for example. We can (1) hit the stop button, (2) record all the bits into long string, (3) agree upon an encoding for the symbols, and (4) generate a Gödel number.

Many macroscopic systems exhibit *chaotic* behavior. In chaotic systems (1) the behavior is infinitely dependent on the initial conditions, and (2) trajectories do not diverge, but are contained within a *strange attractor* that mixes them. Chaos, once exemplified by fluid turbulence and certain chemical reactions, has now been discovered in many biological systems, including the heart and the brain. (See Wiggins 1988.) No one knows the significance of chaos in holding together the bodymind; could it be a significant barrier to moving humans "into" cyberspace or an essential mechanism? Chaos has been discovered in simple cooperative system models. Kephart, Hogg, and Huberman (1989) found that just 300 intelligent agents competing for a finite resource pool can exhibit what looks like chaotic behavior.

Cyberspace will certainly develop over time. *The books* mention the development of "Loas" and "Alephs," and the "evolution" of boppers. There is a contrast here with Gould's picture of the evolution of life on earth. In biological evolution as described in Gould 1989, there were more animal phyla a billion years ago than there are today. There are fewer phyla now because the major forces at work are *decimation* and *diversification*. Decimation, in the form of major disasters, for example, wipes out large populations of animals; then one of the species remaining diversifies to fill in the newly available niches. In his book, Gould emphasizes that evolution is not simply a progression from the "simple" to the "complex."

Simple Darwinian natural selection does not run biological evolution. Hence, computer science models that are based on simple Darwinian natural selection—for example, genetic algorithms—cannot be expected to evolve in the way which life has.

Work in the field of *artificial life* (see Langton 1989) indicates that systems like cyberspace, which have a large number of comparatively smaller interacting elements, exhibit *emergent behavior*. That is, the total behavior may be very much more complicated than the rules of behavior for the elements. To fantasize, "Loas" might be considered to be emergent behavior. CPU run agents may be a lot faster than humans at many complicated projects, which implies that the *time scale* of cyberspace may not naturally fit humans. Actors may fit the description of Rod Brook's robots, "small, fast, and out of control" (Kelly 1990).

Cyberspace may grow in ways quite qualitatively different from life. It will certainly be an A-life zoo. In fact, systematic decimation of A-life forms may be a necessary maintenance function and a spur to the diversification of emergent behavior.

Conclusion

This chapter has described cyberspace largely as an engineering problem and tool. But cyberspace is a tool that may intimately affect our own bodyminds as no other tool ever has. At the end of the twentieth century we cannot pretend to be naive about the consequences of technology. We are drawn to our fates faster and faster.

If I cannot bring a smile, a flower, and a cloud into cyberspace, I will not enter. I will just put it down, as I would any other clunky tool.

References

Agha, Gul. *Actors: A Model of Concurrent Computation in Distributed Systems*. Cambridge: MIT Press, 1986.

Agha, Gul and Hewitt, Carl. "Concurrent Programming Using Actors: Exploiting Large-Scale Parallelism." In *Readings in Distributed Artificial Intelligence*, ed. A. Bond and L. Gasser, 398. San Mateo, California: Morgan Kaufmann, 1988.

Arbib, Michael. *The Metaphorical Brain 2: Neural Networks and Beyond*. New York: Wiley, 1989.

Bond, Alan, and Gasser, Les, eds. *Readings in Distributed Artificial Intelligence*. San Mateo, California: Morgan Kaufmann, 1988.

Chaitin, Gregory. *Information Randomness and Incompleteness*. New York: World Scientific, 1987.

Comer, Douglas. *Internetworking with TCP/IP: Principles, Protocols, and Architecture*. Englewood Cliffs, New Jersey: Prentice Hall, 1988.

Davis, Philip, and Hersh, Reubin. *The Mathematical Experience*. Boston: Houghton Mifflin, 1981.

Davis, Philip, and Hersh, Reubin. *Descartes' Dream*. New York: Harcourt Brace Jovanovich, 1986.

Fennell, Richard and Lesser, Victor. "Parallelism in Artificial Intelligence Problem Solving: A Case Study of Hearsay II." In *Readings in Distributed Artificial Intelligence*, ed. A. Bond and L. Gasser, 106. San Mateo, California: Morgan Kaufmann, 1988.

Gasser, Les, and Huhns, Michael, eds. *Distributed Artificial Intelligence*, Vol. 2. London: Pitman and San Mateo, California: Morgan Kaufmann, 1989.

Gibson, William. *Neuromancer*. New York: Ace, 1984.

Gibson, William. *Count Zero*. New York: Ace, 1987.

Gibson, William. *Mona Lisa Overdrive*. New York: Bantam Books, 1989.

Gould, Stephen. *Wonderful Life: The Burgess Shale and the Nature of History*. New York: W. W. Norton & Co., 1989.

Kelly, Kevin. "Perpetual Novelty: Selected Notes from the Second Artificial Life Conference." *Whole Earth Review 67* (Summer 1990): 20–29.

Kephart, J. O., Hogg, T., and Huberman, B. A. "Dynamics of Computational Ecosystems: Implications of DAI." In *Distributed Artificial Intelligence*, ed. L. Gasser and M. Huhns, 79. London: Pitman and San Mateo, California: Morgan Kaufmann,1989.

Koestler, Arthur. *The Ghost in the Machine*. Chicago: Henry Regnery Company, 1967.

Koestler, Arthur. "Beyond Atomism and Holism: The Concept of the Holon." A. Koestler and J. R. Smythies, eds. *Beyond Reductionism: New Perspectives in the Life Sciences*. Boston: Beacon, 1971.

Lampson, B. W., Paul, K., and Siegert, H. J., eds. *Distributed Systems, Architecture and Implementation*. New York, Springer-Verlag, 1981.

Langton, G. C, ed. *Artificial Life*. New York, Addison-Wesley, 1989.

Minsky, Marvin. *The Society of Mind*. New York, Simon and Schuster, 1985.

Nicolis, Gregoire, and Prigogine, Ilya. *Exploring Complexity: An Introduction*. San Francisco: W. H. Freeman and Company, 1989.

Penrose, Roger. *The Emperor's New Mind*. Oxford: Oxford University Press, 1989.

Prigogine, Ilya. *From Being to Becoming: Time and Complexity in the Physical Sciences*. San Francisco: W. H. Freeman and Company, 1980.

Resnikoff, Howard. *The Illusion of Reality*. New York: Springer-Verlag, 1989.

Rucker, Rudy. *Software*. New York: Avon Books, 1982.

Rucker, Rudy. *Mind Tools: The Five Levels of Mathematical Reality*. Boston: Houghton Mifflin, 1987.

Rucker, Rudy. *Wetware*. New York: Avon Books, 1988.

Schrödinger, Erwin. *What Is Life?* Cambridge: Cambridge University Press, 1944.

Shannon, Claude, and Weaver, Warren. *The Mathematical Theory of Communication*. Urbana: The University of Illinois Press, 1963.

Simpson, Patrick. *Artificial Neural Systems*. New York: Pergamon Press, 1990.

Tenney, Robert, and Sandell, Nils Jr., "Strategies for Distributed Decision Making." In *Readings in Distributed Artificial Intelligence*, ed. A. Bond and L. Gasser, 236. London: Pitman and San Mateo, California: Morgan Kaufmann,1989.

Werner, Eric. "Cooperating Agents: A Unified Theory of Communication and Social Structure." In *Distributed Artificial Intelligence*, ed. L. Gasser and M. Huhns, 30. London: Pitman and San Mateo, California: Morgan Kaufmann,1989.

Wiggins, S. *Global Bifurcations and Chaos: Analytical Methods*. New York: Springer-Verlag, 1988.

Winograd, Terry, and Flores, Fernando. *Understanding Computers and Cognition: A New Foundation for Design*. New York: Addison-Wesley, 1986.

Yonezawa, Akinori, ed. *ABCL: An Object-Oriented Concurrent System*. Cambridge: MIT Press, 1990.

13 *Virtual Worlds: No Interface to Design*

Meredith Bricken

Introduction

In a virtual world, we are inside an environment of pure information that we can see, hear, and touch. The technology itself is invisible, and carefully adapted to human activity so that we can behave naturally. We can create any imaginable environment and we can experience entirely new perspectives and capabilities within it. A virtual world can be informative, useful, and fun; it can also be boring and uncomfortable. The difference is in the *design*.

The platform and the interactive devices we use, the software tools, and the purpose of the environment are all elements in the design of virtual worlds. But the most important component in designing comfortable, functional worlds is the person inside them.

Cyberspace technology couples the functions of the computer with human capabilities. This requires that we tailor the technology to people and refine the fit to individuals. We then have customized interaction with personalized forms of information that can amplify our individual intelligence and broaden our experience.

Designing virtual worlds is a challenging departure from traditional "interface" design. In the first section of this chapter I differentiate between paradigms for screen-based interface design and paradigms for creating virtual worlds.

The engineer, the designer, and the participant co-create cyberspace. Each role carries its own set of goals and expectations, its own model of the technology's salient features. In the second section of the chapter I address these multiple perspectives, and how they interrelate in the cooperative design process.

In conclusion, I consider broader design issues, including control, politics, and emergent phenomena in cyberspace.

Design Paradigm Shifts

We can identify the paradigm shifts between traditional interface design and designing virtual worlds by distinguishing those unique capabilities of virtual world technology that add new elements to the design task.

From Interface to Inclusion

An interface is a surface forming a boundary between two regions. On a computer, this surface is the screen of the monitor, a boundary between the information environment and the person accessing the information. Traditional interface design has concerned itself largely with how that screen can present the most effective indication of a program's scope and functionality, and how to interact with screens in general.

Using a head-mounted display (HMD) and position trackers allows us to move through the screen's barrier, to interact directly with various information forms in an *inclusive* environment. The quality of this distinction between viewing and inclusion can be illustrated with the following analogy. Viewing 3-D graphics on a screen is like looking into the ocean from a glass-bottom boat. We see through a flat window into an animated environment; we experience being on the boat.

Looking into a virtual world using a stereographic screen is like snorkeling. We are at the boundary of a three-dimensional environment, seeing into the depth of the ocean from its edge; we experience being between at the surface of the sea.

Using a stereoscopic HMD is like wearing scuba gear and diving into the ocean. Immersing ourselves in the environment, moving among the reefs, listening to the whalesong, picking up shells to examine, and conversing with other divers, we invoke our fullest comprehension of the scope of the undersea world. We're There.

Is eliminating the screen boundary essential to participation in cyberspace? It depends on what you mean by "cyberspace."

Cyberspace is still in its definitional infancy. The lexicon is in formative turbulence, and the relationship of existing information technology to virtual world implementations is not entirely clear. The necessity of using HMDs is one of many questions. For example, is electronic mail cyberspace? What part do interactive video environments and multi-media play? Must cyberspace be a social experience or can there be personal cyberspaces? Is every three-dimensional computer display cyberspace? Are multisensory systems necessary?

Using the broadest possible definition of cyberspace as any computer-moderated information system, the question of how we connect with this system can be addressed separately. Our participation in cyberspace then becomes a matter of the *bandwidth* of our access.

We can create and enter cyberspace in localized virtual worlds of video, computer graphics, and sound. These interactive environments today run independently, but the goal is to connect them into a networked cyberspace matrix. We can interact with the existing matrix of computerized communications using text and two-dimensional graphics. When we add the capabilities of three-dimensional graphics, we both increase the information density and allow more accurate understanding of what we see. Multisensory interaction adds additional layers of perception and meaning to our experience.

The widest bandwidth of participation in cyberspace is enabled when we pass through the barrier of the computer screen to inhabit, fully sense, and interact directly with people and information. *Inclusion*, this ability to get inside of information, defines a new generation of computing.[1]

The personal impact of inclusion within a virtual world has been documented; in questionnaires given to 300 participants in Autodesk's cyberspace, "being inside" the virtual world was rated as the most compelling aspect of the experience.

An important design consideration stemming from inclusion is that while we interact within a virtual world, we are simultaneously inhabiting the physical world. People are functionally attuned to the earth's gravity and to vertical position. Perceptual conflicts between the virtual and physical worlds cause physical discomfort and feelings of disorientation that can last well beyond the period of inclusion.

We can finesse the problem of conflict using transparent HMDs, which allow natural orientation cues to exist behind superimposed virtual world elements. When using opaque HMDs, orientation cues imbedded into the design of the virtual world are essential.

The fundamental visual feature for vertical orientation is the extended ground plane. Using a clearly defined virtual horizon allows us to engage our natural use of peripheral vision for orientation. The elements of the virtual environment then exist in relationship to that horizon, which serves as a fixed point of reference. Including depth cues in the ground plane is helpful in judging the size and distance of objects: a grid works well, as do elements like train tracks or roads.

Maintaining congruence between the virtual horizon and the perception of our physical position is another design consideration. Even a small disparity between the ground plane of the physical and virtual worlds can throw you off balance because your eyes and your inner ears are getting conflicting messages.

As a point of reference for our own location in cyberspace, the virtual hand (ideally, the virtual body) is both evocative and functional. It is compelling to see your virtual fingers move as you wiggle your hand. Watching your dynamic self-representation within the virtual world is convincing evidence that you're There. And when you're sitting in a chair and simultaneously swooping through a virtual city, focusing on the hand pointing out in front usually eliminates feelings of vertigo.

Slow system response time can also cause dizziness; any lag between turning our head and seeing the world go by is disorienting. This is a design challenge for engineers; no amount of contextual artifice will compensate for lag.

From Mechanism to Intuition

Virtual world technology adapts computers to human functioning rather than training people to cope with interactions based on the computer's mechanism. When we use natural rather than symbolic behaviors, like reaching out to turn an object rather than specifying the view with a coordinate system, we don't have to think about how to do things. We can focus our attention on what we're doing.

It's easy to make mistakes in a symbolic interaction, but turning your head to see what's behind you is inevitably successful. Moving around by pointing in the direction you want to go is equally straightforward, and speaking is the way most people communicate their intentions. These intuitive behaviors are hard to get wrong. There is a natural mapping between intention, action, and feedback.

Visceral access and intuitive interaction evoke our full sensory-cognitive capacity to comprehend. The computational environment becomes a more powerful tool that is also easier to use. New groups of people have access to cyberspace. You don't have to know anything about computers; you don't even have to know how to read.

The task of designing a virtual world, then, does not revolve around helping people interpret what the machine is doing, but on determining the most natural and satisfying behaviors for particular participants, and providing tools that augment natural abilities.

From User to Participant

Among software developers, the term *user* refers to the generic person who, at the end of the programming and interface design process, receives a software application geared to "average" human functioning. *Participants* are active agents. In cyberspace we will use software tools to create our own applications, and to co-create the matrix.

Sensory coupling requires us to regard each participant as an individual, and individuals are highly idiosyncratic. Is tailoring this technology to each person too expensive and time-consuming to consider seriously?

A precedent for fast and easy customization exists in the configuration process of the DataGlove™. Gesture recognition is specific to the size and proportion of each hand, and hands are highly variable. So each person who puts on the glove makes three quick calibration gestures, after which the movement of that particular hand is recognized. It's a piece of code.

However, accounting for physiological differences is not always that easy. For example, an HMD used in the presence of vibration (such as jet cockpits) shifts position on the participant's head as it shakes in response to the oscillation. The display can be dynamically stabilized for a certain range of movement. But people respond differently to vibration, in proportion to their height, their weight, and even their level of muscular tension. Tense people get moved around by vibration more than relaxed people. Tailoring display stabilization to these differences is not a trivial task.[2]

The design of these customized virtual worlds will require our most sophisticated human factors expertise. Reciprocally, it seems inevitable that this new domain will enlarge the scope and diversity of human factors testing. In one Human Interface Technology Lab (HITL) project, we're constructing a virtual airplane for doing research on human fit, view, and reach in cockpit and cabin design.

In the virtual cabin we will be able to move along the aisle, choose a seat, and see what kind of view we get from the window beside our seat; we can stretch our legs to see if they'll fit comfortably, and look to be sure we can see the movie screen. In the virtual cockpit we can sit in the pilot's seat to see if the controls are easy to find and within reach, and if we can see the instrumentation readouts from that position.

By using virtual environments to do ergonomic testing on aircraft and other products during the design phase, the manufactured result will be more comfortable and efficient for us to use. And the manufacturing process will cost less because design mistakes are recognized and changed before they're implemented, saving time and materials.

From Visual to Multimodal

The more sophisticated virtual worlds are "acoustigraphic" environments, achieved by coupling 3-D sound systems with the stereoscopic head-mounted display. Both ambient and localized sounds are coordinated with the graphical representations and with the movements of the participant. We can hear the sounds of traffic in the distance and wind rustling the leaves that move in nearby trees; we can listen to each fish tell a tale as it swims through a musical stream. Today haptic feedback devices are being developed to create the illusion of substance and force within the virtual world; we can feel textures or terrain or the pull of gravity.[3]

All of these capabilities require designers to consider issues of sensory load related to individual learning and performance styles. People who understand what they hear more easily than what they see, for example, could have a world where objects are located by sound and acoustic icons are used.

Creating meaningful and aesthetic combinations of sight, sound, and touch in a virtual world is a complex engineering and design task. One parameter is fidelity between the location of a sound source and the location of its associated visual image. In some applications approximate locations of sight and sound will suffice; in the same way we perceive that a voice is coming from the ventriloquist's dummy, visual cues will override small discrepancies. Other applications, such as concert hall design, will require exacting acoustical models. Ranges of tolerance for mixed-modal operations and for sensory ambiguity and paradox are challenging research questions.

From Metaphor to Virtuality

Metaphors are essential to human understanding[4] and have been thoughtfully employed by interface designers to offer the computer user a clear mental model of what to expect from a particular application. The "desktop" metaphor is one of the most useful and widely known examples of this technique.

Cyberspace participants interact directly with virtuality (rather than reality) to experience the embodiment of the application. This environment is "as if real."

When we use a metaphor to describe an unfamiliar place or process, we think, "it would be approximately like that." When we are inside an environment, however, we think "it is exactly like this." We don't rely on the metaphor of "house" after we've moved into our house; we build a particular cognitive model of our specific domain (to predict where to find our socks, or get around in the dark).

Thus, metaphors serve a different purpose in virtual world design. They become valuable as organizing principles for designers. Theater, architecture, games, and foods have been suggested as useful metaphors for approaching the complex dynamics of cyberspace organization and interaction.[5]

In cyberspace, appearance IS reality. And virtual appearances are completely arbitrary. Despite injunctions not to judge books by their covers, nor to assume that what glitters is gold, we judge real-world things and people by their appearance all the time. Within virtual worlds, we will need alternative criteria for evaluation, especially in our interactions with others. Is there a distinction between the message and the virtual messenger?

Multiple Models of Cyberspace

Virtual world technology provides interactive environments tailored for participants. Hardware and software engineers create the interactive devices and tools; they focus on how systems work. Designers work with engineers to refine prototypes ergonomically and with participants to tailor systems to individual purpose; they focus on the people using the technology. Participants are concerned with costs and benefits of the system, and focus on their particular virtual world.

Because engineers are concerned with execution, designers with evaluation, and participants with function, they each have different conceptual models of the system. To co-create cyberspace requires communication based on some reciprocity of perspective between these groups; awareness of these multiple models is a starting point.

The Engineer's Model

The perspective of engineers is based on implementation. Their goal is to make the technology work. Their model of the system involves the functional intricacies of hardware and software, and their priorities are structured around what is technically possible, or at least feasible.

The technical feasibility of the HMD was demonstrated nearly 30 years ago, and engineers have been refining it ever since. Developments in computer chips and display technology have brought down costs, and now HMDs are economically feasible as well. Engineers are working to provide configurable or optional types of head-mounted display:

• Opaque (for immersion in the virtual world)

• Transparent (for virtual worlds superimposed on reality)

• Acoustically coupled (for 3-D sight and sound)

• Eye-tracking coupled (for control by looking)

• Microphone or ear-mike transducer coupled (for hearing and voice commands)

The obvious problems with the current HMDs are that they're bulky and low in resolution. In search of unobtrusive HMDs, projects that anticipate emerging capabilities in microchip and laser technology, such as the micro-laser scanner, are underway. The tiny HMD microlaser is designed to safely scan pictures directly onto the retina of the eye. This device could allow us to see detailed virtual worlds with the comfort of sunglasses and the clarity of natural vision.

Engineers are also creating and adapting devices for interacting with virtual worlds. There are several kinds of peripherals for this purpose, including:

• Computerized clothing that recognizes physical gestures as commands

• Systems that track the movements of the body

• Trackballs and joysticks that allow movement of perspective

• Devices that allow interaction with 3-D objects such as the bat, wand, and glove

• Feedback devices that use force, pressure, or vibration

• Remote operation systems that translate human movements into the control of machinery

Software engineers are developing new ways for us to represent and interact with information. They are creating the tools we'll use to construct and manipulate virtual world objects and systems, such as:

• Virtual world and matrix operating systems

• Interactive 3-D graphics construction and animation packages

• Specialized information structures and query systems

• Multimodal data visualization and display techniques

• Spatial fields and topologies

• Autonomous agents and entities

Sophisticated software can provide us with dynamic entities, forces, and patterns in even a relatively simple private virtual world. We can have weather, gardens that grow and flower and seed to grow again, and independent creatures that come and go for their own reasons. Within these systems we can slow time or speed it up or run in different time streams. Autonomous entities provide complexity and surprise.[6]

The Participant's Model

Most people are more interested in what they might *do* in cyberspace than in how the system works. The participant's model of the system is based on their experience within it. From the participant's perspective within the virtual world, the environment exists "as if real." The interface technology disappears from view, and the particular acoustigraphic context defines the domain. The participant in the virtual world asks some of the essential questions of humankind:

Where am I? Your primary understanding of the virtual world will derive from what you see and hear when you put on the HMD. You can be nowhere or anywhere, as defined by the acoustigraphic context in which you find yourself. Information forms are utterly malleable; you can be inside worlds patterned after real locations like a city, another planet, or a human body, or you can vary the scale or physics or attributes of the location. You can create your own new worlds to explore, or you can travel through the matrix to visit public virtual environments.

Who am I? What will your virtual self look like? How will you sound? You don't need a body; you can be a floating point of view. You can be the mad hatter or you can be the teapot; you can move back and forth to the rhythm of a song. You can be a tiny droplet in the rain or in the river; you can be what you thought you ought to be all along. You can switch your point of view to an object or a process or another person's point of view in the other person's world.

Assuming multiple perspectives is a powerful capacity; only after young children are developmentally ready to understand that each person sees from a different perspective can they learn to relate to others in an empathetic way. What will we learn from adopting the countless perspectives of cyberspace?

What can I do? There are five general categories of behaviors:

• Relocation: You can move around in your virtual body the same way you do in the physical world, by walking and turning, bending and reaching. (You are presently limited to the range of HMD cables and the field of the position tracker, but less constraining systems are being developed.) You can fly; using a gesture, joystick, or trackball, you can move smoothly with variable speed in any direction. You can jack in to a new perspective, instantly transferring to a new location with or without your virtual body, by naming or pointing to your destination. In cyberspace, the concept of "distance" is optional: relocation is independent of time and space.

• Manipulation: You can move virtual objects with your hands or with your eyes, or you can use a 3-dimensional cursor or wand. You can tell an object to move, or you can program patterns of movement for an object or set of objects.

• Construction: Presently, you build virtual worlds from the outside, and then enter into them. The ability to create and alter these environments interactively is being developed in the form of software toolkits. Your ability to shape and visualize information will depend on the tools you use. Do you prefer virtual palette knives or hammers, lasers or sieves?

• Navigation: Finding your way around in a small virtual world is not difficult; when you can travel through large databases and within the interconnected systems of the matrix, new methods of locating objects, and navigating between domains may be needed. What techniques of wayfinding are most effective in large, complex worlds? Will you create a virtual guide? Or, will you choose an animistic universe in which guidance is embodied in the information form; when you want to find out more about a chair you can ask it, and if you lose it, call its name and it appears. Or you jack in to its location. Do we need maps if there is no distance?

Who is with me? The feasibility of sharing virtual spaces has been demonstrated, but this aspect of cyberspace is now primarily speculative. How will other people be represented in your virtual world; do you control their appearance or do they convey a fixed form to you? The usual concept of "who" people are implies some consistency, but in cyberspace people can adopt an ambiguous form or multiple personae. The meaning of personal appearance changes; in the physical world, you predict certain behaviors from the way a person looks, but what expectations might you have of a winged lobster?

Unique social and cultural forms can emerge in cyberspace as we negotiate the mutual control of a shared environment and decide on conventions for our behavior within it. When we connect with the matrix, how will we choose the company we keep? Can cyberpunks form virtual tribes, and can intruders be deleted?

The Design Model

Designers focus on the way people access and interact with cyberspace. The designer's objective is the creation of comfortable, functional virtual worlds that satisfy the needs and intentions of the participant. Four aspects of this design task are discussed.

1. Designers work with engineers to tailor the technology to people's physical and psychological needs by testing the usability of systems and suggesting refinements.

For example, some of the devices we use to move around inside a virtual world afford more maneuverability than the human body is accustomed to. In one phase of Autodesk's cyberspace project, engineers incorporated a trackball with six degrees of freedom and fast response time for moving around inside a virtual office. In first testing the trackball, we tumbled into perspectives we could barely interpret and careened through the walls and floor. Even after a significant amount of practice, control was elusive.

People found moving through space in this way both frustrating and disorienting. They would lurch into weird positions and then try to reorient toward an upright view. They approached that task by making a series of small adjustments in their perspectives, but each one required conscious interpretation of the new position ("Where am I now?") and planning of the next adjustment ("Where do I want to be?"). The simultaneous changes in pitch, roll, and yaw as well as direction in 3-space was confusing; people are not used to moving without the guiding constraints of ground and gravity. Readjustments were error-prone. Many participants would get three or four of the degrees of freedom right, while spinning out in the others. It was an interesting sensation, but it stopped being fun almost immediately as queasiness set in.

In response to these observations, we constrained freedom of movement. We chose to hold roll constant to the horizon; rolling sideways had the most disconcerting effect on people. This change improved control, and we had fewer complaints of disorientation. Several people did ask to try the unconstrained mode, just for the experience; a few moments of swooping and plummeting satisfied their curiosity, and confirmed the usefulness of the constraint. (Another approach, that we didn't test, would be to slow the trackball's response speed.)

2. Designers work with participants to customize virtual worlds. This requires a dual awareness of the individual needs and preferences of the participant and of the capabilities of the technology.

Eliciting preferences can be done in much the same way an architect works with a client building his home.[8] Physical and perceptual characteristics of the participant will need consideration, such as motion sensitivity, color blindness, handedness and manual dexterity, information mode attention and retention.

The designer can provide participants with information about different cyberspace platforms, software packages, and peripheral devices. Systems provide varying levels of complexity and dynamics in the virtual world.

For example, the PC-based cyberspace system currently limits the graphic image of the virtual world to about 500 flat-shaded polygons in order to achieve a rendering rate of 7 frames per second. (Alvy Ray Smith of PIXAR has estimated that we perceive the equivalent of 80 million polygons at 30+ frames per second when we look at a view of the real world.[9]) People often accommodate to the leisurely frame rate by slowing down their physical movements.

VPL's RB1 system allows models ranging from 1,500 to 15,000 Gouraud-shaded polygons to render at about 20 frames per second, depending on the power of the graphics engine. Elements of the display can be animated, at no cost to rendering speed.

3. Designers compose protoworlds that contain the graphical contexts and interactive possibilities appropriate to particular applications. A protoworld is composed of the default acoustigraphic context, appropriate libraries of objects and sounds, and the vocabulary of interactions available in the virtual world.

We can compare the protoworlds being considered for two HITL projects, *Cyberseas* and *Virtual Mobility*.

The Cyberseas system, intended for shipboard use, will translate real-time sonar and acoustic tomography data into a display of undersea terrain and objects. An opaque HMD configured for three-dimensional sight and hearing will allow the wearer to perceive images of whatever lies in the depths below. The protoworld's default context is a configurable terrain horizon; this will update to reflect the shape of the sea floor as transmitted by remote sensors beneath the ship. The acoustigraphics library will contain objects such as fish that you can hear coming; the pitch of the associated sound becomes higher as the fish approach the participant and lower as they move away. Viewing interaction will allow relocation by virtual movement to any point or perspective within the sensors' range, or the ability to jack in to a remotely operated vehicle (ROV).

The Virtual Mobility project is intended to allow paraplegics to access cyberspace as a working environment for doing computer-aided design (CAD). A transparent HMD allows the wearer to maintain physical-world orientation and awareness. The default context is empty. This protoworld's graphics library will contain CAD objects and menus. Interactivity will be tailored to the individual, since the physical capabilities of paraplegics vary significantly. Pick and place tasks can be performed using voice, eye tracking, and different facial muscles as controls.

4. Designers evaluate effective worlds by observing the learning, accommodation, and performance of participants. The ideal computer domain is fully explorable,[10] allowing the participant to build a specific

cognitive model through experience. A virtual world is inherently information-rich, and if feedback is explicit, the discovery process appears be a relatively efficient way to learn.

One example of this learning process was observed in some detail. Nearly 100 people, including members of the CHI '90 "Local Showcase Tour," explored two virtual worlds at the HITL. Virtual Seattle and Octopus's Garden were designed at the Lab as introductory adventures in cyberspace. They were running on VPL's RB1 system. All demonstrations were videotaped for later data analysis.

Virtual Seattle was a large-scale model centered on Puget Sound; the city contained landmarks such as the Space Needle, the King Dome, and a reasonably accurate skyline. The horizon was defined by mountains. The model was animated; a ferry moved back and forth from a terminal under the Olympics to the docks of Seattle, and an Orca surfaced several times as it moved north in the Sound. This was a viewing and movement demonstration, and the participant could either fly around or jack into the perspective of the ferry as it shuttled between destinations.

Octopus's Garden was an undersea plateau with large rock formations, swaying seaweed, and a school of fish that drifted in and out of sight. The large pink octopus moved about under the control of the system operator using the mouse. There was a treasure chest containing a bag of gold that could be removed and a starfish that could be detached from the rock arch to which it clung. There was also a small ROV. One could jack in to its perspective, or the perspective of the octopus, school of fish, or starfish.

A human guide acted as the kind of query system that might eventually be incorporated into virtual worlds. Participants were introduced into these worlds after a brief description of the gestures that allowed them to fly and to grab objects. Undirected exploration was allowed, and any questions were answered as asked.

Within the first one or two minutes, there were usually three kinds of questions: "Why can't I go there?" (we were testing the concept of a constrained zone of movement in Virtual Seattle; answer: "There are invisible walls around this world"); "Where am I?" (always asked from inside an object such as a mountain or building; answer: "You're inside the <object>"); and "Why aren't I moving?" (preprogrammed glove gestures using thumb-controlled speed demanded more manual dexterity than many participants were capable of; answer: "I'll move your hand into the gesture for flying," followed by the physical action).

After two or three minutes, most people demonstrated a clear understanding of the capabilities and constraints of these worlds. They were fully functional and without further questions. Participants engaged in a wide variety of behaviors. In Virtual Seattle they peered down the Space Needle, chased the Orca, got inside the ferry, flew down the city streets, and looked into the King Dome to see who was playing. In the Octopus's Garden they swam alongside the fish, played tag with the octopus, retrieved the bag of gold, stuck the starfish into the seaweed and explored the inside of the rock pile and the ROV.

These were relatively simple domains. Although few of the participants had experienced the technology before, individual learning time was remarkable short. There were no reports of frustration or feelings of failure; most people volunteered exclamations of exhilaration and enjoyment. This is not the typical novice response to a new computer program.

People really do seem to find virtual worlds easy to figure out. However, a closer look at the design criteria for these particular models is important. All virtual worlds are not equally learnable. How quickly and accurately we build a cognitive model of the environment is influenced by the environment's design. For example, when a virtual world seems familiar to us from some real-world experience, we may accommodate to it more quickly.

Virtual Seattle was approached by participants in much the same way that the actual city would be: key landmarks were quickly identified and located in relation to each other. Remembering where buildings were located did not seem to cause any difficulty. Participants who were local residents gave names to the buildings, identified the particular ferry route, knew where they wanted to go, and spent more time exploring the edges and limits of the domain; out-of-towners wanted to ride the ferry, see inside the Space Needle and watch for the Orca. Although preference for activities varied, adaptation time was consistently short. There was very little disorientation reported by participants in this model.

The Octopus's Garden contained some familiar undersea elements in an unfamiliar context. It generally took longer for people to accommodate to this model. Relocating objects when they were out of direct view caused frequent difficulty. Participants tended to stay within the central area of the defined space, although there were no external constraints on movement in this world. Several people commented that they felt dizzy during or after the experience. Three participants were practiced scuba divers. They were quick to learn their way around and less constrained in their exploration than most; they all reported enjoying the experience very much, and agreed that it was "like really diving."

Virtual Gods: Benign Design

Creating a World: giving form to intention, manifesting a dream, visualizing the unseen . . . this is a job for the gods, is it not? We are only human, but as we develop this technology and build worlds for individual and social use, we assume certain responsibilities. This final section is a personal perspective on subjective design considerations that seem particularly relevant to cyberspace. I address the topics of power, politics, and the undesignable.

Power and Control

There is no doubt that cyberspace and virtual world technology are empowering; but exactly who is being empowered? This is, in part, a design decision. The accessibility of affordable systems, the ease with which people can use them and the degree of control that individuals have over their participation in cyberspace will all be influenced by developers.

The power of the participant begins with access; who gets into cyberspace? The current development of relatively inexpensive systems along with high-end models indicates that the technology will be widely available. Once we are there, who is in control?

Application Control: The design of virtual worlds can induce passivity or facilitate creativity. An application that restricts the interactivity of the medium usurps the participant's power. Passive cyberspace experiences are being considered by some producers in the entertainment industry; their current priority is to reach mass audiences with canned worlds. They are not interested in training novices to become experts at creating their own entertainment; that doesn't sell tickets. Production cyberdramas will probably be enjoyable. But I'm bothered by the notion that we *need* them because it's so hard to make interesting virtual worlds, and most people won't be good at it. Isn't the purpose of designing powerful intuitive tools to make it easy?

Social Control: When cyberspace becomes commonplace in corporations and schools, how will the power of the technology be distributed? The educational applications of cyberspace are stunning in concept, but I've heard teachers respond to this potential in two ways. One response focuses on the new experiences it will afford students, the other centers on the additional control teachers might have over student behavior. Who decides how cyberspace is used in schools?

Personal Control: I watched someone take control from a participant in the middle of a demonstration; he quietly switched power from the participant's glove to a trackball, which he (rather than the participant) controlled. Without warning, he spun the participant's perspective in every direction for about 10 seconds. It had a literally staggering effect on the participant, who emerged pale, dizzy, and visibly upset. I felt like I'd witnessed an assault.

If I had the influence of a science fiction writer, I'd propose two laws of cyberspace:

• Any person can access the cyberspace matrix (voluntary citizenship).

• Each person has full control of his or her interactions within cyberspace (human rights).

Politics of Appropriate Technology

How does cyberspace fit into The Way Things Are? Assessing the long-term political and cultural impact of this technology is difficult, but predicting its immediate impact is almost as hard.

I mentioned the Cyberseas project earlier. The potential applications for this technology include navigation, fish and submarine tracking, undersea resource management, ROV operation, and underwater construction, maintenance and repair. It sounds quite useful.

But, political issues are as numerous as the application areas and include intense competition between various groups of fishermen, Indian rights, military secrets, the activities and attitudes of the Department of Fisheries, and the relative wealth and political power of the fishing industry, shipbuilders, and the Port of Seattle—to name a few.

So, who are we designing for? Who will use this technology, and for what purpose? These issues are inevitable aspects of cyberspace development and will affect the kinds of virtual worlds that come into existence.

The Undesignable: Emergent Phenomena in Cyberspace

Adopting what in Zen is referred to as *beginner's mind* means approaching cyberspace without preconception, resisting the temptation to explain this new technology in terms of previous technology. A car is not simply a horseless carriage; cars have completely changed society. So have movies and TV, and so will cyberspace.

The advantage of this approach is that we're open to surprise. In cyberspace already there are moments we never anticipated and experiences we never imagined. Spontaneous events occur, things that aren't designed, and indeed are undesignable. For example:

Rich Walsh moves forward to enter a virtual world for the first time. His body is paralyzed from his neck downward, and he controls the motion of his wheelchair with a joystick that he operates with his chin. His eyes are sparkling and he expresses his keen anticipation. Because he cannot use his hands, he asks me to wear the glove and fly him around. I assure him that I'll take him wherever he asks to go.

I place the HMD over his eyes and see Virtual Seattle, from his perspective, on a large monitor nearby. I ask where he'd like to go and he responds, "Wherever you like."

"I'll take you into the city," I state confidently, and proceed to veer through the sky with no directional control whatever. I feel utterly awkward, knowing that this person is depending on me to guide him and that I'm getting nowhere. Rich is turning his gaze one way and then another, moving his wheelchair around to get closer to Virtual Seattle, while I'm trying every conceivable orientation of my hand relative to the view I have on the monitor, without successfully steering into the city.

Beginner's mind was easy—I was baffled by the difficulty of the task, until I recognized that we were struggling as separate individuals, ignoring our functional interdependence. We may have understood this at the same instant. I oriented my hand directly over his head, pointing in whatever direction he turned, while we talked about how I couldn't see where to go unless he was looking and he couldn't go unless I could move. We suddenly became a symbiotic organism, functionally fused and operating as smoothly as the way one being coordinates the actions of its hand and eye.

It was an emotional rush, this seamless functional bonding with a stranger during several minutes of elated virtual world exploration. Later, in reviewing the videotape of this experience, I realized that a spontaneous vocabulary shift accompanied the new relationship; after we figured out just who we virtually were together, both sides of our conversation took plural form: "Where will we go next? Can *we* get inside that building?"

This was not an experience I could predict and design for, even in anticipation of Rich's condition. The symbiosis arose from our novel cyberspace interaction, not from his disability or my world building. One day, Rich will steer himself through cyberspace with the same joystick he now uses to control his wheelchair. But the possibility of functional fusion between participants in cyberspace will remain.

This spontaneous aspect of cyberspace reveals its essential vitality: the emergence of the unimagined signals life and growth.

Notes

1. Walker, John. "Through the Looking Glass," Autodesk Internal Paper, 9-1-88.

2. Furness, Thomas. "Helmet-Display Reading Performance during Whole-Body Vibration," Presentation, 52nd Annual Meeting of Aerospace Medical Association, San Antonio, TX, May 1981.

3. Minsky, Ouh-yourn, Steele, Brooks, and Behensky. "Feeling and Seeing: Issues in Force Display," Symposium on 3-D Interactive Graphics '90, Snowbird, Utah, 1990.

4. Lakoff, George, and Johnson, Mark. *Metaphors We Live By*, University of Chicago Press, 1980.

5. Laurel, Brenda. "What to Do in a Virtual World," Panel Presentation, Multimedia & Hypermedia Expo, San Francisco, CA, 1989; Marcus, Aaron. "Beyond the Desktop Metaphor," Panel Presentation, CHI '90, Seattle, WA, 1990; Walser, Randall. "Elements of a Cyberspace Playhouse," Proceedings of the National Computer Graphics Association '90.

6. Bricken, William. "Virtual Reality is Inhabited," Presentation, NCGA '90, Anaheim, CA, 1990.

7. Norman, Donald. "Cognitive Engineering." In D. A. Norman and S. W. Draper, eds., *User Centered System Design: New Perspectives on Human-Computer Interaction,"* Erlbaum Associates, NJ, 1986.

8. Lifchez, Raymond. *Rethinking Architecture*, University of California Press, 1987.

9. Smith, Alvy Ray. Presentation, Autodesk Technology Forum, Sausalito, CA, 1989.

10. Norman, Donald. *The Psychology of Everyday Things*, Basic Books, NY, 1988.

References

Brooks, Fred P. Grasping Reality through Illusion. Proceedings of CHI '88, p. 1.

Buxton, W., and Meyers, B. A. A Study in Two-Handed Input. ACM SIGCHI '86, pp. 231–236.

Ellis, Grunwald, Smith, and Tyler. Enhancement of Man-Machine Communication: The Human Use of Inhuman Beings. Proceedings of the IEEE Comcon '88.

Ellis, Kin, Tyler, McGreevy, and Stark. Visual enhancements for Perspective Displays: Perspective Parameters. IEEE Conference on Systems, Man And Cybernetics Nov. '85, pp. 815–818.

Fuchs, H., Kedem, N. Z., and Naylor, B. Predetermining Visibility Priority in 3-D Scenes (Preliminary Report). ACM pub. 0-89791-004, 1979.

Furness, Thomas. The Super Cockpit and Its Human Factors Challenges. Proceedings of the Human Factors Society Symposium '86.

Furness, T., and Kocian, D. Putting Humans into Virtual Space. Proceedings of the Society for Computer Simulation Aerospace Conference '86.

Kim, W. S., Tendick, F., and Stark, L. W. Visual Enhancements in Pick-and-Place Tasks: Human Operators Controlling a Simulated Cylendrical Manipulator. *IEEE Journal of Robotics and Automation*, Vol. RA-3 No. 5, October 1987, pp. 418–425.

Lakoff, George, and Johnson, Mark. *Metaphors We Live By*. University of Chicago Press, 1980.

Lanier, Jaron. Beyond the Desktop Metaphor. Panel presentation, CHI '90.

Laurel, Brenda. What to Do in a Virtual World. Panel presentation, Multimedia & Hypermedia Expo '89.

Marcus, Aaron. Beyond the Desktop Metaphor. Panel presentation, CHI '90.

Minsky, M., Ouh-yourn, M., Steele, O., Brooks, F. P. Jr., and Behensky, M. Feeling and Seeing: Issues in Force Display. Symposium on 3-D Interactive Graphics '90, Snowbird, Utah, March 1990.

Norman, Donald A. Cognitive Engineering. In D. A. Norman and S. W. Draper (eds.), *User Centered System Design: New Perspectives on Human-Computer Interaction*. Erlbaum Associates, Hillsdale, NJ: 1986.

Norman, Donald A. *The Psychology of Everyday Things*. Basic Books, New York: 1988

Smith, Alvy R. Presentation, Autodesk Technology Forum '89.

Walker, John. Through the Looking Glass. Autodesk Internal Paper, 9-1-88.

Walser, Randall. Elements of a Cyberspace Playhouse. Proceedings of the National Computer Graphics Association '90.

Wise, S. A., Rosen, J., Fisher, S., Glass, K., and Wong, Y. Initial Experience with the DataGlove, a Semi-Automated System for Quantification of Hand Function. RESNA Conference '87, San Jose, CA, 1987, pp. 259–260.

14 Corporate Virtual Workspace

Steve Pruitt
Tom Barrett

The traditional equation of "labor + raw materials = economic success" is rapidly changing as American businesses approach the global, highly competitive markets of the twenty-first century. Strategic advantage now lies in the acquisition and control of information. Capital outlays are increasingly being directed toward investments in information technology. Forward-looking corporations are making these investments to help information workers more effectively capture, analyze, and disseminate information.

Corporations themselves are becoming bewilderingly diverse and geographically far-flung. The ability to bring dispersed assets effectively to bear on a single project or opportunity is becoming increasingly difficult. In order to meet the competitive pressures in the years to come, corporations must better leverage their human resources through fundamental changes in their corporate structures. The lumbering bureaucracies of this century will be replaced by fluid, interdependent groups of problem solvers. Successful organizations in the twenty-first century will be those that can respond most rapidly to opportunities by making the most efficient and creative use of their information.

In this context, we believe that cyberspace technology will be a primary driver toward new corporate architectures. The technology will enable multidimensional, professional interaction and natural, intuitive work group formation. The technology will evolve to provide enterprises with what we call Corporate Virtual Workspaces (CVWs) as highly productive replacements for current work environments. CVWs will be a key factor in the economic success of corporations in the next century. Indeed, CVWs will begin to profoundly change the character of corporations, society, and our economic system.

The following introduction to the Corporate Virtual Workspace is presented in three major sections. At the outset, the reader experiences an early CVW. Since both authors are software engineers, the characters presented in the scenario provide a glimpse of daily life as the inhabitant of a software engineering group housed in a CVW. The CVW described in the scenario is the authors' vision of how cyberspace technology will initially be introduced into corporate America. The two main characters in the scenario show vastly different comfort factors with cyberspace. Over time, cyberspace inhabitants will more fully embrace their alternate reality and much less traditional CVW topologies will evolve. Since the CVW described here is offered only as a glimpse of how cyberspace environments may first appear, the ultimate corporate cyberspace workplace is a subject for exploration in another paper.

With the scenario as a backdrop, attention is then focused on reasons why corporations are naturally evolving toward CVW environments. The third section addresses the ramifications and challenges of widespread use of CVWs from the perspective of the individual, the corporation, and society.

Experiencing the Corporate Virtual Workspace

Meet Austin Curry, a 45-year-old software engineer at Spectrorealm, a worldwide software services corporation. It is 6:00 a.m. and the alarm has just gone off just like it has every workday of Austin's career. As always, his first steps are toward the kitchen to grab a cup of coffee from the pot that faithfully starts the brewing cycle just in time to meet the 6:00 A.M. demand. The dog needs to go out and Austin opens the patio door so Bowser can bound outside. Instead of making those first few steps toward the shower that once ultimately led to the garage and then to the train depot for the long commute to the city, Austin instead shuffles down the hall to his study where he will begin his workday immediately. The shower and breakfast can wait. He has an idea he is anxious to explore.

Entering the CVW

Within 60 seconds of crossing the study's threshold, Austin is at work. By donning his customized computer clothing and logging in to the fiber optic network via his home reality engine, he has attached his

Personal Virtual Workspace (PVW) to his current employer's CVW. His PVW, analogous to the physical cubicle he inhabited back in the old days, provides a personal working environment rich in tools he has developed and collected over the years. These tools greatly extend his productivity, communication skills, and access to knowledge. His home reality engine provides the computing horsepower to sustain his PVW and to manage the I/O interface that connects his PVW to a CVW.

Austin's PVW has been one of the few consistent landmarks in a CVW-based career that has taken him to over 20 projects in seven different CVWs over the past four years. He has been working on his current project for about a month now, a prototype of a data analysis application for an upcoming exploration mission to Mars. He is enjoying this project and his stay in this CVW. The work is stimulating, the environment is rich, and his current colleagues are well respected in prototyping circles.

Structure of the CVW

His current CVW is typical in many respects to others in which he has worked. It utilizes the familiar, old-fashioned office building paradigm to organize the diverse talent required to accomplish a myriad of complex, interwoven activities. As he steps into the CVW, he enters a vast network of interconnected hallways in a bustling virtual corporation. He enters the CVW via his office, number 16 on the red hallway. To his colleagues, it appears as just another PVW accessible along the hallway. A block of offices have been reserved for software engineers involved in the prototyping phase of the data analysis project. Colleagues can readily find Austin by just strolling down the red hallway and knocking on his door, which is rarely closed. On the surface the layout is not too much different from the old days when all the project people were located in the same building. It was easy then and even easier now for Austin and his colleagues to have impromptu idea sessions by simply strolling into a neighbor's office.

Even though Austin's PVW appears to his colleagues as just another office along the red hallway, he has organized *his* perception of it much differently. He perceives it as a control center at the confluence of several hallways. Upon entering his PVW, Austin can look out the window and down about 30 feet to view three hallways converging below him. For simplicity, they are universally referred to in this CVW as the red, blue, and green hallways.

The red hallway contains offices and conference rooms for software engineers from all over the world who are currently involved in the CVW. The talent along the red hallway is vast and varied. Many are highly specialized in certain vertical application areas like CAD/CAM, factory floor control, and business data processing. Others specialize in horizontal areas like user interface design, graphics, and networking. Several offices are reserved for world-renowned experts in particular areas who are kept on retainer by the CVW to provide advice. These experts don't come "in" every day, but a meeting with them in the CVW can typically be arranged in about two hours.

In order to work effectively on his current project, Austin has restricted his view of the red hallway to include only about 20 offices that are essential to his current efforts. By making the other offices invisible to him, he has also filtered out their inhabitants. So, while walking down the red hallway, he only encounters fellow workspacers from the 20 selected offices. Since Austin is currently involved in modeling the data requirements for the Mars application system, he has, at least temporarily, included the offices of experts like Peter Chen, James Martin, and Steve Mellor among the offices that appear to him along the hallway. Just last week, he held a short meeting with Mellor to verify some information modeling concepts.

Although the red hallway is his most frequent stomping grounds, the second hallway Austin can see from the window in his PVW is also quite important to him. This is the blue hallway and it is reserved for client offices (PVWs). Like the red hallway, the blue one has the potential of being very long and densely populated. Austin's corporation works with hundreds of firms and each has one or more representatives accessible along the hallway. During his prototyping project, however, Austin is only interacting with a subset of representatives from one particular corporation. As a result, he has constrained his view of the blue hallway to include only these client CVW inhabitants and their offices.

For security reasons, a client's view of the CVW is limited. They need to be able to stroll down the blue hallway to visit colleagues. In addition, they are free to access the red hallway, and to come by Austin's office and the offices of other software engineering team members. However, they are not permitted to scamper at will through all CVW hallways. Furthermore, much of the content of the hypermedia bulletin boards that adorn the hallway walls is automatically rendered invisible to a client. As a result of security requirements, the client sees a much smaller subset of the CVW than does Austin or one of his colleagues.

Conference rooms are liberally scattered along both the red and blue hallways. They make an excellent place to bring clients and software engineers together to discover and refine specifications for a system. Specialized tools are available in each conference room to facilitate communication and creativity. For example, 3-D drawing and painting tools are commonly used. Large vertical expanses that are practically infinite in dimension are available on each wall. These vertical expanses take the place of the old chalkboards and whiteboards many workspacers used earlier in their careers. Chalk and markers are replaced by a light-emitting stylus capable of rendering images on the vertical expanse from a palette of 16 million colors. Sophisticated pan and zoom features allow the presenter to quickly traverse the immense surface area of an expanse. If the meeting is to be held in an atmosphere where formal walls are impractical like in a forest, or on a beach or in Earth orbit, the vertical expanses are supported by three-legged easels or any other metaphor deemed appropriate by the meeting organizer.

The meeting organizer can quickly customize the shape and size of a conference room to match the meeting requirements. For example, a broad rectangular-shaped room with a high ceiling provides a fitting location for lectures. A small, circular configuration with a stained-glass skylight has proven popular for intimate group discussions. The presenters use the sophisticated lighting and sound equipment to set the mood for meetings and for calming things down when tempers flair.

The last hallway visible to Austin from his PVW window is the green hallway. It supports a collection of resource centers. Resource centers are organized around particular subject areas like modeling, data base design, data communications, and simulation. It is a place where the software engineering staff can go to do research and grow their skills. Vast collections of trade journals and technical publications are available. These documents are no longer published in printed form, but are distributed as mini cyberspace experiences that the reader enters and interacts with.

The centers also offer tutorial sessions covering various key concepts and techniques. Thousands of topics are supported by recorded memories of lectures, discussions, and demonstrations done in the CVW. Since everything that happens in the CVW is under computer control, any series of events may be stored as a recorded memory and be relived later with all the fidelity of the original experience. Last week, for example, Austin brushed up on the latest object-oriented data base techniques by replaying a recorded memory of a symposium conducted earlier in the year by a consulting firm. The traditional field of Computer-Assisted Instruction (CAI) has evolved to make good use of the capabilities that recorded memories provide. In contrast to the CAI Austin had once endured in the learning center at his old office building, modern CAI allows him to interact visually, aurally, and tactilely with objects in the lessons. As a result, the learning experience is much more effective and stimulating.

Daily Chores

Quickly, Austin checks his VMAIL (Virtual Mail) system for messages received while he was away in the physical world. The highest priority item is a request from his manager to evaluate the resume of a software engineer who has expressed a desire to join the project. The evaluation will take some concentration and Austin is distracted by the traffic and conversations going on in the CVW hallway. To filter out the distraction, Austin gestures to display an environmental control panel before him at eye level. By manipulating the switches and dials, he reduces the visual, aural, and tactile acuity of his computerized clothing to ignore all but close-range external stimuli. With another feature on the control panel, he temporarily switches his view to his study's physical environment in order to locate his misplaced coffee cup. After switching his field of vision back to the cyberspace environment, Austin determines that he would like a green tint to the light in his PVW. With the turn of another dial on the control panel, his PVW is now bathed in a relaxing green hue.

With the proper conditions set for concentration, he calls up the resume his manager forwarded to him. It is not the typical hard copy resume that Austin evaluated in the old days. This resume is a multimedia extravaganza that takes full advantage of CVW capabilities. The introduction looks more like another *Star Wars* sequel than a summary of the candidate's rationale for leaving academia to pursue a career in the commercial sector. By using some of the hypermedia links embedded in the resume, Austin is able to wade through much of the fluff to reach a section where the candidate offers a demonstration of some object-oriented design work he did on his last project. Given his current interest in object-oriented approaches, Austin can't resist the temptation to step inside the demonstration and interact with it. About 15 minutes later, Austin exits the demonstration eager to report a favorable first impression of the candidate. Austin logs on to the VMAIL system, records his recommendation that the candidate be brought into the CVW for an interview, and sends the VMAIL message to his manager.

Next, Austin consults two intelligent agents that have been busy working for him during the night. Both manifest themselves in his PVW as somewhat comical looking, pint-sized robots. Most of the time, they are extending Austin's capabilities by interacting autonomously in cyberspace. Austin calls the first robot Phil. Phil stands guard just inside the door of the PVW and is responsible for filtering all incoming interactions with Austin's PVW. For example, mini virtual realities are sent to cyberspace inhabitants much like junk mail found its way to Austin's old office in the physical world. Phil filters the junk mail, forwards some to other team members, or simply disposes of it. It was Phil who realized the significance of the resume evaluation request from Austin's manager and logged it as high priority on the VMAIL system.

Austin refers to his PVW's second intelligent agent as George. George is rarely in the PVW since Austin keeps him busy constantly scouring for knowledge in the vast interconnected web of PVWs, CVWs, UVWs (University Virtual Workspaces), PDVWs (Public Domain Virtual Workspaces), and GVWs (Governmental Virtual Workspaces). This network of cyberspace environments is illustrated in Figure 14.1.

By supplying George with a profile of current topics of interest, Austin keeps abreast of a wide range of intellectual endeavors and opportunities. George gets the credit for alerting Austin to the availability of the object-oriented data base tutorial that he enjoyed last week in the CVW's resource center.

Finally, Austin checks his daily calendar to review commitments. His daily calendar is really two separate calendars since there are demands on his time from both the virtual and physical worlds. In the CVW, he

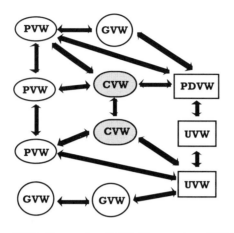

PVW - Personal   CVW - Corporate   GVW - Government
PDVW - Public Domain   UVW - University Virtual Workstation

Figure 14.1
Overview of virtual workspace domains.

notices that his only commitment today is for a short status meeting with his client late in the day. The physical world is also rather uneventful today since his only appointment is to take Bowser to the vet right after lunch. It appears that he has plenty of time to solicit opinions from a few of his colleagues on a particularly nagging problem he has been facing during his modeling sessions with his current client. Hopefully, he can bring some new perspectives to the meeting this afternoon. With a clean conscience that he is on top of his daily chores, Austin opens the door of his PVW and begins strolling down the red hallway.

Strolling the CVW Hallways

Several of his colleagues are also up early this morning. Four of them are gathered around one of the hallway bulletin boards viewing and discussing the highlights from the latest SIGGRAPH conference. They have used the bulletin board's hypermedia links to traverse the conference proceedings and have displayed the benchmarks for a new graphics engine unveiled at the conference. There is considerable discussion on the potential the new engine may have for enhancing capabilities available inside PVWs and CVWs.

Austin has never physically met most of these virtual colleagues. They live in places like Japan, Germany, Guam, and the Bahamas. His newest colleague; Samuel Jamison, lives and works in an A-frame in Aspen. Samuel and several of the others have just joined Austin's CVW and are only expected to remain until the application prototype is presented to the client. Then, they will be off to another CVW and another prototyping project. If the opportunities look enticing, and they usually do, Austin will follow.

The session at the bulletin board is engaging, but this morning Austin is particularly interested in taking his problem to one of his new colleagues, Johann Grumman, a user interface design specialist from Bonn. Johann has not only brought solid academic credentials to the project, a Ph.D. in cognitive psychology, he has also exhibited a creative flair for getting the most out of CVW capabilities. Austin knows that Johann has hung his shingle in the 84th office along the red hallway. He is in a hurry by now and elects not to walk down the rest of the hallway to reach Johann's office. So, he gestures with his right hand to display a control panel at eye level. He toggles the navigator controls to "Red 84" and claps his hands. The control panel disappears and he is whisked down the red hallway. Deceleration brings him smoothly to the door where the name and title of Johann Grumman are proudly displayed in shimmering purple and green neon. Johann is a creative sort.

The PVW Environment

The door is partially open, so Austin slowly enters Johann's PVW hoping not to startle him. Johann has apparently stepped out of the workspace for a moment to grab a glass of milk from his refrigerator in Bonn, but a note jotted on the wall indicates that he will be back shortly. Austin decides to wait.

The office is quite dramatic. Johann is obviously an avid mountain climber. One entire wall of the eight-sided office sports a lifelike panorama of El Capitan from Austin's home state of California. At first glance, it appears that the wall is just a still photograph of the majestic climbing magnet, but as Austin investigates it more closely, he notices that the leaves on the trees in the foreground are fluttering as if the wind were blowing. Upon still closer inspection, he notices brilliantly colored objects about halfway up the side of El Capitan. By invoking the zoom feature with a gesture toward a telescope icon in the lower-right corner of the wall, Austin brings the scene progressively closer. With another gesture toward a stop sign icon, he stabilizes the picture at about 50 yards from the band of hearty climbers that are making their ascent.

The event is visually exciting and so is the audio content. By gesturing again at the wall's icon controls, the sounds of the scene are now audible via the directional sound receptors in Austin's computerized clothing. He hears the calls of "belay on" and "on belay" as the second climber clad in a burnt orange rugby shirt, dark brown knickers, and sky blue climbing shoes makes his way toward the lead climber positioned some 25 feet above at the next pitch. The wind whistles behind Austin as he hears the calls of a distant songbird.

Suddenly, Austin's serenity is partially broken, as Johann reenters the CVW and appears between him and the El Capitan experience. Austin quickly gestures at the wall's control panel to return the settings to their original position and greets Johann by extending his right hand. Austin notices the firm handshake that confirms his earlier guess that Johann is a youthful, athletic sort. Austin introduces himself as the leader of the prototype team and asks Johann if he has some time to critique an idea from a human interface perspective. Johann acknowledges and the discussion formally begins.

Problem Solving in the CVW

Austin starts the discussion by turning around and opening up the small backpack that followed him into Johann's office. The backpack contains communication aids that Austin wants to have readily available wherever he goes in the CVW. The backpack automatically follows him like a shadow. In fact, Austin has thought about representing it as a shadow, but remains comfortable with the idea of carrying his mental supplies around with him in a backpack. Familiar metaphors die slowly, at least for middle-aged software engineers. Even after years in the CVW, Austin needs anchoring points like the backpack in what still seems a bit like a brave new world to him. He continues to endure the kidding from his younger colleagues, who feel completely at ease in the CVW, as they call his faithful backpack "Bowser" in honor of his faithful canine companion on the "physical side."

Austin reaches into the backpack and pulls out an entity relationship diagram (ERD) for the prototype project. It's not just a piece of virtual paper with ERD symbolism sketched on it. It is a large interconnected network of individual nodes—squares representing entities and arcs denoting relationships. He hoists it high enough into the air so that the bottom nodes clear the zipper on his backpack. He leaves it suspended in the middle of the office for easy viewing by Johann and anyone else who might wander in during the discussion.

Since Austin is a software engineer, he has frequently used techniques like ERDs and data flow diagrams to collect requirements for application prototypes. The requirements gathering sessions that Austin has been leading have made substantial use of ERDs and Austin is disturbed that this particular client is not fully comprehending the semantics embodied in the diagram's symbolism.

Using the ERD that he pulled out of his backpack as an example, Austin introduces Johann to the basics of entity relationship diagramming. Most of the client representatives are comfortable with the concept that objects of interest to their business (entities) are expressed as squares on the ERD, but it is in the relationships, represented by labeled arcs connecting the cubes, where the confusion begins. A small section of a larger ERD is illustrated in Figure 14.2.

Relationships on an ERD have characteristics like cardinality and total participation. Cardinality expresses the ratio of occurrences between two entities like "sensor" and "test," for example. If one sensor can be responsible for more than one test and a single test is carried out by only one sensor, the cardinality of the relationship is one-to-many. As illustrated in Figure 14.2, double arrowheads (indicating many) appear at the end of the arc near the test entity's square. The absence of arrowheads on the other end of the arc near the sensor entity denotes a cardinality of one.

The total participation characteristic of a relationship determines if every occurrence of an entity must participate in a particular relationship. For example, if information cannot be stored about a sensor until it is associated with at least one test, the sensor entity totally participates in the relationship with the test entity. As illustrated above, total

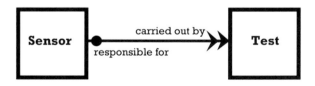

Figure 14.2

participation is denoted by a dot near the sensor entity along the relationship arc.

The ERDs have been done in CVW conference rooms along the blue hallway, but Austin feels that the imagery employed during the modeling sessions hasn't taken full advantage of the capabilities of CVW. As a result, the client is still struggling with the symbolism in spite of the enhanced communications potential of the CVW.

In order to provide Johann with a firsthand impression for how modeling sessions have been conducted thus far, Austin reaches into his backpack and grabs what looks like a cassette tape and a boom box tape player. Once he loads the tape and presses the play button, voices are heard from across the PVW and a 3-D recorded memory of last Tuesday's modeling session appears in a replay cube. To save space in Johann's PVW, the scene is being replayed in miniature in a cubic area about six feet on a side. Within this replay space, the 3-D images of Austin and three client representatives appear.

In the replay space, Austin is leading the session and is armed with a light-emitting stylus. He is standing in front of a large vertical expanse. As the discussion leads to the identification of another entity or relationship, he draws the appropriate 2-D symbolism on the vertical expanse where it can be evaluated by all the client participants. Features like pop-up annotation windows, infinite zoom, and rapid panning have been well-utilized during the sessions, but there is still something missing. The clients seem confused with the symbolism. Since the ERD must capture the data requirements upon which the prototype is developed, the risk of incomplete specifications increases with the fogginess of the symbolism.

After only a few moments of viewing the recorded memory, Johann has some ideas on how some additional virtual tools could improve the modeling sessions. To illustrate his ideas, he invites Austin to enter the 3-D recorded memory with him. Once they have entered the replay space, Johann decides there is little need to continue viewing the entire memory from last Tuesday's modeling session. Johann does want to keep part of the memory active and reuse it. He is particularly interested in the vertical expanse and the diagram that Austin has sketched with the stylus. The replay space's control panel appears in response to one of Austin's gestures. By selecting various editing tools, Austin is able to quickly remove the client representatives and himself, as the leader, from the recorded memory.

Once the replay space has been reconfigured, Johann, who is bubbling with enthusiasm by now, comments that Tuesday's modeling session showed a classic example of a physical side technique that has not yet evolved to take full advantage of the CVW capabilities. Johann emphasizes three points in particular. First, the sessions seem needlessly boring and he is puzzled by why the 3-D capabilities of the light-emitting stylus were not employed to spice up the sessions. Second, he is dubious of diagramming techniques that use symbols having little correlation to their physical counterparts. For example, in traditional ERDs, even though sensors and tests are dramatically different concepts, they are both represented as squares. Only the labels in the squares distinguish one entity from another. Third, Johann feels that ERDS always seem to show too much information at one time. This results in cognitive overload for new modeling participants.

Johann begins to explore tailoring the ERD technique to more fully utilize the capabilities inherent in the CVW. He appears to be genuinely enthused not just in providing some immediate advice to Austin, but in crafting a new virtual tool that could be sold as a product in the CVW to assist other software engineers with similar frustrations. At the very least, the new tool will become a part of his own PVW.

To illustrate a possible solution to his first two concerns—lack of pizzazz and all entities represented by squares—Johann gestures by touching the top of his head. Following a flash of lightning and a loud clap of thunder, a small fluffy white cloud appears over Johann's head. The cloud reminds Austin of the comic strips he used to read as a kid. Thought clouds like this appeared above the comic strip characters when they were thinking. True to his creative bent, this is where Johann stores his mental tools. Austin watches as Johann peers into the cloud and pulls out a tool, called the Concept Cannon, that he developed for another project. The cannon is about six inches long and four inches high and rolls around on two tiny wheels. It was designed to render images onto a vertical expanse, and Johann developed it for the marketing folks in the CVW as a sales presentation tool. By combining 3-D shapes and stereophonic sound outputs with each cannon burst, it proved to be an exciting means of displaying concepts to a group of potential customers assembled in front of a vertical expanse. It may be a bit wild for Austin and his situation thought Johann, but it would surely get the creative juices flowing.

Before demonstrating the Concept Cannon, Johann decides to enlarge it. By directing a gesture at the muzzle of the cannon, a control panel appears. Johann adjusts the dimensions and the cannon instantaneously re-forms three times as large. With the control panel still open, Johann shows Austin that the cannon can be configured to lob numerous styles of 2-D and 3-D objects onto the vertical expanse. Johann recommends that the cannon be set so that the objects it renders on the canvas display still 3-D images when viewed from the front and optional, dynamic 3-D images (movies) when viewed from the back or rotated 180 degrees.

Johann closes the control panel, loads a concept cannonball and fires his first volley onto the vertical expanse. It is a 3-D image of a sensor much like the one carried on the Martian probe. Via a sequence of gestures, Johann guides Austin and himself up to the surface of the vertical expanse and inside the sensor itself to roam among its component parts.

The cannon can deliver a concept to any expanse within 50 meters of the cannon. Once an image has been fired to the expanse, an entire group may view and interact with the object under the control of the cannon operator as Johann had demonstrated by their quick voyage into the sensor. The operator can also fire the cannon directly at a particular individual. Upon "striking" the individual, a personal vertical expanse is automatically created in front of the individual and the object appears in such a form that the individual can view and interact with it privately under his own control. At the cannon operator's discretion, the changes that the individual makes to the private copy of the objects can also be reflected on any public copies displayed on the public vertical expanse.

Austin sees that the Concept Cannon could be an exciting replacement for his relatively primitive light-emitting stylus. So, instead of drawing 2-D square entity symbols on the vertical expanse, he can display 3-D versions that look like the objects they represent. Then, under his control, he can guide the clients through an interaction with the entities. If someone wants to further analyze a particular entity, he can make a personal copy of the object available to the individual client by firing the Concept Cannon at them.

Johann sees that Austin appreciates the power of the Concept Cannon to help the client experience the semantic content of the entities in the ERD. So, he now moves on to some ideas to help reduce the confusion over the relationship characteristics like cardinality and total participation that Johann feels clutter up the diagram. Johann suggests that the key might be the perspective from which relationship characteristics are viewed. He goes on to explain that rather than showing them all at one time for the entire diagram, it may be more useful to have relationship information appear only when viewing the diagram from the perspective of a particular entity. For example, in exploring the cardinality between two entities like sensor and test, the session leader can bring the participants, as a group, inside the sensor entity object. Inside the entity, relationships could appear as portholes. Inside the sensor object, they could open the cover over a porthole representing a relationship, for example, and look through the porthole towards the test object. If the cardinality is one-to-many, the viewers would see multiple test objects. From the test perspective, where a particular test is typically carried out by only one sensor, the group would see a single sensor object through the corresponding relationship porthole. Thus, this would obviously be a one-to-one relationship.

As for the total participation characteristic of a relationship, if a sensor must be associated with a particular test when it is created, this could be designated by a coverless relationship porthole inside the sensor entity. The analogy would suggest to the viewer that since the porthole cannot be covered up, there must always be a relationship through which to peer. Conversely, portholes having covers would represent relationships without total participation specified.

Sensing that Austin might be feeling a bit overwhelmed by the possibilities, Johann suggests that they exit the replay space and return to his PVW to summarize some of the key ideas covered thus far. By pointing upward with both hands, they float out of the replay space and appear in front of the El Capitan expanse where the conversation began some 45 minutes earlier. Austin is flabbergasted by the brief discussion. He collapses the replay space into his backpack. He will reexpand it back in his own office later today. He can't wait to experiment further with the Concept Cannon.

Exiting the CVW

Austin thanks Johann and strolls back out into the red hallway with his trusty backpack close behind. It is now almost 7:30 and Austin feels like taking a shower and grabbing some breakfast. He returns to his office to drop off the backpack and leave a note that he will return in about an hour. Then, taking off his computer clothing, his attention turns to the physical world where Bowser is more than ready to come in from the patio.

Why the Corporate Virtual Workspace?

The Corporate Virtual Workspace (CVW) experienced by Austin and his colleagues is a response to several forces clearly visible in today's physically based corporations. The following discussion briefly examines three of these forces: the evolution of software technology, complexity of large-scale projects, and demands for higher productivity.

Evolution of Software Technology

Software technology has inevitably evolved towards higher end-user interaction. One useful measurement of this evolution is software's increasing accessibility. From the early arcane machine code to today's high-level, fourth-generation languages (4GLs) and PC-based productivity tools, software has become increasingly easier to conceptualize, build, and use. All signs indicate that this trend will not only continue, but will increase as hardware becomes more powerful and software grows even more intuitive and powerful.

The latest driver behind this trend toward interactiveness and accessibility is object-oriented programming, a conceptual paradigm where computerized objects are viewed as independent actors in a complex web of interaction. Each actor embodies its own behavior and, if relevant, its own interaction with the outside system.

Object-oriented programming will flower in the rich bed of cyberspace systems. The promise of visual programming will be powerfully realized. Nonprogrammers will interact with purely software objects with the same familiarity with which they have manipulated nonsoftware objects. Whereas nonsoftware objects in the "real" world play out roles prescribed by nature's rules, software objects will act out roles in compliance with rules enforced by their hosting cyberspace. Cyberspace software objects will represent themselves polymorphically to their users through a combination of visual, aural, and tactile sensations. Where, traditionally, computer access has almost been purely nonsensorial, cyberspace objects will behave and interact through an array of sensory metaphors.

Large-Scale Projects

In addition to the demands for greater end-user interaction, the scale of problems being tackled by software engineers has dramatically increased. The ability to build large-scale software systems economically has become a daunting challenge for both managers and individual contributors in most large corporations. Contemporary programming-in-the-large is a complex undertaking consisting of both serial and concurrent tasks. Each task is typically attacked by a small team whose results are on the critical path on the project schedule.

In addition to in-house talent, there is increasing use of outside consultants. Just as Johann Grumman provided Austin with advice on human factors, consultants can bring a narrow focus of skills and participate briefly on a project. They may also enjoin the project for longer durations with broader skills and wider responsibilities.

Over the life cycle of a software engineering project, it is not uncommon to form a variety of teams with well-defined objectives. For example, at the outset of the project, an analysis team typically performs the formal problem definition and requirements gathering. Next, a prototyping team is assembled to verify the analysis by constructing and implementing a preliminary design. As the life cycle continues through formal design, implementation, and maintenance, other teams are formed to focus the appropriate specialized skills on the tasks at hand.

There are two inherent technical and management challenges stemming from these team activities: interconnectedness and transience. Both challenges require a tremendous degree of interpersonal communication that is not fully facilitated by contemporary communication systems.

Regarding interconnectedness, every effort is associated with one or more teams. A large degree of connectivity means each effort must be deftly coordinated with one or more dependent steps. The output from one effort is input to another, and each major task takes its turn being on the critical path.

Each effort is connected through personnel. An engineer may serve on more than one team, serially or in parallel. "The ability to wear more than one hat becomes a very valuable skill." Many tasks are multi-disciplined and require that personnel connect with experts in several areas in order to augment skill sets and acquire domain knowledge. Often these experts are not part of the project per se, but may exist nearby on related projects, at company headquarters in the research labs, or out in the field at a branch location thousands of miles away. Historical experts may be brought in on a consultative basis.

In addition to the interconnectedness of large-scale projects, they are often characterized by transience. Expertise flows in and out on a strategically timed basis. Teams may be initially infused with expertise from transient outside consultants. This expertise is critical to starting activities off properly and in employing the latest techniques. For example, the analysis team's work may be launched with sessions on the latest methodology, while the prototyping team is trained in cutting-edge prototyping environments. As consultants come and go, it is the project's responsibility to capture their contributions so that the consultants' knowledge is available long after they have gone on to the next client. Austin's CVW addresses the need for storing valuable consulting expertise by making seminars available as recorded memories in the resource centers in the green hallway. In addition, Austin has access to several experts via offices set aside for them along the red hallway.

Productivity

As if the demands for greater software interaction and the complexity of large-scale projects weren't vexing enough, productivity is a serious problem. Today, enhancements in productivity are not keeping up with the demands for new software solutions. System engineers in large corporations can't write application systems fast enough to keep up with the needs of their corporations. Knowledge workers in these corporations have a steadily growing backlog of unaddressed software system needs. Some industry analysts estimate that this backlog represents at least 60 man-months of software engineering effort. In addition, there is undoubtedly an even larger invisible backlog of system solutions that aren't even communicated by knowledge workers due to the obvious inability of software engineers to address their needs.

In order to provide knowledge workers with the software solutions demanded by the twenty-first century's competitive environment, the application development backlog must be addressed. This will require tremendous enhancements in software engineering productivity. Enriched CVW environments facilitate the communications necessary for efficient team interaction. In addition, the CVW provides exciting potential for heightened productivity through use of virtual tools like Austin's intelligent agents, vertical expanse, light-emitting stylus, and recorded memories.

Ramifications

Widespread use of PVWs and CVWs will have profound effects on the individuals and the corporations they inhabit.

For the Individual

Despite advances in telecommunications and the loosening of traditional hierarchical power structures in the workplace, there has been no mass movement toward the long, predicted electronic commute. Our freeways are still jammed with frustrated commuters. This is because most knowledge-intensive jobs require personal interaction. Much of this interaction is in the form of nonverbal communication. Physical cues are needed for additional information transmission and for spawning spontaneous creativity. No current means of electronic communication fully facilitates physical communication; consequently, workers are required to physically transport themselves and remain constrained by space and time.

Cyberspace will free an individual from space and time constraints. An individual will be able to instantly link up "physically" much the same way people link up via telephone today. Electronic commuting will become commonplace, as cyberspace workers like Austin commute by connecting their PVW to other PVWs or to their employer's CVW.

The PVW will become a cyberspace worker's most valuable economic asset. His or her career and experience will be reflected in their PVW. More than a personal workspace, each PVW will house a cyberspace worker's accumulated personal tools, references, and productivity artifacts. Indeed, the value of a new prospective team member will be measured by the richness of his PVW in addition to the knowledge and experience he has garnered on past projects.

Personal tools developed in past positions will no longer be relegated to disk or tape obscurity. Instead, they will be visible and active inhabiters of the PVW just like Austin's intelligent agent robots Phil and George. Easily available and malleable, personal tools will live long and productive lives. Time spent on their creation and modification will increase in productivity value as these tools evolve along with the cyberspace worker.

Over time, a cyberspace worker will collect and maintain a set of colleagues who stay in touch throughout careers or lifetimes. These relationships will remain as strong as the relationship between next-door neighbors with common interests who are accessible only to the owner of the PVW. Despite job or career changes, these neighbors will remain lifelong collaborators. Such a relationship might be forming between Austin and Johann.

And yet, for all this closeness, the freedom found in cyberspace will endow cyberspace workers with tremendous job mobility. No longer bound to a physical "company town," changing jobs will be as easy as disconnecting a PVW from one CVW and connecting into another. Job mobility may reach a level where a large number of cyberspace workers become entrepreneurs, either contracting their services to companies, or starting their own companies.

Along with increased job mobility, cyberspace workers will find that the environment offers increased educational mobility. Opportunities for lifelong learning will abound as most universities and companies specializing in job skill seminars will offer ongoing education accessible via a cyberspace classroom connected to the PVW. As an adjunct to the PVW, the cyberspace classroom will increase the capability for transferring knowledge from the classroom to an ongoing work assignment. A current job assignment could be integrated into the participant's classroom exercises, thus facilitating the transfer of knowledge. These educational opportunities might also be available as standard components of a CVW, as they are for Austin in the resource centers located along the green hallway.

Increased job mobility and easily accessible education will mean smoother career changes. As new knowledge is garnered from the virtual resource centers, the cyberspace worker can carefully guide job selection and incrementally bridge from one career to another. Abrupt dislocations caused by returning to school can be eliminated if the individual wishes to remain economically active.

The knowledge, experience, and contacts a cyberspace worker gains and accesses through cyberspace will augment their skill and ability far beyond what a noncyberspatial worker can manage. The productivity and worth of cyberspace workers will entice corporations to rehire them.

Electronically nomadic, cyberspace workers will live wherever the electronic cyberspace infrastructure reaches. Physical locations with high physical appeal will become popular. Areas once deemed strictly vacation spots or resorts will be available for permanent habitat. Separated from the stress caused by crowded and polluted urban areas and able to instantly turn to their real environments for recreation and exercise, cyberspace workers will lead highly productive and healthy lives.

For the Corporation

The advent of the PVW and the CVW will produce a new classification scheme for corporations. One group will include those corporations whose primary goods and services are physical reality-based. They will provide manufacturing, health care, transportation, construction, and agriculture, for example. The other group will include corporations whose marketplace offerings are cyberspatially based. For example, education and training, software, entertainment, and financial services will be provided by these virtual reality-based endeavors.

Advent of the Cyberspace Corporation   Precursors to the cyberspace corporation exist today. Their goods and services are inherently information or knowledge. However, the true cyberspace corporation is more than just a grouping of goods or services.

Having no need for physical facilities other than the system hosting the CVW, the cyberspace corporation will exist entirely in cyberspace.

With little need for start-up capital, cyberspace corporations will form quickly around an individual or group of individuals who have identified an opportunity and formulated a market plan. Additional cyberspace workers will quickly be gathered from previous endeavors or new talent will be recruited. Profit shares will be apportioned across participating members.

The cyberspace corporation may provide a single product or service and then disband, or it may be formed with a longer-term vision and remain to serve the product's market. Other cyberspace firms may specialize in assuming ongoing maintenance of products if the developer decides to pursue other market opportunities.

Cyberspace corporations will be very fast-acting and transient. They will be composed of bright, creative, high-tech nomads who will coalesce into work units for dynamic market opportunities. Personnel turnover will be high as tasks are completed and cyberspace workers decide to migrate to other opportunities. As new corporations form, cyberspace workers may find themselves working periodically with the same people. A very productive informal network will form as cyberspace workers leverage their rich set of experiences and contacts.

Adoption by the Traditional Corporation   Cyberspace will impose powerful pressures to change the traditional corporate structure and how products are created and marketed. Corporations will respond by using cyberspace for conducting both internal and external business. Communication and control, inherent to all corporate structures, will evolve with the cyberspace medium.

All forms of communication will be conducted via cyberspace, introducing new opportunities for customer and vendor relations, employee relations, supervisor/subordinate interaction, and intergroup communications. These traditional communication forms will become interactive visual representations, providing a richer bandwidth.

Jargon-loaded and abstract documents will transform, becoming dynamic actors in cyberspace. For example, contracts will be a composite cyberspace object that the reader can step inside of and in which he can visit precise interactive objects that simulate the product or service that is being contracted. Delivered goods could then be compared with the contract representation of the ordered good. The contract parties and contract author could include interactive representations of themselves to answer specific questions or conduct demonstrations of the contracted goods or services. Austin encountered this technique in the resume he evaluated when he interacted with the demonstration that the candidate had included.

Personal and group communications could also take the form of interactive cyberspace objects. As an appendage to an individual's or a team's ongoing work, an interactive cyberspace progress report would include a work simulation open to asynchronous visitation by any authorized party. Status tours with the responsible party could be arranged for detailed demonstration.

Interactive cyberspace objects will give managers and users unprecedented intimate access to the life cycle of a system or product. Their changes or comments, easily annotated, could be instantly included by the responsible parties. Liaison personnel following ongoing projects will be able to integrate developments faster.

On a larger scale, the CVW will provide upper management with unprecedented ability to move about and observe corporate activities. CVW tours will give upper management easy access to ongoing work either internal or external to the corporation. The resultant lessening in the need for extensive reporting chains will flatten bureaucratic hierarchies.

Perhaps the most dramatic opportunity cyberspace provides for a corporation is a dynamic and exciting public face. Public visits to the corporation will provide intimate interactions with company products either simulated or real. The information flow will not be one way. A corporation will dynamically gather customer impressions of new or planned products for instant measurements of acceptance.

As an extension to the CVW, a cyberspace showcase will be maintained with changing product updates. A visitor could arrive for product announcements or merely go window-shopping. Products will be demonstrated in a context-specific environment for customer feedback and possible immediate reconfiguration to meet the customer's requirements.

The cyberspace showcase will make most forms of advertisement obsolete. For narrow target audiences, interactive cyberspace objects for specific products can be sent to identified individuals. Of course, the receiver would have no obligation to open these objects.

An individual will not be limited to a specific showcase. He will be able to dynamically assemble his own shopping mall containing the mix of products he is interested in. Again, interactive demonstrations will enable the shopper to try, firsthand, the simulated or actual products. By gathering products from the various showcase inhabitants, the shopper will find trial and integration easier. A shopper will be able to piece together anything from a new wardrobe to a fully equipped factory.

Underlying the showcase will be an ongoing market analysis. No longer limited to statistical studies or contrived product prototype trials, a company will maintain a real-time data base of customer demands as they try out and respond to specific products. On-site product analysis and feedback will enable a company to immediately respond with a highly dynamic product mix.

The showcase medium will grow so powerful and useful that a corporation may disassociate itself entirely from middlemen distributors. Nearly instantaneous market analysis feeding highly automated factories that operate under just-in-time inventories will require minimum warehousing. The economic gains from dealing directly with customers may more than offset the increase in the distribution investment.

In addition to showcasing their products, a company will showcase themselves and opportunities for working in its CVW. The personnel department will maintain a cyberspace structure enabling prospective applicants to tour ongoing projects and current employment opportunities. Co-workers may be easily reached for interviewing, answering questions, or demonstrating unclassified projects. While visiting a personnel showcase, an individual's resume could be explored as an interactive cyberspace object, from which personnel could tour an individual's career history. Work detailed in the resume may provide an ideal demonstration of past accomplishments.

Challenges

As discussed above, the potential positive impacts of cyberspace on the individ-ual and the corporation are substantial. However, the transition to cyberspace will not be easy. Individuals, corporations, and our society all need to come to grips with major challenges.

To the Individual
The dynamics of cyberspace will challenge the individual professionally and personally. Lifelong learning will become an economic necessity. Old skills will become obsolete as new technology is developed, or supply may exceed demand as many people will have access to excellent cyberspace-based education. As skills become useless, so may entire career fields. An individual will need to be adept at moving into fields which hold long-term promise and demand.

If the individual takes full advantage of cyberspace and becomes an independent economic unit, he will need to become more self-reliant. Health care and financial cushions may become the responsibility of the individual.

To the Corporation

Cyberspace will present major challenges to existing corporations. Some will lead in the adaptation process, some will follow, and many will disappear into history. The resulting competitive pressures will require companies to fully embrace cyberspace. Learning to survive in cyberspace will make old ways of doing business obsolete and require new creative replacements.

Perhaps the greatest challenge in surviving in the cyberspace era will be personnel management. Armed with cyberspace mobility, cyberspace workers will find it much easier to change companies. Adapting to rapid turnover will require innovative approaches.

To begin, a corporation must perceive the advantages that highly connected cyberspace workers bring with them. The more projects, either internal or external to the company the worker is involved in, the more experienced he or she becomes and the richer his or her PVW. Many workers may eventually become stagnant remaining at the same company; careers may hit a dead end. By releasing workers with no repercussions, a company may find itself rehiring a rejuvenated and enriched worker.

To protect themselves, companies may enforce length of stay contracts to ensure a return on investment for training and education expenses. They may also require nondisclosure contracts.

Paralleling the new realities of worker turnover, the corporate reward system will need to be overhauled. Reward systems based on employee tenure and loyalty will be harder to justify. Benefits such as medical insurance, vacation, and profit sharing may disappear as universal offerings. They may instead be reserved for enticing long contracts and rewarding extended service.

Manufacturing-based corporations may face a unique challenge. Daunted by the expensive task of maintaining a competitive manufacturing base, the manufacturer may find competing in the fast-moving and dynamic cyberspace marketplace prohibitive. Except for a few megacorporations, many manufacturers may let fast-acting cyberspace corporations discover the product opportunities and assume a proportional share of the risk. These manufacturers may become foundries for cyberspace corporations and concentrate their resources on maintaining excellent manufacturing capabilities.

To the Economy and Society

As the cyberspace economy overlays the traditional economy, its infrastructure will become critical to world economic health. Governmental regulations may become necessary to ensure stability by detecting attempts to subvert it. Unquestionably, cyberspace bureaucracies will emerge.

As traditional economic infrastructures wither from disuse, many economic structures will disappear. The potential job dislocation will be immense and expensive. For instance, cyberspatial commuting will drastically reduce the load on transportation systems. Entire industries, built upon commuting and shopping traffic patterns, will find their sustenance severely threatened. Fast-food restaurants, convenience stores, and gasoline suppliers all depend on large commuting traffic. Shopping centers compete against direct sales outlets. By eliminating or drastically reducing their markets, these industries will become inconsequential. Cyberspace educational resources must be made available and the cyberspace economy will have to expand fast enough to assimilate the resultant unemployed.

Directly related, the value of large centralized business districts will be highly suspect. Currently, vast amounts of capital are invested in existing business districts. An effort to protect these investments may be mounted by companies who will not offer CVW facilities. But, as the cyberspatial economy grows, more workers will cast off their old commuting ways. The recalcitrant companies will find themselves unable to attract quality employees. As the centralized business districts become irrelevant, financial bridges will be needed for accommodating the lost fortunes.

Nationalities will find that national borders will be irrelevant in the cyberspace economy. Citizens will be able to work without regard to national borders. While this may threaten sovereignty in many instances, it will provide an excellent avenue for those individuals who are stuck in a stagnant local economy. The need for some form of cyberspace "green card" may arise, but it will be exceedingly hard to police.

The dynamics and transience of the cyberspace corporation will mean that existing corporate laws will need to be revamped. The responsiveness of the current bureaucratic means of registering a corporation must be improved to accommodate quick formation of corporate entities. The bureaucracy must also be able to deal with the short lifespans of many of the corporations that are formed. Taxation policies will be challenged by a cyberspace corporation having no true state of registration.

Conclusion

A major effect of technology has been the acceleration of change. Cyberspace technology will provide twenty-first-century corporations with revolutionary capabilities. Successful corporations will employ the technology to compete in emerging marketplaces, both physical and virtual, where opportunities present themselves suddenly and last only as long as it takes a fast-moving supplier to respond. Instantaneous interaction on a national and global level will make the window of opportunity smaller and smaller. Corporations will succeed based on their ability to quickly identify market needs, marshal the resources, develop innovative designs, and deliver high-quality products. Short product lead times will be critical.

Cyberspace offers a rich medium for thriving in the hypercompetitive economy of the twenty-first century. Individuals will have unprecedented resources from which to build prosperous multicareer lives. Modern corporations will arise and flourish by embracing the advantages inherent in a cyberspace economy. Although cyberspace will be a harbinger of certain economic problems and dislocations, it will also supply many of the solutions.

15 Making Reality a Cyberspace

Wendy A. Kellogg
John M. Carroll
John T. Richards

Introduction

In the field of human-computer interaction, the 1990s are already being described as the decade of multimedia. On the leading edge of interactive multimedia systems is cyberspace and the development of three-dimensional virtual world technology. The evolution and application of this technology is already occurring and it is incumbent upon us now to begin to envision and understand cyberspace as such: what people might do with it and in it, its potential and limitations in furthering human goals. For to be believable, useful, and usable, cyberspaces will have to support existing practices of individuals and groups; their activities, purposes, modes of perceiving and knowing, to name a few. If cyberspace is effectively to *lead* human practice, by the same token, it will have to make contact with these practices as it enables new ones.

In this chapter, we will take the established vision of cyberspace to be that represented in the science fiction literature (for example, Gibson 1988), the beginnings of which are apparent in current "goggle + dataglove" implementations of virtual realities.[1] In this vision, cyberspace is an independent artificial reality, accessed individually by any number of users who themselves have some representation within the world. Both real-world objects and data objects are accessible in this envisioned cyberspace such that a user's virtual actions might have consequences both in the physical world and in cyberspace.

As psychologists interested in human-computer interaction, our attention was readily drawn to the unique characteristics of cyberspace as a "user interface." In particular, our own work had considered how the characteristics of particular user interface designs can help as well as hinder the activities people want to engage in with those artifacts (see Carroll and Kellogg 1989; Carroll, Kellogg, and Rosson 1991; Kellogg 1990). We were thus struck by the idea of how cyberspace might replace the desktop metaphor, an idea that arose in some of the early discussion of virtual realities. We imagined secretaries, data gloved and goggled, doing their jobs in cyberspace. We wondered what possible advantages there might be in a 3-D virtual desktop as opposed to current "flat" desktops. Perhaps there would be memory benefits in an enhanced spatial environment for storing and retrieving office documents. But what would happen when the phone rang, or the boss stuck her head in the office to ask not to be disturbed for an hour? For much of a secretary's activity, the potential disadvantages of isolation from normal reality seemed to outweigh the potential advantages of immersion in a virtual space.

We imagined ourselves conducting our research in cyberspace; for example, accessing and searching a bibliographic database. Here, perhaps because there is less demand for research activity to remain immersed in normal reality, it was easier to see how overall advantages might accrue. Searching bibliographic databases through logical and syntactically well-formed queries is notoriously difficult, and keywords are not particularly efficient for sorting the relevant from the irrelevant. But we could envision a cyberspace affording direct manipulation query and 3-D representations of semantic relationships (amount of overlap in citations between retrieved documents and a model defined by the user) for improving the task of finding relevant publications. However, we also realized, as we thought about our own work, that a significant and perennial problem that cyberspace could *not* address was retrieving relevant information from the publications *physically* contained in our own offices!

These initial considerations led us to reframe the design problem of envisioning and understanding cyberspace. Rather than working with the assumption of an enclosed, independent virtual world, we opted to consider how a virtual reality might be created within the context of the real world. This strategy essentially inverts the design problem: rather than asking how cyberspace might be *realized*, it asks how the ordinary practices and objects of reality might be *cyberized*. We propose a *distributed, augmented reality* rather than an *enclosed, simulated reality*, a reality in which cyberization is integrated seamlessly into people's everyday activities, and real-world objects take on virtual attributes and behaviors that support and enhance those activities. The essence of this proposal is to bring cyberspace to the people, rather than the people to cyberspace.

As a complementary research strategy for discovering the potential of cyberspace, the strategy of envisioning cyberized realities may offer unique leverage. Creating distributed cyberspaces in real-world objects may be an easier, more tractable design problem than creating full-blown enclosed virtual worlds. In some cases, distributed and enclosed cyberspaces will have different limitations and constraints; a strategy encompassing both offers the advantages of a more broadly based approach to the development and application of new technology. Building distributed cyberspaces may provide a medium for developing, prototyping, and testing ideas about enclosed cyberspaces. Not all of these possibilities need be the case for a dual strategy to be valuable; if any of them are demonstrable, then a complementary strategy is desirable.

Considering virtual realities as *part* of reality as opposed to *apart from* reality expands the virtual world design space. These different senses of cyberspace could turn out to be complementary in an even stronger sense than merely being a good research strategy on the way to convivial virtual worlds. They may both be necessary components of an ultimate cyberspace. As we examine different realms of human activity as candidate applications for virtual world technology, we will no doubt discover certain domains, or activities within domains, particularly suited to each kind of cyberspace, and yet others where distributed and enclosed cyberspaces will cooperate to produce a viable artificial reality. Developing an understanding of the unique characteristics of augmented and simulated realities and how they may be employed to extend and improve current practice is a necessary prerequisite for capitalizing on the entire design space.

For the remainder of this chapter, we focus on developing such an understanding of augmented realities. First we consider an example of an augmented reality, then we turn to an elaboration of characteristics that seem to be critical for creating meaningful augmented realities, and finally, in a second example, we suggest how augmented and simulated realities might be combined to take advantage of the characteristics of each.

An Augmented Reality: Jack's Kitchen

A distributed cyberspace fashioned from real-world objects with virtual attributes and behaviors will be the preferable alternative for some activities. Domains where there is a high need for the person to remain integrated with normal reality will be natural candidates for augmented rather than simulated realities. In addition, ideal domains for augmentation will have the characteristic that parts of the activities the person wishes to engage in can effectively be delegated to objects in the environment. The example we give to illustrate distributed cyberspace —an augmented kitchen—has both of these characteristics. In particular, we consider the scenario of planning and preparing a meal for a dinner party.

A Scenario

Jack belongs to a "wine and dine" group of six people that meets once a month to enjoy an unusual wine or wines with an appropriate accompanying meal. This month is Jack's turn to host, which means that he is responsible for selecting the wine and planning and preparing the meal.

Jack has entered this event on his social calendar with a notation that he is responsible for doing the dinner. The calendar has responsibility for "remembering" that Jack is in charge and taking appropriate and timely action. One week before the dinner, the calendar initiates planning for the meal and reminds Jack of the upcoming event. The calendar sends a request to Jack's Dinner Party Planner to initiate planning of a "wine and dine" meal.

Since this dinner party centers around the wine, the Dinner Party Planner begins by querying Jack's wine cellar for suggestions. Other dinner party plans might begin from the type of party (casual, elegant), the guests, or the food available in Jack's kitchen and/or garden. The wine cellar knows the wines it contains, knows which are unusual or interesting for some reason, and which are ready to be consumed. The wine cellar responds with several possibilities, returning names and descriptive information. The Dinner Party Planner posts the suggestions by placing a miniature of each label on the kitchen bulletin board, and posts a request to Jack to choose the wine or wines he wants to serve. When Jack gets home that night, he looks over the wine cellar's suggestions and finds over 20 bottles suggested. Not wanting to consider so many options, Jack asks the Dinner Party Planner to limit selections to French wines. The number of labels shrinks to about a dozen, and Jack selects a twenty-year-old Chateau Pétrus Bordeaux and a Sauternes for dessert.

The Dinner Party Planner incorporates the two wines as topmost constraints on the meal to be planned and sets about inquiring after other constraints. As the "descendant" of the more general Meal Planner, the Dinner Party Planner always respects certain constraints it has inherited—for example, constraints on constructing balanced and nutritious meals. Other constraints are relevant only to this particular meal being planned and are discovered by querying objects with potentially relevant information.

For example, Jack's social calendar has responsibility for archiving information about past meetings of the wine and dine group and for knowing about the group itself. The Dinner Party Planner can query the social calendar with the name and date of the event, and consider what this group has prepared in recent months (the Planner maintains a bias for new dishes) and the dietary constraints or preferences of the guests attending this meal. For example, Kristin dislikes onion, particularly raw onion, Jeff prefers low-cholesteral meals, and Bob is allergic to mushroom. Other kinds of constraints may show up based on the state of Jack's kitchen itself: for example, the Meal Planner will try to use perishable food by placing a higher priority on the food's use in the meal as its "use before" date approaches.

Jack's Dinner Party Planner accesses a gourmet recipe database to which Jack subscribes and begins to construct three complete "views" into the recipe database, where each view represents a complete meal proposal. The Dinner Party Planner generates the following suggestions[2] through its interaction with the recipe database, which it posts to the message board with a request to Jack to evaluate the suggestions:

salmon caviar and white bean salad
ruby venison ragout
bulgar pilaf with green peppercorns
creamy fennel puree
maple hazelnut mousse
german chocolate lace cookies

broiled oysters with arugula purée and champagne sabayon
foie gras sautéed with raspberries
parslied rack of lamb
pear and parsnip puree
arugula and chèvre salad with trastevere dressing
chocolate chestnut torte

zucchini-watercress soup
beef carbonnade
parslied butter noodles
sweet potato and carrot purée
arugula salad with vinaigrette
apple tart

Jack gets home too late to look at the message board in his kitchen. The next day when he returns home, the message board is flashing to get his attention. He looks over the suggestions made by the Dinner Party Planner but doesn't like any of the meals as they are. He picks up the light pen and circles the dishes that interest him from the three meals. He then sends a request to the Dinner Party Planner to formulate a fourth suggestion utilizing the selected items if possible.

Within a few minutes, the Dinner Party Planner posts a new suggestion to the message board. Jack reviews it briefly and indicates his acceptance. The

Dinner Party Planner sends a request to the Shopper to organize the required grocery shopping expedition, and another request to the Chef for coordinating a plan to guide Jack in preparing the meal.

The Shopper has several responsibilities based on the meal plan. It must query the kitchen objects that store food (refrigerator, cupboards, etc.) to see what needs to be purchased for preparing the meal. It must send a request to these objects to put food that is presently in the kitchen and needed for the meal on "reserve" so that the Shopper will be informed if the availability of these items changes before shopping day. It must integrate the special items needed for this meal with other routine shopping needs (and/or other planned meals). Since Jack normally shops at one of three grocery stores, the Shopper is able to put the final shopping list in an order reflecting the layout of any of these stores. When Jack tells the system he is ready to shop and where he wants to go, the Shopper queries the store to make sure that critical items are available and to warn Jack if current conditions would make shopping difficult (for example, if the store is unusually crowded). In addition, once shopping time has arrived, the Shopper informs the storage devices of food in the kitchen with "reserve" status that reserved items can no longer be replaced; accordingly, the storage devices will protest the removal of these items until preparation time.

The Shopper organizes a shopping list and posts a request to the message board for Jack to tell it when he will shop, and whether he will shop at Gelson's, the store Jack customarily uses for dinner parties. Jack ignores the Shopper's request. When the day of the dinner party arrives, the Shopper queries all three stores, and finds that all can satisfy the shopping list. It then waits for Jack to request the shopping list for the store he is going to. The shopping list will plug into his shopping cart at the store to carry out additional responsibilities, like pointing out items on sale from the shopping list, or issuing coupons available for items on the list. When Jack gets to the store, he realizes he has forgotten to bring the shopping list. Fortunately, though, for a nominal fee one of the store's grocery carts can contact Jack's Shopper via phone lines for remote access of the shopping list.

After Jack has shopped, the time to prepare the meal arrives. Jack's Chef has responded to the request from the Dinner Party Planner to guide preparation of the meal. Accordingly, it has assembled all of the recipes to be displayed on a portable recipe holder, has integrated the steps and time needed to prepare the various components of the meal, has made recommendations for things that can be done in advance, and has assembled a suggested order for preparation of each part of the meal. Since Jack is an experienced cook, he usually generates his own plan for preparing the meal, which he then checks against the Chef's recommendations for anything he might have forgotten. This afternoon, Jack finds his own plan in harmony with the Chef's suggestions, and he brings off an excellent meal without a hitch.

The activity of hosting a dinner party consists of two major separable objectives: planning the meal, and preparing the meal. These objectives entail distinctly different types of activity, with concomitantly different kinds of involvement on the part of the augmented kitchen. Planning, for example, is primarily an information transaction, relying on gathering and taking into account various constraints in generating a solution to the problem of what to offer for dinner. The salient point here is that Jack is interested in solving the problem of what he should do for dinner. This makes him interested in suggestions his kitchen might be able to make based on the information it knows how to consider, but does not make him interested in understanding how the kitchen is arriving at its suggestions or in getting into a dialogue about how to decide. Of course, in some instances Jack may prefer to get the Meal Planner to critique a meal he proposes or to plan a meal around a particular dish he wishes to prepare. But no matter how the planning phase is initiated, or which constraints will apply in the current case, Jack is primarily interested in the *results* rather than the *process* of developing suggestions. This sets up an ideal situation for delegation; Jack can effectively off load some of his commitments, at least temporarily, to the augmented objects of his kitchen.

A different picture emerges in the domain of preparing the meal. Here Jack is interested in *both* process and results. Preparing a meal involves continuous interaction with the (physical) environment; that is, it is primarily action-oriented rather than information-oriented. Since a main goal of dinner party hosting is to have each part of the meal ready to be served at the appropriate time, and a single chef can only perform one action at a time, a feasible process plan is necessary for a good result. For this reason, the kitchen recommends a preparation plan but does not attempt to intervene in or manage the preparation activity (prompting Jack to cut the tomatoes *now*). Further, Jack enjoys gourmet cooking, especially for knowledgeable and appreciative guests. Unlike in the planning task, then, Jack's kitchen must play a strictly supportive and passive role in meal preparation to preserve the quality of Jack's experience in cooking.

Finally, Jack's delegation of responsibilities to objects in his kitchen on any particular occasion is optional. The augmented kitchen can function perfectly well when Jack does not behave as it "expects" or requests. If Jack leaves the house to shop without declaring his intentions to his Shopper, it is of no consequence. If he forgets to bring along his shopping list, he can shop without it as easily as retrieving it electronically. The use or nonuse of the virtual properties of his kitchen objects is seamless and smoothly integrated with Jack's ordinary practices; there exists a kind of "graceful degradation" of the full functionality of which the augmented kitchen is capable. Augmented reality easily and gracefully collapses to ordinary reality in direct response to Jack's behavior.

Emergent Virtual Reality

One way to view distributed cyberspace is as a way to improve on "flawed" design in ordinary reality. One of the most fundamental ways in which reality is flawed is that the objects in our environment are largely "dead" to distinctions we care about. Television sets and stereo systems are socially insensitive: they do not turn themselves down when we talk on the phone. A second way in which reality is flawed is that physical objects often make too little contribution in carrying out activities where it would be appropriate and most welcome. Many tasks that we want or have to do require significant amounts of tedious busywork, left to the human only because a more sensible alternative does not exist. By way of an example, the task of getting an appliance under warranty repaired is discussed below. Addressing such concerns as a design problem is made more difficult by the fact that the world in which our concerns arise is dynamic: even if a distinction and our objectives remain constant (desiring to be safe from pollutants in the home), what these mean, specifically, in terms of the world may change over time (microwave leakage isn't a problem until there are microwaves; radon gas is a relatively recent worry, having arisen as a concern in part as people responded to another concern—increased energy efficiency—by reducing air leaks in the house).

The "distinctions" people care about can be viewed as virtual worlds, or as we prefer, information webs. Culture, religion, social customs, academic disciplines, to name a few—all are familiar examples of information webs that represent systems of distinctions that matter to various groups of people. In the augmented kitchen, distinctions that matter to Jack are represented by the kinds of information the objects of the kitchen make use of (the gourmet database that Jack subscribes to) and by the assistance the kitchen provides in accomplishing tasks that Jack wants to do. In the augmented museum, described in the next section, an information web of stories, issues, arguments, and intrigues about the subject matter of the museum *re*presents distinctions that matter to natural historians.

With the idea of a distributed cyberspace as an augmentation of ordinary reality, we pursue the concept of virtual reality in a different sense than with the notion of an enclosed cyberspace. Augmenting reality (rather than replacing it with a simulated reality) means asking how real-world objects, inherently dispersed and disconnected, can be made sensitive to personal or cultural distinctions, in a way that can be directly experienced or acted upon by a person. Creating an augmented reality means giving real objects attributes and behaviors that reify or respond to concerns that emerge from the worlds we already inhabit. In analogy to object-oriented program design, we may consider various real-world objects and how they (currently) relate to things we care about, and ask what "responsibilities" involving those concerns they *ought* to have; we may engage in "object-oriented" (re)design of reality.

On a trivial interpretation, *any* object with computational capabilities might be said to constitute an augmented reality. Perhaps so, but at the very least, augmentations of reality will vary in their ability to support important distinctions and to do so through coordinated and autonomous collaboration with other objects. We may not be able to exclude a clock radio as a denizen of distributed cyberspace, but at best it will be an uninteresting and degenerate case, constituting a limited set of attributes and behaviors (having responsibility for waking its owner, displaying the time, or playing music) and operating in isolation from other objects and tasks in which it might logically play a role (informing the coffee maker, climate control system, and water heater what time the alarm is set for).

In contrast, a more typical case of a distributed cyberspace will involve many objects acting in concert to support a rich set of tasks via the responsibilities they have and the actions they take. We propose that realization of at least the following principles will be necessary for creating significant and usable distributed cyberspaces: *richness, connectivity, persistence,* and *direct interaction.*

Richness refers to the user's experience of objects: not only what can be perceptually experienced (visually, aurally, tactilely, etc.), but also their significance—both practical (what can be done with and through the object) and emotional (what the object represents to a particular person). Objects in the world are already rich in many of these senses. Cyberizing real-world objects offers an opportunity to enhance their inherent richness. As an object's ability to support human tasks or to embody information is extended, the user's experience of it will accordingly tend to become more multidimensional.

For example, Jack's experience of the objects in his kitchen is richer because of the variety of Jack's commitments in which they can participate and the delegation they afford. Some responsibilities—for example, the warranty, proof-of-purchase, and repair information associated with a kitchen appliance—Jack is content to delegate unconditionally. If the coffee maker breaks, Jack is happy for it to maintain whatever information the manufacturer needs to certify his ownership and date of purchase and to tell him where to send it for repair. By taking responsibility for at least part of this concern, Jack's direct experience of the coffee maker is enriched.

A second critical attribute exemplified by Jack's kitchen is the *connectivity* of objects. Connectivity refers to an object's ability to communicate with and be influenced by its environment, including other objects relevant to some activity, the distinctions represented in virtual information webs of various sorts, and the characteristics of its users. Connectivity ameliorates the inherent discontinuity of real-world objects and allows coherent groups of objects to interact and cooperate in support of an activity: it forges logical paths among objects where none existed before.

Connectivity is a means by which richness is achieved, and it takes a variety of forms. When objects are augmented with certain responsibilities they may need to tell or request things of other objects in carrying out those responsibilities. Similarly, objects may need to interact with information webs that exist outside of themselves (the gourmet database) or that they build up as part of their persistent memory (the social calendar tracking past meals or the dietary constraints of Jack's friends). Finally, objects are also connected with their users, both in the sense of registering user characteristics that affect their responsibilities and in the sense of supporting a rich and diverse set of ways for users to access and interact with them in pursuit of their activities.

Persistence refers to endowing objects with the ability to maintain and utilize historical and state information about user interactions and other events pertinent to their responsibilities. Persistence is necessary because users' activities span time, and often what has happened in the past is relevant to what is happening (or what *ought* to happen) now (past meals of Jack's group influence the planning of the current meal). Persistence is a second critical means by which the richness of objects is enhanced: it means, for example, that what a user requests of an object today will be "remembered" tomorrow when the user interacts with that object again. The kind of subjective, personalized experience of physical objects that is intrinsic to the notion of augmented realities is not achievable if objects lack a persistent memory.

Finally, a fundamental characteristic of successful distributed cyber-spaces will be affording *direct interaction*. The gains of augmenting reality must outweigh the costs. For example, the storage devices in Jack's kitchen have responsibility for knowing what they contain, but this capability ought not be achieved by asking Jack to encode each item from the grocery store as he puts it away. Rather, the storage device ideally gains its knowledge as an indirect side effect of Jack's normal actions in putting away food; in this case, perhaps machine-readable bar codes on the food that are read as the food is put away.

The kinds of interaction problems that are ideally addressed by direct interaction (or unobtrusive side effects of natural activity) include at least the following user concerns: (1) how to indicate interest in an object; (2) how to select from multiple paths (of tasks, or kinds of virtual information) running through an object; (3) how to provide feedback to objects where appropriate; and (4) how to maintain the fidelity of objects' representations of the real world with the actual state of the world.

Although these concerns apply to the general case of human-machine interaction, achieving these tasks through direct interaction is particularly important to augmented realities, where users are carrying out activities not through a machine (as in using a word processor to write a letter), but directly in the world.

Augmented and simulated realities represent different approaches to interacting with virtual worlds. Rather than a virtual world created out of visual illusion, augmented realities are created out of the stuff of the real world. Rather than being immersed in a perceptually omnipotent simulation, the user of an augmented reality remains in contact with the ordinary world. Rather than concentrating responsibility for action in the human agent, augmented realities inherently provoke distributed responsibility across human and nonhuman agents. Rather than an abrupt boundary between being in or not being in the cyberspace, augmented realities can exhibit graceful degradation, merging smoothly with ordinary reality.

The differences between augmented and simulated virtual realities are just that; one kind is not inherently better than the other. Each form of reality will find domains of application in which it is unsurpassed. For example, distributed cyberspaces have many potential applications in information and action domains, where objects can be given responsibility for maintaining information in order to change the nature of human action required to carry out a task. Simulated realities, on the other hand, are perhaps ideally suited for activities where the primary goal is providing a milieu in which the user can *experience* something inaccessible in ordinary reality.

Despite, or perhaps because of, different strengths and uses, augmented and simulated realities will often be complementary, combining to create rich composite cyberspaces that can address information, action, *and* experience domains. The design of an appropriate virtual reality depends on the situation. We turn now to a situation in which a cyberspace incorporating both genres of virtual reality can accomplish more than could either an augmented or a simulated reality alone.

A Natural History Museum Cyberspace

Traditionally, museums have had the primary purpose of archiving objects of interest and supporting the research of curators and scientists. Now, however, the role of the museum as an exploratory, educational, public environment is seen as central. Here the objective is "not to create subject-matter mastery, but to accomplish certain appreciative goals—to awaken interest, to broaden perspectives, to induce deeper understanding, to enrich aesthetic sensitivities, and so forth" (Lee 1968).

Museums are ideal candidates for hybrid cyberspaces. Artifacts of interest are distributed throughout the building. Connected with these artifacts are traditional "virtual" worlds, for example, of natural history and paleontology, comprising (among other things) a collection of stories for which the artifacts have significance and issues that they raise. Of course, there are severe limitations on what might be accomplished by directly augmenting museum objects, and authentic artifacts can afford little direct interaction with visitors. As a result, there is a great opportunity for simulated realities, dispersed throughout the museum, to be associated with particular artifacts, offering direct interaction and experience *in* a virtual world (an enclosed cyberspace) of a virtual world (natural history).

In the following scenario, a distributed system of enclosed cyberspaces cooperates with special augmented objects and museum visitors to create trails through an augmented museum. Each enclosed cyberspace is an environment in which a museum artifact and its role in natural history can be explored.

A Visit to the Natural History Cyberspace Museum

A crowd of Saturday visitors stands around a display of Archaeopteryx, the first bird, at the Natural History Cyberspace Museum in Washington, D.C.[3] Most of the visitors look at the exhibit for a few moments and move on, but a few of them, wearing special glasses, earphones, and sheer gloves of translucent material, stay longer. Charles, a 10-year-old with an abiding interest in dinosaurs, who has visited the museum many times, has come directly to the exhibit of Archaeopteryx, as he often does when mulling over the question of whether the dinosaurs should be considered extinct. He knows that some scientists believe that Archaeopteryx descended from a group of dinosaurs called coelurosaurs, but his fondness for dinosaurs makes him reluctant to accept this small bird as legitimate progeny of the dinosaur.

Charles takes one of the empty seats associated with the exhibit in order to enter its habitat cyberspace. A monitor at the back of his seat will show curious onlookers what he is seeing. Activating the cyberspace, as he has done many times before, the goggles Charles is wearing opaque for a moment before he is suddenly swept into the virtual landscape of Archaeopteryx. As he orients and looks around, Charles sees one of the birds sitting on a low-hanging branch in a nearby tree. He approaches the tree and reaches out a virtual hand to touch the bird. As he makes contact, Charles's perception shifts as he takes on the perspective of the bird and surveys the landscape from his perch. Presently, he spots what he is looking for: he sees a coelurosaur on a distant ridge. He jumps down from the tree and runs towards the ridge. Confronting the beast, Charles again reaches out a virtual hand to touch it. His perception shifts again as he "becomes" the dinosaur. The bird stands nearby.

Though Charles is familiar with some of the arguments supporting an evolutionary relationship between Archaeopteryx and the coleurosaurs (he has heard the bird and the dinosaur discussing clavicles, feathers, and warm-bloodedness in previous visits to this cyberspace), he has developed his own logic about the plausibility of this proposition. Charles reasons that if the bird is truly related to the coleurosaur, then he should be able to find similarities between them that he can feel as he becomes each in turn or that he can see in their environments. Charles lingers in the habitat cyberspace, alternating his perceptual experience between the bird and the dinosaur, seeking his experiential link.

Laura, a 16-year-old high school biology student, is preparing a term paper on the taxonomic classification of life and has come to the museum to investigate two current proposals from scientists for changing the existing taxonomy. She has just come from a cyberspace exhibit of prokaryotes and eukaryotes, fundamentally different kinds of single-celled organisms that have been proposed as kingdoms in a five-kingdom taxonomy to replace the current two-kingdom taxonomy of plants and animals. Laura has collected information on the new taxonomy in a virtual backpack she carries with her. She now stands before the exhibit of Archaeopteryx ready to explore a second proposal, of creating a new vertebrate class called Dinosaurus, containing both dinosaurs and birds.

Laura takes a seat next to Charles. She enters the habitat cyberspace and sees a bird in a nearby tree. Glancing over the landscape, she sees another bird and a dinosaur in the distance—Charles doing his exploration. But Laura has something specific in mind; she turns her attention away from the ridge and makes a gesture with her gloved hands, turning them palm up and flat, as if asking a question. The landscape quite suddenly is filled with virtual people— mostly scientists with some kind of interest in Archaeopteryx. Each scientist, or sometimes a group of several scientists, is standing around a blackboard. Approaching them makes the work on the blackboard visible.

Laura knows that the proposal for Dinosaurus was made by two paleontologists, Bakker and Galton, and begins to look among the blackboards for those with two people. Like the other habitat cyberspaces in the museum, the cyberspace Laura now inhabits has responsibility for incorporating and responding to what she has already seen and done today in her experience here. Accordingly, it has arranged the blackboards so that the more related the story or conversation taking place around the blackboard is to what Laura has already collected in her virtual backpack, the closer to her they appear. Spotting one close by, she travels nearer and sees "Dinosaurus" near the top of the blackboard. She stops in front of the blackboard and touches it with a virtual hand. As she makes contact, she hears the two scientists discussing the advisability of putting birds and dinosaurs in the same class.

One of the scientists points out that Dinosaurus will be accepted by the scientific community if it can be shown that birds inherited the primary characteristics underlying their evolutionary success, namely, feathers and warm-bloodedness, from dinosaurs. The other scientist agrees and details reasons to believe that dinosaurs were warm-blooded, including dinosaur bone structure (typical of warm-blooded rather than cold-blooded animals), geographical distribution (some dinosaurs lived too far from the equator to have been cold-blooded), ratio of predators to prey in dinosaur populations (estimated to be that typical of warm-blooded rather than cold-blooded populations), and dinosaur anatomy (suggesting that dinosaurs were built much like modern mammals).

As the second scientist finishes, another appears and takes up the story. He introduces himself as Dr. Ostrom and says that he too believes that the proposed Dinosaurus class should be accepted, based on a recent anatomical reexamination of Archaeopteryx he has completed. He says he believes that the coleurosaurs had feathers to help keep them warm, and that Archaeopteryx was a flightless bird, who used feathers, like its dinosaurian ancestor, for warmth, and possibly for capturing small flying prey. He goes on to say that his detailed comparisons of the anatomy of Archaeopteryx and the coleurosaurs show remarkable similarities, so many similarities, in fact, that it seems unlikely that their common features evolved independently in each species. Rather, he believes, they must indicate common ancestry.

Laura listens avidly to the scientists' conversation, transfixed by the rapid flow of ideas. She watches the scientists' arguments get posted to the blackboard as they speak. When the conversation wanes, Laura opens her virtual backpack to collect the contents of the blackboard. She backs off from the scientists and makes a fist. The scientists and blackboards disappear from the landscape. When she is finished for the day, Laura will pick up a diskette bearing the information she has collected during her visit and marking the trail she has created through the museum on this visit.

Despite the sweeping changes in computer technology in the last two decades, the practice of visiting a museum, and the nature of museums themselves, with a few exceptions, have changed little from descriptions given in the proceedings of the 1968 conference "Computers and Their Potential Applications in Museums."[4] For example, Lee notes that museum visitors are "diverse in terms of age, background, and the orientations and attitudes they bring with them," that they "usually come to the museum in small groups, continually interacting with each other as well as with the exhibits" (1968, 375). He stresses that visitors are free explorers of the museum environment. "The visitor comes to the museum for reasons of his own; he explores the environment at his own pace and in his own terms. His motivation is intrinsic—he is interested in furthering his own self-development in a way that is enjoyable to him. When he no longer finds the experience meaningful and pleasing, he disengages and leaves the museum."

The nature of a museum as an educational environment differs from the more typical educational environment of the school. "In contrast to the public school . . . the museum provides a rich opportunity for direct meaningful experience with things. It is a setting where experiential learning can take place. For example, one can truly get a feeling for the massive size, the power, and the essentially social character of a herd when confronted with an entire herd of stuffed elephants at the

American Museum of Natural History. As compared with books, lectures, or pictures, exhibits can offer the drama of firsthand contact, of a vivid, immediate experience, and of the reality of a phenomenon" (Lee 1968, 374). Lee concludes that the nature of learning possible in the museum environment can render it "deeply personal, rich, and highly rewarding."

One of the foremost appreciative goals of the museum should be to exploit visitors' curiosity: for example, to generate interest in some question of natural history, and then use the artifacts and experiences available to the visitor to convey aspects of the question. Of course, *some* appreciative goals are served merely in virtue of the variety of creatures and specimens on exhibit. Following the principle of "graceful degradation," the augmentation of the museum should not interfere with visitors who simply want to browse the exhibits in the traditional way (existing practice). Rather, the goal is to *lead* the existing practice of museum visits by "leveraging" off certain exhibits and allowing interested visitors to go beyond the artifacts they see into an exploration and understanding of the "virtual" scientific and cultural worlds of information that are tied to museum objects: for example, issues in natural history, means of deciding issues, the current status of issues, etc. The hope is to afford a far richer experience by allowing visitors to make contact with the "web" of human knowledge surrounding the artifacts of the museum in the open-ended, exploratory manner in which they currently view exhibits passively.

A second goal is to increase the richness of the experience by *personalizing* it. Both within and across visits, the museum should be sensitive to characteristics of individual visitors. For example, visitors might enter enclosed habitat cyberspaces in which they can experience an animal's environment by becoming that animal virtually. Or a visitor might enter a data space to gain access on demand to particular kinds of information about exhibits they are viewing. The museum might provide guidance for a visitor's activities: for example, bringing an activity to the visitor's attention (something available in a hands-on exploratory room) if it is related to other activities she has engaged in today. Across visits, the museum might have responsibility for knowing when a visitor last visited, what has already been seen, the kinds of things particularly of interest, etc., and make recommendations about what to do during the current visit.

Summary and Conclusion

We argued initially that exploring the notion of distributed cyberspace might be a good research strategy for the development and application of virtual world technology, and further, that augmented realities might be complementary to simulated realities in the construction of an "ultimate" cyberspace. Augmented realities differ from simulated realities in at least the locus of their implementation: simulated realities are created within a single object (a computing system), augmented realities are created in distributed, discontinuous real-world objects. They will certainly differ in the kinds of psychological consequences they have for users, and perhaps in the kinds of activities each can appropriately support. For these reasons, part of the research effort in virtual world technology should be directed toward considering and implementing distributed cyberspaces.

Precedents for Augmented Realities

Like the current implementations of three-dimensional virtual worlds as precursors to an established, consensual cyberspace, the notion of augmented realities can be seen to be "under construction" today in several guises. Perhaps the most familiar technological precursor is the international system of automated teller machines (ATMs). Although limited in richness, ATMs do have connectivity (with the information web of users' bank accounts), do support a distinction that matters to people (having cash in one's possession), and have a limited degree of persistence (in the per diem limits on cash withdrawals). Automated teller machines form a distributed, augmented reality: they have virtual behaviors and attributes (the ability to determine whether a requested transaction is acceptable vis à vis cultural conventions). Their use is voluntary and is integrated seamlessly into their users' everyday activities. ATMs have effectively (and swiftly) changed the everyday consumer practice of banking.

Science fiction literature also offers examples—not in its depiction of the established view of cyberspace per se, but in the "background" characteristics of the world in which stories are set. One example is the Maas-Neotek biochip unit described in Gibson's 1988 *Mona Lisa Overdrive*: a hand-held and activated "ghost" that produces a virtual companion that only the user sees and hears, and that provides conversation and all manner of cultural and pragmatic information relevant to the user's current concerns and welfare. Gibson's device is easily as far beyond current technological capabilities as his vision of the cyberspace matrix, but it exhibits the critical characteristics of richness, connectivity, persistence, and direct interaction.

Finally, some of the MIT Media Lab's projects might also be seen as precursors to augmented realities, for example, the "personal newspaper" (see Brand 1987 for a description). In this example, a computer system monitors various print and television news sources, and then constructs a personalized version of the day's news according to the user's preferences. The personal newspaper demonstrates adaptation of an important information web (the news media) to an individual user's perspective and preferences.

Distributed Cyberspace as a Complementary Research Strategy

While Jack's kitchen may not be built tomorrow, other examples of augmented realities demonstrate that this approach can be more tractable than creating full-blown simulated realities for the same purpose. For example, a newspaper article discussing potential applications of virtual reality mentioned the idea of electronic cadavers for medical training.[5] Such an enclosed cyberspace could be used by medical students to learn anatomy or to develop skill in surgical or physical examination procedures.

As it turns out, though, there already exists an example of an augmented reality for a precisely similar purpose: the upper gastrointestinal endoscopy simulator created by David Hon of IXION, Inc.[6] While arguably not *simple* to build, Hon's system is already built, and provides both richness and connectivity. Hon's system is implemented in a physical object—a manikin—where physicians learning the endoscopy procedure carry out realistic, free form examinations. The manikin is instrumented to coordinate immediate visual feedback on endoscope movements, as well as other kinds of information available to learners on request. Physical barriers within the manikin create realistic resistance to the movements of the endoscope, further enhancing the richness of the simulated experience. The system exhibits connectivity through its use of a "web" of information regarding anatomical landmarks, endoscope position, and motion video of the human upper gastrointestinal tract.

The integration of the physical simulation and the instantaneous, coordinated access to the information context of the endoscopy procedure in Hon's system creates what he calls a "visual feel" that would be a difficult experience to produce, at least at present, in an enclosed cyberspace. Whereas some of the informational aspects of the endoscopy procedure could surely be addressed by an enclosed cyberspace, producing the *feel* of an actual procedure in an electronic simulation is beyond current capabilities. One way to view what Hon has created is as a *virtual* world of upper GI endoscopy that is experienced by the user both as an activity in the *real* world (practicing carrying out the physical procedure) and as an exploratory environment for building skills for controlling and interpreting the procedure (visual recognition of physical anomalies, deciding when enough information has been gleaned from the procedure).

Distributed and enclosed cyberspaces have different limitations and constraints. Enclosed cyberspace will excel in its unfamiliar kind of perceptual immediacy and in its ability arbitrarily to render objects. The drawbacks of these very advantages are those associated with being *disconnected* from ordinary reality and objects. Distributed cyberspaces have the advantages of maintaining continuity with the real world and the inherent richness of real objects. The difficulties of distributed cyberspaces are the inherent limitations of certain objects and the interactions they afford (some objects are simply too small, too fast, etc., to support interaction). However, the augmented museum, when contrasted to a purely virtual museum, demonstrates how a balance of the two might work, and the chief advantage of building cyberspaces into the physical world.

Finally, distributed cyberspaces may provide a medium for prototyping enclosed cyberspaces. For example, suppose the books and research papers in our offices were electronically sense-*able* and could support identifying and locating currently relevant information. Creating such an augmented reality could yield insight into strategies people use for searching and retrieving information that might be of use in the design of an enclosed cyberspace of bibliographic references. For example, the fact that when searching one's office one is searching a universe of information that is already and perpetually filtered for general relevance to one's interests may lead to considering how enclosed cyberspaces could generate perpetually individually filtered versions of public databases.

Distributed Cyberspace as a Component of the "Ultimate" Cyberspace

People live in the world, and their practices have evolved in the world. Virtual worlds do not exist solely in some enclosed cyberspace: they exist in human culture, knowledge, and values as well. The image of cyberspace "jockeys" in a trance, hands flying over a console, bodies slumped in a chair, while careening through matrixes of corporate information, is a bizarre one, to say the least, when applied to human activities more mundane than fictional information system espionage. Designers and implementers of enclosed cyberspaces for everyday human practices, no less than designers of spreadsheets or word processors, will have to take care to make contact with the real world and existing nontechnological virtual worlds if their creations are to be successful. When (and if) the ultimate cyberspace is created, distributed, augmented realities may well comprise an important part.

Acknowledgments

We thank Eric Gold, John Thomas, and members of the User Interface Theory and Design group for discussion of the ideas raised in this chapter.

Notes

1. Numerous examples of these systems now exist, including work at the University of Washington, NASA Ames, the University of North Carolina, and at a variety of industrial research labs.

2. The recipes in this scenario are adapted from Rosso and Lukins 1982, 1985.

3. The issues and surrounding "stories" that visitors in this scenario explore are based on Gould 1980.

4. The Metropolitan Museum of Art Presentation, "Computers and Their Potential Applications in Museums." (New York: Arno Press, Inc., 1968). A conference sponsored by the Metropolitan Museum of Art and supported by a grant from the IBM Corporation, April 15–17, 1968, New York, New York.

5. Frederick Case, "Virtually Real," Seattle Times, February 26, 1990.

6. Upper G.I. Endoscopy Simulator. David Hon, President, IXION, Inc., 4335 N. Northlake Way, Seattle, WA.

References

Brand, Stewart (1987). The media lab: Inventing the future at MIT. New York: Viking Penguin.

Carroll, John M. and Kellogg, Wendy A. (1989). Artifact as theory-nexus: Hermeneutics meets theory-based design. In K. Bice and C. Lewis (eds.), Proceedings of CHI '89 Conference on Human Factors in Computing Systems. New York: ACM.

Carroll, John M., Kellogg, Wendy A., and Rosson, Mary Beth (1991). The task-artifact cycle. In J. M. Carroll (ed.), Designing interaction: Psychology at the human-computer interface. New York: Cambridge University Press.

Gibson, William. (1988). Mona Lisa overdrive. New York: Bantom Books.

Gould, Stephen J. (1980). The panda's thumb. New York: W. W. Norton & Co.

Kellogg, Wendy A. (1990). Qualitative artifact analysis. In D. Diaper, D. Gilmore, G. Cockton, and B. Shackel (eds.), Human-Computer Interaction: Interact '90. Amsterdam: North-Holland.

Lee, Robert S. (1968). The future of the museum as a learning environment. In Metropolitan Museum of Art (ed.), Computers and their potential applications in museums. New York: Arno Press, Inc.

Rosso, J., Lukins, S., and McLaughlin, M. (1982). The Silver Palate cookbook. New York: Workman Publishing, Inc.

Rosso, J., Lukins, S., and Chase, S. (1985). The Silver Palate good times cookbook. New York: Workman Publishing, Inc.

Contributors

Tom Barrett

is a research and development specialist at Electronic Data Systems in Richardson, Texas. Currently a member of EDS's Research and Advanced Development Group, he is also a "technology evangelist" for EDS's Technical Development Division in computer assisted systems engineering, artificial intelligence, hypermedia, and virtual reality.

Michael Benedikt

is Professor of Architecture at the University of Texas at Austin. He is the author of *For an Achitecture of Reality* and *Deconstructing the Kimbell: An Essay on Meaning and Architecture*. He is also President and CEO of Mental Technology Inc., of Austin, Texas, a software design consultancy.

Meredith Bricken

is a research scientist at the Human Interface Technology Laboratory at the University of Washington and is the director of its education program. Her current work is in designing and building virtual worlds, with particular emphasis on virtual prototyping (for Boeing), on curriculum modules for teaching world-building techniques, and on educational applications in general. Previously, she was at The Research Lab at Autodesk.

John M. Carroll

is manager of User Interface Technology at the User Interface Institute, IBM Thomas J. Watson Research Center, in Yorktown Heights, New York. His research is in the analysis of learning and problem-solving capacities that underlie human interaction with computers. His most recent books are *The Nurnberg Funnel: Designing Minimalist Instruction for Practical Computer Skill* and *Designing Interaction: Psychology at the Human-Computer Interface.*

F. Randall Farmer

was operator and coauthor of Lucasfilm's Habitat. He is now at The American Information Exchange Corporation, where he is helping bring free markets to cyberspace.

William Gibson

is the author of *Burning Chrome, Neuromancer, Count Zero, Mona Lisa Overdrive,* and most recently, with Bruce Sterling, *The Difference Engine.* He lives and works in Vancouver, B.C.

Michael Heim

is the author of *Electric Language.* He was a Fulbright Scholar for three years and lectures in philosophy at California State University at Long Beach. He is a consultant to the computer industry and a frequent contributor to the debates surrounding virtual worlds and cyberspace.

Wendy Kellogg

is a research staff member at the IBM Thomas J. Watson Research Center, in Yorktown Heights, New York. Her current work involves design analysis of computer artifacts and the use of analysis in the design of applications employing emerging technologies such as multimedia and virtual reality.

Tim McFadden

is a staff software engineer at Acer/Altos Computer Systems in San Jose, California. He is currently working on TCP/IP, the "plumbing" of cyberspace, as well as studying neural nets for input modes to virtual realities.

Chip Morningstar

was a project leader for Lucasfilm's Habitat. Currently he is Vice President for Software Development at The American Information Exchange Corporation in Palo Alto, California.

Marcos Novak

is Assistant Professor of Architecture at The University of Texas at Austin and a Ph.D. candidate at the University of California at Los Angeles. His research is in algorithmic composition, cyberspace, and the relationship of architecture to music.

Steve Pruitt

is a software engineer working on the design of the next generation computer integrated manufacturing system for Texas Instruments' semiconductor fabrication facilities. His current assignment is designing an integrated factory simulator. He is also studying TI's possible entry into the field of cyberspace and virtual reality.

John T. Richards

is the senior manager at the User Interface Institute, IBM Thomas J. Watson Research Center, in Yorktown Heights, New York. His research interests include user interface design, rapid prototyping tools and techniques, and object-oriented programming.

Nicole Stenger

is a computer animation artist. From her practice in Paris she became a Research Fellow at MIT's Center for Advanced Visual Studies and is currently working on an extended virtual world project, using animation techniques, at the Human Interface Technology Laboratory at The University of Washington in Seattle.

Allucquere Rosanne Stone

is a visiting professor in the Department of Sociology at the University of California in San Diego. She has done research for The National Institutes of Health and was a member of the Bell Telephone Laboratories Special Systems Exploratory Development Group. She produces the Monterey Symphony radio broadcast series and is the director of the Group for the Study of Virtual Systems at the Center for Cultural Studies at the University of California in Santa Cruz.

Carl Tollander

is Project Scientist with the Cyberspace Project at Autodesk, Inc., in Sausalito, California. He is currently involved with issues of representation and distributed simulation for near-future, commercial cyberspace products.

David Tomas

is an artist and anthropologist teaching in the Department of Visual Arts at the University of Ottawa in Ontario. He is currently writing a book on the history of the Western technologies of observation and representation on the Andaman Islands. He has published numerous articles on ritual and photography, including one on technicity in William Gibson's novels (*New Formations*, Issue 8, 1989).

Alan Wexelblat

is a staff systems engineer with Bull Worldwide Information Systems in Billerica, Massachusetts. He has been involved in work on artificial realities since 1986 when he worked at MCC. Currently, he is trying to understand the underlying mechanisms of cyberspace and to translate those into a commercial product for Bull.